D1222675

Antiproton-Nucleon and Antiproton-Nucleus Interactions

ETTORE MAJORANA
INTERNATIONAL SCIENCE SERIES
Series Editor:
Antonino Zichichi
European Physical Society
Geneva, Switzerland

(PHYSICAL SCIENCES)

Recent volumes in the series:

Volume 38 **MONTE CARLO TRANSPORT OF ELECTRONS AND PHOTONS**
Edited by Theodore M. Jenkins, Walter R. Nelson, and Alessandro Rindi

Volume 39 **NEW ASPECTS OF HIGH-ENERGY PROTON–PROTON COLLISIONS**
Edited by A. Ali

Volume 40 **DATA ANALYSIS IN ASTRONOMY III**
Edited by V. Di Gesù, L. Scarsi, P. Crane, J. H. Friedman, S. Levialdi, and M. C. Maccarone

Volume 41 **PROGRESS IN MICROEMULSIONS**
Edited by S. Martellucci and A. N. Chester

Volume 42 **DIGITAL SEISMOLOGY AND FINE MODELING OF THE LITHOSPHERE**
Edited by R. Cassinis, G. Nolet, and G. F. Panza

Volume 43 **NONSMOOTH OPTIMIZATION AND RELATED TOPICS**
Edited by F. H. Clarke, V. F. Dem'yanov, and F. Giannessi

Volume 44 **HEAVY FLAVOURS AND HIGH-ENERGY COLLISIONS IN THE 1–100 TeV RANGE**
Edited by A. Ali and L. Cifarelli

Volume 45 **FRACTALS' PHYSICAL ORIGIN AND PROPERTIES**
Edited by Luciano Pietronero

Volume 46 **DISORDERED SOLIDS: Structures and Processes**
Edited by Baldassare Di Bartolo

Volume 47 **ANTIPROTON–NUCLEON AND ANTIPROTON–NUCLEUS INTERACTIONS**
Edited by F. Bradamante, J.-M. Richard, and R. Klapisch

A Continuation Order Plan is available for this series. A continuation order will bring delivery of each new volume immediately upon publication. Volumes are billed only upon actual shipment. For further information please contact the publisher.

Antiproton-Nucleon and Antiproton-Nucleus Interactions

Edited by

F. Bradamante

University of Trieste
Trieste, Italy

J.-M. Richard

University of Grenoble
Grenoble, France

and

R. Klapisch

CERN
Geneva, Switzerland

Plenum Press • New York and London

Library of Congress Cataloging-in-Publication Data

International School of Physics with Low Energy Antiprotons on
Antiproton-Nucleon and Antiproton-Nucleus Interactions (3rd : 1988 :
Erice, Sicily)
Antiproton-nucleon and antiproton-nucleus interactions / edited by
F. Bradamante, J.-M. Richard, and R. Klapisch.
p. cm. -- (Ettore Majorana international science series.
Physical sciences ; 47)
"Proceedings of the third course of the International School of
Physics with Low Energy Antiprotons on Antiproton-Nucleon and
Antiproton-Nucleus Interactions, held June 10-18, 1988, in Erice,
Sicily, Italy"--T.p. verso.
Includes bibliographical references.
ISBN 0-306-43501-2
1. Antiprotons--Congresses. 2. Nucleon-antinucleon interactions-
-Congresses. 3. Nucleon-nucleon scattering--Congresses. 4. Proton
-antiproton annihilation--Congresses. 5. Quantum chromodynamics-
-Congresses. I. Bradamante, F. II. Richard, J.-M. (Jean-Marc),
1947- III. Klapisch, Robert. IV. Title. V. Series.
QC793.5.P72I57 1988
539.7'212--dc20 90-6767
CIP

Proceedings of the Third Course of the International School of
Physics with Low Energy Antiprotons on Antiproton–Nucleon and
Antiproton–Nucleus Interactions, held June 10–18, 1988,
in Erice, Sicily, Italy

A Division of Plenum Publishing Corporation
233 Spring Street, New York, N.Y. 10013

Printed in the United States of America

PREFACE

The third course of the International School on Physics with Low Energy Antiprotons was held in Erice, Sicily at the Ettore Majorana Centre for Scientific Culture, from 10 to 18 June, 1988.

The School is dedicated to physics accessible to experiments using low energy antiprotons, especially in view of operation of the LEAR facility at CERN with the upgraded antiproton source AAC (Antiproton Accumulator AA and Antiproton Collector ACOL). The first course in 1986 covered topics related to fundamental symmetries; the second course in 1987 focused on spectroscopy of light and heavy quarks. This book contains the Proceedings of the third course, devoted to the experimental and theoretical aspects of the interaction of antinucleons with nucleons and nuclei.

The Proceedings contain both the tutorial lectures and contributions presented by participants during the School. The papers are organized in several sections.

The first section deals with the theoretical aspects of $N\bar{N}$ scattering and annihilation, and the underlying QCD.

The experimental techniques and results concerning $N\bar{N}$ scattering are contained in Section II.

Section III contains theoretical reviews and contributions on antiproton-nucleus scattering and bound states.

Section IV is devoted to the experimental results on the antiproton-nucleus systems and their phenomenological analysis.

Finally, some possible developments of the antiproton machines are presented.

We should like to express our gratitude to Dr. Alberto Gabriele and the staff of the Ettore Majorana Centre who provided for a smooth running of the School and a very pleasant stay. We are particularly grateful to Mrs Anne Marie Bugge for her efficient help during the preparation and running of the School and for the editing of these Proceedings.

F. Bradamante, J.M. Richard, R. Klapisch

CONTENTS

I QCD AND THEORETICAL DESCRIPTION OF $N\bar{N}$ INTERACTION

II EXPERIMENTAL ASPECTS OF $N\bar{N}$ SCATTERING

III ANTINUCLEON-NUCLEUS: THEORY

IV ANTINUCLEON-NUCLEUS:
PHENOMENOLOGY AND EXPERIMENTAL RESULTS

V MACHINE DEVELOPMENTS

LATTICE QCD : WHY AND HOW

André Morel

Service de Physique Théorique - CEN Saclay
91191 Gif-sur-Yvette cedex

1. INTRODUCTION

These lectures aim at giving a general survey of what means "Lattice Quantum Chromodynamics". Not only all technicalities will be systematically avoided, but also many interesting aspects, both conceptual (connections with the theory of phase transitions in statistical mechanics, for example) and practical (numerical analysis, computer science, quantitative results), will be at best outlined. A more substantial account of these subjects can be found in lectures given by E. Brézin, E. Marinari and myself at the Ecole de Gif-sur-Yvette 1986, with the additionnal advantage that two thirds of the notes are written in French.

We shall concentrate on two questions about lattice QCD : *why* and *how* ? The answer to the first question will essentially consist in explaining that the non-abelian character of the colour SU(3) gauge group implies that the low energy features of strong interactions (hadron spectrum, chiral symmetry breaking) *cannot* be accounted for by perturbative quantum field theory. This part of the lectures will end up with the Wilson's proposal that discretizing space-time (lattice) realizes a non perturbative regularization of the theory. The second part will describe how lattice QCD hopefully is, or seems to be, an adequate manner to predict quantitatively the long distance properties of *hadrons* : the building of the partition function associated with the QCD Lagrangian, the appropriate way of approaching the continuum limits and the steps to be performed in order to extract the quantities of physical interest will be outlined. A qualitative account of the present situation will close these notes.

Antiproton-Nucleon and Antiproton-Nucleus Interactions
Edited by F. Bradamante *et al.*
Plenum Press, New York, 1990

2. LATTICE QCD : WHY

2.1 - Quantum Chromodynamics : a gauge theory based on the SU(3) group

It is convenient to distinguish, among the QCD fields, matter fields (the quarks) and gauge fields (the gluons). They play the role respectively of the electrons and of the photon of Quantum Electrodynamics (QED). Like in QED, matter and gauge fields interact via a current-gauge field local coupling described by the following interaction Lagrangian

$$\mathcal{L}_I(x) = \frac{g}{2}\,\bar{q}(x)\;\gamma_\mu\;\Lambda.G^\mu(x)\;q(x) \quad . \tag{1}$$

Here g denotes the strong interaction coupling constant (in QED $g = 2 \times e$, e = electron charge). The notation $\Lambda.G^\mu$ stands for

$$\Lambda.G^\mu(x) = \sum_{a=1}^{8} G_a^\mu(x)\;\lambda_a \quad , \tag{2}$$

with λ_a representing 8 independent, hermitean, traceless 3×3 matrices, a basis for the SU(3) algebra. To each of these matrices is associated one gluon field G_a^μ, where μ = 0, 1, 2, 3 is the Lorentz index (the gluons are massless vector fields). The γ_μ matrices act in Dirac space, the λ_a matrices in colour space, so that the quark fields $q(x)$, and their Dirac conjugate $\bar{q}(x)$, carry Dirac indices α = 1, 2, 3, 4 and colour indices i = 1, 2, 3, in addition to the relevant flavour indices which will be understood.

Note that the 8 λ matrices play the role of the (unique) charge operator Q of QED. *Unlike* QED (U(1) group), QCD corresponds to a *non-abelian* group, characterized by non trivial commutation relations among the generators :

$$[\lambda_a, \lambda_b] = \sum_{c=1}^{8} f_{abc}\;\lambda_c \quad . \tag{3}$$

The constants f_{abc} are the SU(3) structure constants.

This non-abelian character of the QCD group has drastic consequences, on which we want to elaborate further.

2.2 - The QCD group is non-abelian : so what ?

While Eq.(1) describes the interaction term of the gluons with the matter currents, the full QCD Lagrangian also contains, in addition to the free quark Dirac Lagrangian $\mathcal{L}_D$, what is called the "pure gauge" term $\mathcal{L}_G$. This term generalizes the $F_{\mu\nu}\,F^{\mu\nu}$ contribution of the photon A_μ in

QED, where the QED field strength $F_{\mu\nu}$ is given by

$$F_{\mu\nu} = \partial_\mu A_\nu - \partial_\nu A_\mu \tag{4}$$

In this case, $(F_{\mu\nu})^2$, quadratic in derivatives of the photon field is purely kinetic, and in absence of charged matter fields a photon moves freely. In the QCD case, the field strength tensor for each $a = 1, \ldots, 8$ contains, in addition to a similar kinetic term, a new term quadratic in the gluon fields :

$$F^a_{\mu\nu} = F^{a[\mathrm{kin}]}_{\mu\nu} + ig \sum_{b,c=1}^{8} f^{abc}\, G^b_\mu\, G^c_\nu \quad . \tag{5}$$

Accordingly, the pure gauge term proportional to $F^a_{\mu\nu}\, F^{\mu\nu}_a$ not only contains a kinetic term, but also terms of third and fourth order in the gluon fields, with respective couplings g and g^2. This feature, which follows from first principles as soon as one decides to consider a non abelian gauge theory, is of fundamental importance. It means that in even in the absence of matter (quark) fields, there exists non trivial interactions between gluons. Among the (expected) drastic consequences of this property, let us quote :

- existence of glueballs (hadrons with no valence quark, made of "pure glue"),

- confinement (free quarks and gluons are not observable, whatever the observation scale is),

- spontaneous breaking of chiral symmetry (with its standard manifestations, e.g. light pions, current algebra, non-vanishing proton mass in the chiral limit where the quark mass tends to 0).

Although these properties are not proven, there are good evidences, both from experiment and theory, that they hold. If they do, it is not difficult to convince oneself that they are *non-perturbative* consequences of QCD, as we shall argue shortly when discussing hadron masses. But let us before outline what we mean by non-perturbative property, by comparing with QED, once again.

Consider the binding energy E_B of, say, the hydrogen atom. We find in text books

$$E_B = -\frac{1}{2}\, m\, \frac{e^4}{\hbar^2} + O(e^6) \tag{6}$$

with

$$m = \frac{m_{e\ell.} \times m_{\mathrm{proton}}}{m_{e\ell.} + m_{\mathrm{proton}}} \quad .$$

We immediately notice two basic facts :
(i) - this physical quantity can be reached by perturbation theory (expansion in the coupling constant),
(ii) - the energy scale is set *by the masses of the constituants.*

This clearly contrasts with what we are going to find in QCD, when we consider the mass m_k of standard hadrons (quark bound states) :

$$m_k = c_k \Lambda_{QCD} + O(m_{quark}) \ . \tag{7}$$

Here the energy scale is some (unknown) constant Λ_{QCD}, in units of which any physical quantity can be measured. The numerical constant c_k is the value of the mass of k in units of this energy scale. In the limit of small quark masses, there is a correction to this chiral limit, proportional (at least in general, see below) to m_{quark}. So the challenge to the theorist here is *to compute* c_k from first principles, and possibly the quark mass corrections, or in other words to compute *ratios* of hadron masses since Λ_{QCD} is a priori unknown. Note also that in (7) there is no mention of the coupling constant, the only free parameter of the original theory in the chiral limit. One sometimes speaks of this property as "dimensional transmutation", the unknown dimensionless parameter g being replaced by the unknown energy scale Λ. This will hopefully become clearer from our discussion of renormalization group. Let us also at this point clarify the vocabulary concerning what is called quark mass. In the above, and in the Dirac part $\mathcal{L}_D$ of the Lagrangian, the notation m_{quark} represents *the current quark mass*, a free parameter for each quark flavour. This mass is the one which governs the chiral limit (no $m\bar{q}q$ term in $\mathcal{L}_D$), and enters standard relations of current algebra such as PCAC (the axial current is conserved up to terms of order m_{quark}). In contrast, the notion of *constituant mass* is purely phenomenological with respect to QCD theory : roughly speaking, since a proton is made of three quarks, one may think of it as a non-relativistic bound state of massive quarks with (constituant) mass $\simeq m_{proton}/3$. A glance at Eq.(7) clearly shows that the constituant mass has nothing to do with the current mass, at least in the subspace of light (current) quarks. In fact, the constituant mass represents the effective result of the complex, non perturbative, mutual interactions of gluons and current quarks.

There is one important exception to Eq.(7) in the hadronic world, which concerns the flavour non-singlet pseudo-scalar sector, pions, kaons, ... There the right equation is

$$m^2_{\pi,K} = B \times m_{quark} + \text{smaller terms} \quad , \tag{8}$$

where B is a non-perturbative constant computable within QCD. Eq.(8) is a direct manisfestation of spontaneous breaking by QCD of the chiral symmetry which the Lagrangian possesses at zero quark mass. It is based on the Goldstone theorem, which states that there are as many zero mass bosons (Goldstone bosons) as there are symmetry generators broken in the process : in the (u,d) quark world, the broken generators are the three axial generators of the chiral group $SU(2)_{right} \otimes SU(2)_{left}$, hence there are 3 pions, massless at $m_u = m_d = 0$. That spontaneous breakdown can only be a non perturbative feature can be understood if one remembers that in a renormalizable field theory all the symmetries of the Lagrangian are preserved order by order in perturbation (with the exception of the so-called "anomalies"), as a consequence of the associated Ward identities : all the counterterms which the perturbation series build have the same symmetries as the original terms.

2.3 - The renormalization group and the non perturbative nature of hadron masses

Let us start with a renormalizable field theory such as QCD, treated in perturbation. There is no scale in the Lagrangian, at least in the chiral limit $m_{quark} = 0$. In particular, the coupling constant g, as for any gauge theory, is dimensionless in dimension $d = 4$. However, a scale μ *has to be introduced* in the regularization-renormalization scheme, otherwise the Feynman integrals generated by the perturbative expansion have just no meaning at all (ultra violet divergences in the integral over the loop four-momenta). An example of such a scale is the one introduced by dimensional regularization : logarithmically u.v divergent Feynman integrals are made convergent by lowering the dimension d of space time from $d = 4$ to $d = 4-\epsilon$. So doing, dimensional counting shows that the gauge coupling is no more dimensionless : the interaction $g_\epsilon\, j_\nu\, A^\nu$ has the proper dimension $4-\epsilon$ in energy (in such a way that $\int dx^{4-\epsilon}\, \mathcal{L}(x)$ has no dimension in units of $\hbar$) provided that

$$g_\epsilon = (\mu)^{\epsilon/2}\, g\ . \tag{9}$$

Here g is dimensionless and μ is an *arbitrary* energy scale.

With the lattice technique, Feynman integral convergence is automatic since the integration over any internal momenta k is restricted to any interval of length $2\pi/a$, for example

$$-\frac{\pi}{a} < k_\nu \leqslant \frac{\pi}{a}\ , \tag{10}$$

where a is the lattice spacing. Then there is again a scale introduced by the regularization, $\mu = a^{-1}$, and all physical quantities will be measured in units of it.

In all cases, the scale μ introduced is *arbitrary*. Nevertheless, different physicists choosing *different* scales should eventually find the *same* physics. Proving that a theory is renormalizable actually proves that it is so and leads to the so-called renormalization group equations (RGE). Here we take renormalizability of QCD as granted (see standard textbooks), and just draw a few simple consequences.

Let Q be a physical quantity and suppose it has been computed perturbatively. One thus has chosen some particular value of a scale parameter μ, defined a renormalized, dimensionless, coupling constant at that scale, $g_R(\mu)$ and obtained some result

$$Q = Q[g_R(\mu),\mu] \ . \tag{11}$$

If this final result is independent of μ, then

$$\mu \frac{\mathrm{d}}{\mathrm{d}\mu} Q[g_R(\mu),\mu] = 0 \ , \tag{12}$$

which is called the renormalization group equation for Q. Let us now specialize to, say, a hadron mass in QCD, in the chiral limit. By dimensional analysis, μ being the only scale and g_R dimensionless, one necessarily has

$$m_k = \mu \, f_k[g_R(\mu)] \ , \tag{13}$$

so that the R.G.E. for m_k reads

$$f_k(g) + \beta(g) \frac{\mathrm{d}f}{\mathrm{d}g} = 0 \quad , \tag{14}$$

with the definition

$$\beta(g_R) \equiv \mu \frac{\mathrm{d}}{\mathrm{d}\mu} g_R(\mu) \ . \tag{15}$$

So the function $\beta(g)$ contains the information on how the renormalized coupling constant depends on the (arbitrary) regularization scale μ. Suppose now that $\beta(g)$ is known. Then Eq.(14) can be solved, leading to

$$m_k = \mu \, f_k(g_o) \, \exp\left[- \int_{g_o}^{g_R(\mu)} \frac{\mathrm{d}g}{\beta(g)}\right] \ . \tag{16}$$

This expression, together with the definition (15) of $\beta(g)$, expresses that m_k is a physical quantity, independent of μ. The constant g_o can be thought of as the value of g_R at some value $\mu = \mu_o$ of the scale.

Why is this useful in QCD (and in many other contexts) ? Although $\beta(g)$ is unknown, it can be computed perturbatively, and the result to lowest order in g is :

$$\begin{aligned} \beta(g) &= -\beta_o \, g^3 + O(g^5) \\ \beta_o &= \frac{1}{48\pi^2} [33-2n_f] \end{aligned} \tag{17}$$

In the expression of β_o, n_f is the relevant number of flavours; the constant 33 would be $11N$ for a gauge theory based on SU(N). This result has two major, and related, consequences

(i) - Asymptotic freedom

Let us solve Eq.(15) when $\beta(g)$ is truncated according to (17). We obtain

$$\frac{dg_R(\mu)}{g_R^3} = -\beta_o \frac{d\mu}{\mu} \tag{18}$$

which leads to

$$\frac{1}{g_R^2(\mu)} = 2\beta_o \ \ell n \ \mu + \text{cst.} \tag{19}$$

Since in our case β_o is *positive* for $n_f \leqslant 16$, we learn that as $\mu \longrightarrow \infty$, $g_R^2(\mu) \longrightarrow 0$ logarithmically. So, there exists a range of μ sufficiently *large* where the renormalized coupling constant actually is *small*, which justifies a posteriori the truncation of the perturbative series giving $\beta(g)$. This is the celebrated property of QCD to be an *asymptotically free* theory : at large enough scale, strong interactions effectively weaken. The coupling $g_R^2(\mu)$ at large μ is known as the "running" coupling constant. This first consequence leads to many applications where perturbative QCD yields predictions directly comparable with experiment (e^+e^- annihilation at high energy, heavy Drell-Yan pairs, large transverse momentum production, deep inelastic lepton scattering ...).

(ii) - The non-perturbative character of hadron masses

Restricting $\beta(g)$ in Eq.(16) to its first term in (17), we find for the hadron mass

$$m_k = \text{cst.} \ \mu \exp\left[-\frac{1}{2\beta_o \ g_R^2(\mu)}\right], \tag{20}$$

a result which is valid in a range of large μ where the expansion (17) is accurate enough. In this range, the independence of m_k upon μ is of course insured by (19), an equation which we can also write

$$\mu = \Lambda \exp\left[\frac{1}{2\beta_o \ g_R^2(\mu)}\right], \tag{21}$$

where Λ, an unknown fundamental parameter of QCD, sets the scale for all physical quantities. Indeed, with this definition a hadron mass reads from Eq.(20) :

$$m_k = c_k \ \Lambda, \tag{22}$$

where c_k is a numerical constant. From this it follows that the challenge for theorists is to compute all hadron masses in the same unit Λ, that is the corresponding numerical constants c_k. In turn, expression (20) tells us that there is no way to compute m_k through a perturbative expansion around $g_R^2 = 0$ since m_k has an essential singularity there.

One may wonder why these conclusions do not apply to QED. The reason is that the first term in the expansion (17) of $\beta(g)$ for QED is positive ($\beta_0 < 0$) so that Eq.(19) does not makes sense at large μ.

2.4 - Discretizing space-time : the Wilson's proposal

As we have seen, following the standard perturbative treatment of QCD leads nowhere as far as quantities such as hadron masses are concerned. The origin of the whole story lies in the ultraviolet structure of the theory (signalled by the divergence of Feynman integrals at large values of internal loop momenta), or equivalently, in the existence in space-time of short distance singularities. This equivalence can be understood from the properties of Fourier transforms : if $\tilde{A}(k)$ is ill-behaved at large k values, its Fourier transform $A(x)$,

$$A(x) = \int \frac{d^4k}{(2\pi)^4} \tilde{A}(k)\, e^{ikx} \quad , \tag{23}$$

is singular at small x values. In other words, quantities (propagators, amplitudes, in general Green's functions) relating fields at vanishingly distant point x and y of space time are singular in $(x-y)$. Wilson proposed in 1974 to realize the short distance regularization of the theory by replacing continuous space-time by a discrete lattice, where by construction the minimum distance $|x-y|$ between two points is a, the lattice spacing.

From now on, then, we consider that the fundamental fields can leave only on the sites (or the links) of a regular lattice :

$$x_\mu(\text{continuum}) \longrightarrow x_\mu = a\, n_\mu \quad , \tag{24}$$

where n_μ is any integer.

The Fourier transform of a function $G(x)$ is

$$\tilde{G}(k) = \sum_{n_\mu} e^{-ia\, n_\mu k^\mu}\, G(x) \quad , \tag{25}$$

and the values of k_μ can be restricted to any interval of length $2\pi/a$ because $\tilde{G}(k)$ is obviously periodic (invariant under $k_\mu \longrightarrow k_\mu + 2\pi/a$).

So, a lattice formulation presents the following features

(i) - there is a scale $\mu = a^{-1}$ introduced by the regularization,

(ii) - the scale μ acts as a cut-off on momenta,

(iii) - if the lattice is finite (L sites in each of the d directions), the world is reduced to a box of volume $(La)^d$, the system is finite and looks like a statistical system whose dynamical variables are fields rather than spins, magnets, or atoms, etc.

In order to make the technique operational, the steps to be taken are

(i) - to define on the lattice an action derived from the continuum action one starts with,

(ii) - to treat the statistical system by suitable methods, in particular numerical simulations,

(iii) - to take the appropriate limit towards the continuum : naively speaking it consists in taking the limit $a \longrightarrow 0$; we will see what it means in practice.

3. LATTICE QCD : HOW

In this second part, we examine more practically how the QCD theory can be transcribed on a lattice (we will first discuss the simpler case of a scalar theory), how physical quantities are extracted from numerical simulations, and what the general results obtained so far look like.

3.1 - The scalar theory case

The continuum Lagrangian for a scalar theory with $\lambda\varphi^4$ interaction reads

$$\mathcal{L}(x) = \sum_{\mu=0}^{3} \frac{1}{2} (\partial_\mu \varphi(x))(\partial^\mu \varphi(x)) - \frac{m^2}{2} \varphi^2(x) - \lambda\varphi^4(x) \tag{26}$$

Here we start with the Lorentz metric $x^\mu x_\mu = x_0^2 - \vec{x}^2$, and $\partial_\mu \equiv \partial/\partial x^\mu$. The field $\varphi(x)$ represents a scalar particle of mass m, with a self interaction specified by the coupling constant λ. The standard perturbative field theory makes use of the *continuum action* :

$$S = \int d^4x \; \mathcal{L}(x) \; , \tag{27}$$

out of which the Feynman amplitudes are generated through the expansion of

$$z = \exp \left[i/\hbar \; S \right] \; , \tag{28}$$

considered as a functional of the *field configurations* $\{\varphi(x)\}$ (a configuration is a set of values of $\varphi(x)$ at *all* points x of space time).

It is convenient for our purpose to next define the Euclidean (or "Imaginary time") version of the theory. It is obtained by changing the time-variable x_0 according to

$$x_0 \longrightarrow -ix_4 \ , \tag{29}$$

so that the Minkowskian scalar product and the integration measure are changed according to

$$\begin{aligned} x_\mu\, x^\mu &\longrightarrow -x_E^2 = -\sum_{\alpha=1}^{4} x_\alpha^2 \\ d^4x &\longrightarrow -i\, d^4x_E \end{aligned} \tag{30}$$

where x_E^2 and $d^4\, x_E$ are the Euclidean product and measure. In these new variables, the functional z reads

$$z=\exp(-S_{E/\hbar}) = \exp\left\{-\frac{1}{\hbar}\int dx_E^4\left[\frac{1}{2}\sum_1^4 (\partial_\alpha\varphi(x))^2+\frac{1}{2}m^2\varphi^2(x)+\lambda\varphi^4(x)\right]\right\}, \tag{31}$$

which defines a *positive* Euclidean action S_E. It is instructive to view z as the Boltzman factor associated with the configuration $\{\varphi(x)\}$. This comparison is made more transparent by rescaling the fields,

$$\varphi^2(x) \longrightarrow \varphi^2(x)/\lambda \ ,$$

which leads to the writing of z in terms of the new fields as

$$z \equiv \exp\{-\beta \times H\} \ , \tag{32}$$

with $\beta = \dfrac{1}{\hbar\lambda}$,

and

$$H = \int d^4x_E\left[\frac{1}{2}\sum_\alpha (\partial_\alpha\varphi)^2 + \frac{1}{2}m^2\varphi + \varphi^4\right] . \tag{33}$$

Noting that H is independent of λ, one has a strict analogy with a statistical system defined by the (pseudo) hamiltonian H, at a (pseudo) temperature T such that

$$\beta = \frac{1}{kT} = \frac{1}{\hbar\lambda} . \tag{34}$$

The term pseudo- refers to the facts that : *(i)* - here we work in *4* dimensions (not 3 or less as in realistic statistical physics); *(ii)* - the points at which the dynamical variables (here $\varphi(x)$) are located are not (yet) discrete points; *(iii)* - the temperature defined has nothing to do with any *physical* temperature. In particular, it should not be mistaken for the temperature introduced in the QCD context to discuss the possible existence of a deconfined phase (liberation of quarks and gluons as free quanta), hopefully reached, or reachable, in very high energy heavy ion collisions and, according to the Big-Bang model, in the primordial Universe. This interesting subject extends well beyond the purpose of these lectures.

With this analogy in mind, we now define the *lattice Euclidean version* of the scalar theory. Variables φ of the system are restricted to live on the sites x of a regular lattice

$$x_\alpha = a n_\alpha \quad ; \quad \alpha = 1,2,3,4 \tag{35}$$

where a is the lattice spacing and n_α an integer. One then has to choose a definition for the lattice derivative in order to replace $\partial_\alpha \varphi(x)$. A *particular* choice which gives $\partial_\alpha\varphi$ in the limit $a \longrightarrow 0$, provided $\varphi(x)$ is sufficiently well behaved, is

$$\frac{\partial \varphi(x)}{\partial x_\alpha} \longrightarrow \frac{1}{a} \left[\varphi(x+a\hat{\alpha}) - \varphi(x)\right] , \tag{36}$$

Other choices can be made with the same classical limit, which differ from the latter one by positive powers of a. They should lead to the same continuum physics. According to the above analogy, we now have a 4-dimensional problem in statistical physics, governed by the "Hamiltonian"

$$H = a^4 \sum_{\text{sites } i} \left[\frac{1}{2a^2} \sum_\alpha \left[\sigma(i+a\hat{\alpha})-\varphi(i)\right]^2 + \frac{1}{2} m^2 \varphi^2(i) + \varphi^4(i)\right] . \tag{37}$$

Note that the only term which connects variables at different sites is the crossed product of the lattice derivative squared : were it absent, one would have a (trivial) problem of independent objects. This crossed product,

$$-a^2 \sum_\alpha \varphi(i+a\hat{\alpha}) \varphi(i) ,$$

is a "nearest neighbour" interaction between site variables, exactly as in common spin problems in statistical mechanics.

3.2 - Statistical mechanics and Feynman paths: the need for stochastic methods

In this section we continue to describe analogies between statistical mechanics and field theory, in a totally heuristic manner, that is without proving any statement about the fact that a statistical treatment of the lattice action obtained is actually equivalent to standard quantum field theory. These analogies will nevertheless help us to introduce the quantities of interest and to understand how these can be measured.

From now on, we consider the functional

$$z = \exp[-\beta H] \ , \tag{38}$$

with β and H defined as above, as the Boltzman weight associated with some lattice configuration of the field $\{\varphi(i)\}$. Then one introduces the corresponding *partition function*

$$Z = \sum_{\text{all}\{\varphi\}} \exp[-\beta \ H(\{\varphi\})] \ , \tag{39}$$

so that the *probability density* of a particular configuration $\{\varphi_0\}$ is

$$\mathscr{P}(\{\varphi_0\}) = \frac{1}{Z} \exp\ [-\beta \ H(\{\varphi_0\})] \ . \tag{40}$$

It is clear that, at least in principle, $\mathscr{P}$ can be computed for any configuration, in a finite time for a finite lattice. Hence any question about averages over all configurations of arbitrary functions of the fields can be answered. For example, the φ propagator (which answers the question : what is the probability amplitude to have $\varphi(i)$ at site i given $\varphi(j)$ at site (j)), is nothing but the 2-point correlation function

$$\Delta_{ij} \equiv \langle \varphi(i)\varphi(j) \rangle_{\mathscr{P}} = \frac{1}{Z} \sum_{\text{all}\{\varphi\}} \varphi(i)\varphi(j) \ \exp[-\beta H(\{\varphi\})] \tag{41}$$

Let us make the meaning of $\sum_{\{\varphi\}}$ more precise. In the present case (real scalar field), at each of the L^4 sites of a lattice of length L along each direction, $\varphi(i)$ represents a number which may take any real value. Hence for any functional X

$$\sum_{\text{all}\{\varphi\}} X \equiv \prod_{i \in \text{lattice}} \int_{-\infty}^{+\infty} d\varphi(x_i) \ X(\{\varphi\}) \ . \tag{42}$$

So we are faced with the task of performing integrals over L^4 continuous variables when our world is restricted to a box of linear size L ! The only way out of this terrific challenge is to use stochastic methods. The general method is to prepare a large number N of configurations distributed according to the probability distribution Z^{-1} $\exp(-\beta H)$. If $\{\varphi\}_c$,

$c = 1, \ldots, N$ is such a set, then any average can be evaluated using

$$\langle A \rangle = \sum_{\text{all}\{\varphi\}} A(\{\varphi\})\ \mathscr{P}(\{\varphi\}) \simeq \frac{1}{N} \sum_{c=1}^{N} A(\{\varphi\}_c) \ . \tag{43}$$

If the N configurations are statistically independent, then the error on this stochastic measure of $\langle A \rangle$ is of order $1/\sqrt{N}$.

The lattice formulation of a field theory, which has just been outlined in this section on the scalar example, can be viewed as a *definition* of its formulation through functional integrals (or Feynman paths). It actually gives a meaning (but this statement requires several chapters in textbooks) to the partition function in the continuum and Minkowsky space, commonly written as

$$Z^{\text{cont.}} = \int \left(\prod_x \mathrm{d}\varphi(x) \right) \exp\left[\frac{i}{\hbar} \int \mathrm{d}^4 x \ \mathscr{L}(\varphi(x)) \right] \ . \tag{44}$$

Clearly in this expression $\prod_x \mathrm{d}\varphi(x)$ is ill-defined when x belongs to the continuum, and this disease is at the origin of all short distance (or u.v.) problems of the standard approach. Of course the use of Eq.(39) as a definition of the partition function does not make these short distance features irrelevant, but it constitutes a workable starting point, from which one then has to reach the continuum limit. Naively speaking, it "just" means to let a, the lattice spacing, go to zero. In practice, "It's a long way to go" (see below) !

3.3 - QCD on the lattice

As we know, there are two kinds of fields in continuum QCD : matter fields, the quarks and gauge fields. There are as many Dirac fields q and Dirac conjugate $\bar{q}$ as there are colours and flavours. Here we concentrate on colour degrees of freedom and write the quarks $q_m(x)$, where $m = 1, 2, 3$ and where x denotes its location in space-time. When discretizing space-time by a lattice, one would naively expect, by reference to the scalar case, that restricting $q_m(x)$ to live on the sites is the right thing to do. This however leads to serious problems, known generically as the "species doubling" problem. In fact, under a few general conditions specified in theorems due to Nielsen and Ninomiya, there is an unavoidable conflict between saving chiral symmetry on the lattice and fixing the number of flavours. This problem is handled in most lattice calculations either by recovering chiral symmetry only in the continuum limit ("Wilson fermions") or by keeping only part of it and tolerating a residual quark species degeneracy ("Kogut-Susskind fermions"). The rather cumbersome technicalities associated with these features will be omitted, the above considerations being just an introduction to a few keywords of the specialized literature.

So we consider that some way out of the lattice fermion puzzle has been chosen, so that, on the lattice, there are objects representing the quarks, and hence carrying a colour index. Let us recall that a *local gauge transformation* belonging to colour SU(3) acts as follows on the quarks (and their conjugate)

$$
\begin{aligned}
q_m(x) &\longrightarrow \sum_n V_{mn}(x)\; q_n(x) \\
\bar{q}_m(x) &\longrightarrow \sum_n \bar{q}_n(x)\; V^\dagger_{nm}(x) \;,
\end{aligned}
\tag{45}
$$

where $V(x)$ is an arbitrary, x dependent, 3×3 unitary matrix. We have

$$
V^\dagger(x)\; V(x) = 1 \;,
$$

so that a mass term $\bar{q}q(x)$ is obviously gauge invariant, whereas a lattice term written

$$
\bar{q}\;\Delta\; q = \bar{q}(x)\;\gamma_\alpha[q(x+a\hat{\alpha})-q(x-a\hat{\alpha})]/2a \tag{46}
$$

is not invariant because $V^\dagger(x)V(y)$ is not unity. Since the kinetic term connects fields situated at the end of lattice links $[x,\; x \pm a\hat{\alpha}]$, the desired invariance can be restored by introducing the gauge degrees of freedom through *link variables*. The covariant derivative on the lattice can indeed be written

$$
\bar{q}\;\Delta\; q = \frac{1}{2a}\,\bar{q}(x)\;\gamma_\alpha\Big[U_\alpha(x)\;q(x+a\hat{\alpha}) \;-\; U^\dagger_\alpha(x-a\hat{\alpha})\;q(x-a\hat{\alpha})\Big] \;. \tag{47}
$$

Here $U_\alpha(x)$ represents an $SU(3)_c$ group element attached to the link $[x,x+a\hat{\alpha}]$, $U^\dagger$ its hermitean conjugate. One immediately verifies that this derivative is invariant under the general local gauge transformations of Eq.(45), provided under such a transformation one imposes

$$
U_\alpha(x) \longrightarrow V(x)\;U_\alpha(x)\;V^\dagger(x+a\hat{\alpha}) \;. \tag{48}
$$

Note that the gauge fields enter the lattice action through the *group elements*, whereas in the continuum the covariant derivative involves the *group algebra*. Finally, we have to transcribe on the lattice the "pure gauge action", written for the continuum in Eq.(5). The most commonly adopted lattice gauge action, proposed by Wilson, is

$$
L_G^{\text{Latt.}} = \frac{1}{g_L^2} \sum_{\text{plaquettes}} (\text{Tr}\; U_P + \text{h.c.}) \;. \tag{49}
$$

The lattice gauge coupling constant is g_L, a plaquette P is an elementary square of the lattice made of four links. The plaquette located at x in the plane $(\hat{\alpha},\hat{\beta})$ consists of the links

$$[x,x+a\hat{\alpha}],\ [x+a\hat{\alpha},x+a\hat{\alpha}+a\hat{\beta}],\ [x+a\hat{\alpha}+a\hat{\beta},x+a\hat{\beta}],\ [x+a\hat{\beta},x].$$

The 3×3 unitary matrix U_P is the product over these links of the link matrices defined above. Taking its trace in colour space insures gauge invariance, due to (48), (46) and the cyclic invariance of a trace

$$\mathrm{Tr}\left[V^{+}U_P V\right] = \mathrm{Tr}\left[U_P V V^{+}\right] = \mathrm{Tr}\ U_P \quad .$$

That the form (49) gives back the pure gauge term of Eq.(5) in the limit $a \longrightarrow 0$ can be checked by explicit computation. One uses that in this limit, and with the notations of Eqs.(1) and (2), a group element $U_\alpha(x)$ and the associated algebra element $\Lambda.G_\alpha(x)$ are related through

$$U_\alpha(x) \simeq \exp[-ia\ \Lambda.G_\alpha(x)] \ . \tag{50}$$

In this computation, of course, the non-abelian character of the group must be carefully taken care of.

According to this discussion, the full lattice QCD action looks like

$$S = \sum_{x,\alpha} \bar{q}\ \gamma_\alpha\ \Delta_\alpha\ q + \sum_x m\bar{q}q(x) + \frac{1}{g_L^2} \sum_P \mathrm{Tr}(U_P + \mathrm{h.c.}) \quad , \tag{51}$$

and the corresponding partition function is written :

$$Z_{\mathrm{QCD}}^{\mathrm{latt}} = \int \prod_{\mathrm{sites}} \left[\mathrm{d}q\ \mathrm{d}\bar{q}\right] \prod_{\mathrm{links}} [\mathrm{d}U]\ \exp\ [-S] \tag{52}$$

The following remarks are in order.

(i) - the integration measure $\left[\mathrm{d}q\ \mathrm{d}\bar{q}\right]$ over the fermion fields is special. Indeed, in this framework, $q(x)$ represents a so-called Grassmann variable, not a c-number : note that due to the anticommutation relations for fermions, one must have $q^2(x) = 0$. This leads to special rules for derivation, integration etc ... (Grassmann algebra). For our purpose, we just quote the following result. Let M be any c-number $n \times n$ matrix, q_i and $\bar{q}_i$, $i = 1, \ldots, n$ $2n$ such Grassmann variables, then

$$\int \prod_{i=1}^{n} \left[\mathrm{d}q_i\ \mathrm{d}\bar{q}_i\right]\ \exp\ \left[\bar{q}\ Mq\right] = \mathrm{cst.} \times \det M \quad . \tag{53}$$

det M is called the "fermionic determinant" in the literature. In our case, M is defined by the quadratic form

$$\bar{q}\, M\, q = \bar{q}\, \not{\Delta}\, q + m\, \bar{q}\, q$$

appearing in the action. The index i of M is a multi-index including all the labels of a field q : site, colour, possibly flavour and Dirac index.

(ii) - $[dU]$ represents an invariant measure (under gauge transformations), called the Haar measure of the group. So the integration measure of the gauge fields does not spoil the invariance of the action.

(iii) - From the expression of the partition function, one sees that g_L^2 plays the role of a temperature, at least for the pure gauge part of the action. Here again one should not be confused by this terminology ; it is *not* the temperature relevant to the possible phase transition of QCD.

According to the remark *(i)* above, the partition function we have to deal with is

$$Z_{\mathrm{QCD}}^{\mathrm{latt.}} = \int \prod_{\mathrm{links}} [dU] \exp\left[- \frac{1}{g_L^2} \mathrm{Tr}\left(U_P + U_P^{\dagger}\right) + \ln \det(\not{\Delta} + m) \right] . \qquad (54)$$

To go on with the terminology employed in the literature, neglecting the fermionic determinant in (54) is called the *quenched* approximation of QCD. It amounts to omit the feed back of the quarks on the gauge field dynamics. In the perturbative language, it also corresponds to the neglect of all closed quark loops in all Feynman diagrams. One also speaks of *static* quarks. By contrast, the case of full QCD as in (54) is called *unquenched*, or *annealed*; the quarks are *dynamical* quarks.

The integrand in Eq.(54) constitutes the Boltzman weight relevant to the computation of any average value relative to correlation functions of interest. Finally, after making such averages, one still has to take the appropriate continuum limit.

3.4 - The steps of a lattice QCD numerical simulation

Given the partition function of Eq.(54), or its quenched approximation, one performs the following steps.

(i) - Choose a set of parameters for the simulation. These parameters are the QCD parameters, quark mass value(s) and coupling constant g_L, and the lattice parameters, typically the linear box size L (it may be useful in some applications to take different L values in different directions). We will discuss the choice of g_L and L later on, in connection with the continuum limit to be taken.

(ii) - Generate a large enough number of statistically independent gauge field configurations (that is of link variables), distributed according to the Boltzman weight. This can be done by various methods, Monte-Carlo generation (Metropolis algorithm), stochastic evolutions (Langevin equation) for example, or combinations of these techniques. A notorious problem in practical calculations is the evaluation of the fermionic determinant, which consumes an enormous amount of computer time. Note that the size of the fermion matrix is proportional to the box volume since its indices include the lattice site label. This is why the quenched approximation has been most preferred till now.

(iii) - Take averages, over the configurations selected , of well chosen operators. Mass spectrum calculations provide a typical example. One chooses a local operator $\Gamma(x)$ with the quantum numbers of a given particle (π, ρ, nucleon, ...). For example, one may consider the correlation function of

$$\Gamma_\pi(x) = \bar{q}(x)\ \gamma_5\ \vec{\tau}\ q(x) \quad , \tag{55}$$

and measure the average over configurations

$$\Delta^\pi_{xy} = \left\langle \Gamma_\pi(x)\ \Gamma^\dagger_\pi(y) \right\rangle \quad , \tag{56}$$

which gives information on the π-state propagation(*). More precisely, a state at rest of mass m_k evolves with Minkowski time x_0 proportionally to $\exp\,[i\ m_k\ x_0]$ and hence with the lattice Euclidean time an_4, proportionally to

$$C(n_4) = \exp[-an_4\ m_k] \ . \tag{57}$$

So extracting a mass from a hadron propagator Δ_{xy} amounts to analyze it in terms of exponential decay. Because for given quantum numbers not only the ground state, but excited states of larger mass and 2-particle states may contribute, it is wise to look at the *large* n_4 behaviour of the correlation function Δ_{xy}. Another reason to use large lattices.

(*) Since Γ involves the quark fields, this average is to be taken with the form (51-52) of the partition function, and the integration over the quark fields performed according to the proper rules for Grassman variables. As a result one finally gets an average, with the weight (54), of functions of the gauge fields only.

(iv) - As illustrated above by Eq.(57), the number which can be measured in a lattice simulation is expressed in units of the lattice spacing. In the spectrum example, one obtains

$$m_k = a^{-1} \times (\text{measured number})_k \quad . \tag{58}$$

It is thus time to remember that we have modelized by a discrete lattice a world which looks to us as continuous at all observed scales : we do have to take the limit $a \longrightarrow 0$. How can Eq.(58) be useful in this limit ? The answer has in fact been already given in Section (2.4). The depicted simulation is supposed to be made at some value g_L of the lattice gauge coupling, so that fixing m_k at its physical value means fixing the relationship between g_L and a. The renormalization group equation for QCD tells us that this is indeed possible, and that the continuum limit of lattice QCD is approached by letting g_L go to zero (not *be* zero ! Freedom is only *asymptotic* !) according to (see e.g. Eq.(21)) :

$$a^{-1} = \mu = \Lambda_L \exp\left[\frac{1}{2\beta_0 \; g_L^2(a)}\right] . \tag{59}$$

(v) - The task of the QCD simulator will be achieved provided he proves that the "measured number" behaves with g_L according to the "scaling behaviour" predicted by the renormalization group. This behaviour is known as the "asymptotic scaling" behaviour (strictly speaking, this term applies to a more precise version of the R.G.E. which includes the effect of the term of order g^5 in $\beta(g)$ of Eq.(17), leading to a power law correction g^{γ} in Eq.(21) or (59)). A weaker form of scaling is the statement, or the check, that the *ratio* of two physical quantities of the same dimension becomes independent of the coupling constant.

Lattice artefacts and Finite Size Effects (FSE)

The dimensionless correlation length associated with the existence of a state of mass m_k is

$$\zeta_k = (m_k a)^{-1} \tag{60}$$

Under the above (necessary) conditions for a reliable numerical simulation of QCD, this correlation length goes to infinity as one approaches the continuum, scaling behaviour. Physically, it means that one works in a situation where the details of the lattice (at the scale of a few lattice spacings) become unimportant. This situation is the same as that of problems of statistical mechanics close to a second order transition : there one knows that systems with different short distance interactions have the same long distance behaviour (critical exponents, universality); these systems are said to belong to the same universality class. From this point of view, lattice QCD, with the mentioned analogy between g_L^2 and a temperature, is a statistical system which has a continuous transition at the critical temperature $T_c = g_L^2 = 0$. Hence the absence of lattice artefacts (dependence over the "details" of the discretization) requires large correlation lengths in units of the lattice spacing.

There is an obvious drawback to this nice situation permitted by asymptotic freedom : looking for large correlation lengths implies working on large lattices. The absence of lattice artefacts *and* of finite size effects (FSE) requires that the following conditions are fulfilled

$$1 \ll \zeta \ll L \ , \tag{61}$$

where ζ is the largest physical correlation length (the inverse of the lightest particle mass in lattice units) and L the linear size of the box. It happens however that FSE may be used when the upper bound in Eq.(61) is not fully respected. This refers to the property known as Finite Size Scaling (or Fisher scaling) which states that, when measured in a box of size L, the correlation length ζ_L is related to its infinite volume limit through

$$\zeta_L/\zeta = f(\zeta/L) \ . \tag{62}$$

The function f is of course unknown, but the scaling property (dependence only on the *ratio* ζ/L) may help in analyzing data taken at various g_L's and medium size L's, and in performing extrapolations to the infinite volume limit if scaling is actually observed.

Problems associated with the chiral limit of QCD

It is well known that at least two of the quarks in nature, u and d, have very small masses compared to a standard hadronic scale (hundreds of MeV). The chiral limit is the limit where quarks are massless, a limit where the QCD Lagrangian recovers an additional, global, symmetry, the chiral symmetry. Actually, arbitrary unitary transformations in flavour space can then be performed independently on the left and right components $q_{L,R}$ of the quark fields

$$q_{\substack{L\\R}} = \left(\frac{1\mp\gamma_5}{2}\right) q \ . \tag{63}$$

That a mass term breaks this symmetry follows from the remark that

$$m\bar{q}q = m\left(\bar{q}_R q_L + \bar{q}_L q_R\right) \tag{64}$$

is not invariant under the above transformations. The reason why chiral symmetry is an interesting symmetry of QCD is that it appears that, *even for massless quarks*, one finds

$$\left\langle \bar{q}q \right\rangle \neq 0 \ . \tag{65}$$

This is an example of spontaneous breaking of a continuous symmetry in particle physics : although the Lagrangian is symmetric, the corresponding spectrum is not. Experimental confirmation of this fact is provided by the absence of degeneracy of states with opposite parities. More than that, in agreement with the so-called Goldstone theorem which states that any spontaneous breaking of a continuous symmetry is associated with massless particles (Goldstone bosons), the pions are found very light compared with other hadrons. They are not strictly massless because m_u and m_d are not strictly zero.

The non-vanishing of $\langle \bar{q}q \rangle$ in the chiral limit is an accepted fact, well confirmed by many numerical simulations of QCD. Note however that this very interesting feature is especially hard to study on a finite lattice : since the phenomenon is accompanied with Goldstone bosons, there exists in the system a correlation length $\zeta_\pi = 1/(m_\pi a)$ which becomes large not only because of aymptotic scaling ($a \longrightarrow 0$) but because m_π itself vanishes in the chiral limit. The unavoidable FSE's which follow constitute one reason why actual calculations are not performed for realistic values of the u,d quark mass, but extrapolated down from larger masses. There are systematic errors associated with this procedure.

There is another reason why simulations avoid the use of small quark masses. In the course of the numerical calculation, one has to compute the quark propagator $(\not{D} + m)^{-1}$ and the fermionic determinant, for many field configurations. These quantities are difficult to compute because the matrix $(\not{D} + m)$ has vanishing eigen values when $m \longrightarrow 0$, a feature which costs much computer time.

3.5 - A summary of results

There is an enormous amount of results accumulated from numerical simulations made possible all over the world by the appearance of large scale supercomputers. Plenary reports in general conferences, and proceedings of the specialized meetings held during the few past years give a full account for the quantitative progresses made (the numerical methods for QCD started to be developped around 1980). Here we limit ourselves to a few basic questions and propose a general (and personal) point of view on the tentative answers brought to them up to now.

Confinement - String tension

Confinement is the property attributed to QCD not to allow quarks and gluons to propagate freely in the vacuum, at least at moderate (physical) temperature and matter density. More specifically, only those objects, the hadrons, which are colour singlets may exist as asymptotic states. This situation which contrasts with that of QED is interpreted as a consequence of the non abelian character of the $SU(3)_c$ group (see sections 2.1 to 2.3). Needless to recall that the absence of free quarks and gluons is compatible with all reproducible experimental data.

On the theory side, a criterion for confinement of quarks and gluons in QCD has been proposed by Wilson. He showed that the energy $V(R)$ necessary to separate two colour charges at distance R (static quark and antiquark) could be evaluated by measuring on a lattice the average value of a suitable loop operator L_{RT},

$$\langle L_{RT} \rangle = \exp\,[-TV(R)]\ . \tag{66}$$

This result will not be shown here. The loop L_{RT} is a rectangle drawn on a lattice, of sides R along a spatial direction and T along the Euclidean time direction. The operator L_{RT} is the trace over the colour indices of the product along the loop of the gauge group elements attached to each link :

$$L_{RT} \equiv Tr\left(\prod_{\text{rectangle}} U_{\text{link}} \right)\ . \tag{67}$$

In QED, the separation energy is finite because the Coulomb potential vanishes at large distances. In QCD, it was anticipated by Wilson that one should find

$$\underset{R \longrightarrow \infty}{V(R)} = \sigma R\ , \tag{68}$$

at least in the absence of dynamical quarks (quenched case). The parameter σ, with the dimension, carried by a^{-2}, of an energy squared, is called the string tension. The separation costs an infinite energy, hence no free colour charge exists. Taking Eq.(66) for granted, the situation (68) implies

$$\ell n \,\langle L_{RT} \rangle = -\sigma RT\ , \tag{69}$$

that is an "area law" (as opposed to a perimeter law $(R+T)$ in a deconfined case) for the logarithm of the Wilson loop operators. The area law has been probed in extensive numerical simulations, leading to the widely accepted result that there exists a non-vanishing string tension σ, scaling as expected like a^2 when $g_L^2 \longrightarrow 0$. Such measurements can be used to fix the energy scale a^{-1} in a given "experiment" by relating σ to the slope $\alpha' \simeq 1\ \text{GeV}^{-2}$ of the linear trajectories formed by the hadrons observed in nature. This relation, inferred from the free string theory (Nambu-Goto action), is

$$\sigma = \frac{1}{2\pi\alpha'} \simeq (400\ \text{MeV})^2 \tag{70}$$

so a^{-1} is given by 400MeV/(σ in lattice units).

There are other ways to measure the same quantity, in particular via the so-called "Polyakov loops", and they lead to similar results. In the presence of dynamical quarks, one expects the large distance potential $V(R)$ to be screened by colour charges created (in pairs) from the vacuum. Then $V(R)$ should tend to a constant at infinity. This situation anyway leads to the same result as far as the absence of free charge is concerned : the quark and antiquark of a pair recombine with quark and antiquark of other pairs to produce the colour singlet hadrons observed. For example, jets are produced in e^+e^- collisions from $e^+e^- \longrightarrow qq^c$ at short distance, followed by $qq^c \longrightarrow$ hadron jets as soon as q and q^c start to leave each other. This screening by pair creation is actually visible in measurements of $V(R)$ in numerical simulations involving dynamical quarks.

Glueballs

We have seen that, because it is non-abelian, QCD theory is non-trivial even with no quarks (pure glue theory). Then "pure glue" hadrons, or glueballs, may exists. There is strong evidence for them in lattice investigations, an no doubt at all concerning the existence of a 0^{++} state around 1 GeV. The situation however is far from clear concerning two points.

(i) The status of glueballs of higher spins : in particular, there is a debate about the 0^{++}-2^{++} splitting which is not yet solved.

(ii) What is the fate of the glueballs when quarks are there, which is undoubtly more realistic ? Certainly one expects their mixing with more conventional states having the same quantum numbers. So, what should be a clear experimental signal for the glueballs ?

Hadron spectroscopy

The method for extracting hadron masses from suitable two-point correlation functions has been explained in Section 3.4 The exploration until now has mainly concerned the lowest lying mesons $0^{\pm}$, $1^{\pm}$ and baryons (nucleons, Δ, ...). It is usually performed in association with a study of chiral symmetry properties. From this point of view, the evidence for

- the non-vanishing of $\langle \bar{q}q \rangle$ as $m_q \longrightarrow 0$,
- the corresponding vanishing mass of the pion,
- the non-vanishing of the other masses ($\rho, p, \ldots$) is convincing, so that the general pattern is the expected one. This is especially so in the quenched approximation where large statistics can be collected, which allows for good (statistical) accuracy.

For reasons which are best summarized by inequalities (61), the ideal simulation which requires m_{quark} and g_L going to zero costs an infinitely growing amount of computer time. This is why it has not yet been shown quantitatively

(i) that all masses in lattice units scale according to the RGE,

(ii) that, e.g., the ratio m_N/m_ρ is what is observed in nature

The problem of the ratio m_N/m_ρ is of fundamental importance : since the u,d quark masses are very small with respect to the QCD scale, this pure number is a *zero parameter* prediction of QCD. Because very low quark masses lead to a large correlation length associated with the π mass, they require *very large* lattices (unrealistic yet). So, people perform simulations at higher $m_{u,d}$, with the hope of extrapolating the measured ratio down to $m_{u,d} \sim 0$. Until now, this method *has not succeeded* to give the right answer (neither to disprove it).

Finite temperature simulations

There are many systems of statistical mechanics for which an order-disorder transition appears when they are heated. For example, at low temperature, all the spins of a ferromagnet tend to be aligned along the same (arbitrary) direction even in the absence of magnetic field. Above some temperature $T = T_c$, thermal agitation wins and destroys the order. In QCD, similar transitions are expected concerning chiral symmetry an confinement. If this is true, there should exist temperatures T_χ and T_D at which the order parameter for chiral symmetry $\langle \bar{q}q \rangle$ and the string tension σ respectively vanish. Field theoretic methods allowing for non-zero temperature calculations have been developped. They will not be reviewed here. Applied on the lattice, they give good indications that transitions occur at $T_\chi \simeq T_D \simeq 100$ to 300 MeV. Here again there are difficulties in getting good data and in interpreting them. One should indeed remember that, strictly speaking, there is *no phase transition* in a *finite* system, because the energy barrier between different ordered phases is finite, being proportional to the volume of the box. So, although there is no doubt that $\langle \bar{q}q \rangle$ and σ *are seen to nearly vanish*, there remains uncertainties, in particular on the values of the transition temperatures and on their equality. Also, the order of the deconfinement transition is still under debate, although first order is theoretically favoured.

These questions are far from being academic, at least two physical applications being currently discussed. One is of cosmological nature. In the Big-Bang model, the early universe was hot, at a temperature well above T_D. When cooling down during the expansion, the universe was submitted to a phase transition from a plasma state of quarks and gluons (deconfined phase) to the present phase where only colourless objects are seen to exist. Important issues such as the observed isotropy of our universe or the history of nucleosynthesis depend on properties of the deconfining transition (its order in particular).

A second application concerns efforts made to produce in the laboratory the conditions of the transition from cold matter (nuclei) to hot matter (quark-gluon plasma). This is tempted in the ultrarelativistic heavy ion collisions undertaken at CERN : according to the theoretical evaluations made of T_D, may be these experiments provide high enough energy densities to deconfine the quarks and gluons. A major problem is then to design clear-cut signals of a transition.

Weak matrix elements

The standard model of electroweak interactions allows to predict (perturbatively) any weak process at the quark level. In order to predict the corresponding hadron properties (e.g. weak decays, CP violation in the K_o-$\bar{K}_o$ system, ΔI =1/2 rule) one must take into account that the quarks are not free, but confined inside particles, a non perturbative feature. This is a very difficult subject, whose study has been undertaken later than spectrum investigation. A few results only have come out, not yet stabilized. The origin of the ΔI = 1/2 rule remains obscure.

Lattice simulation for other purposes

To stay within the domain of particle physics, there are a few other subjects which can be investigated using lattices. Let us mention the computation of electroweak form factors, and, as an especially outstanding problem, the study of the Higgs mechanism. One interesting result of lattice simulations is the existence of an upper bound for the Higgs mass in the standard model, consistently found by several groups to be of the order of 700 MeV.

Acknowledgments

I thank the organizers of the Erice School of Physics with Low Energy Antiprotons for their invitation to give these lectures. I am very grateful to J. Ergotte for her prompt and careful typing of the manuscript.

Bibliography

Field theory

C. Itzykson, J.B. Zuber : "Quantum field theory", Mc Graw-Hill (1980).

Ta-Pei Cheng, Ling-Fong Li : "Gauge Theory of Elementary Particle Physics", Clarendon Press, Oxford (1984).

P. Ramond : Field Theory - "A modern Primer", Frontiers in Physics, Benjamin/Cummings Publishing Company (1981).

M. Le Bellac, "Des phénomènes critiques aux champs de jauge", Inter éditions / Editions du CNRS, Paris 1987.

Lattice - Non perturbative QCD

K.G. Wilson, Phys. Rev. D10 (1974) 2445.

L.P. Kadanoff : "The application of renormalization group techniques to quarks and strings", Rev. of Modern Physics 49 (1977) 267.

J.B. Kogut ; "The lattice gauge theory approach to quantum chromodynamics", Rev. of Modern Physics 55 (1983) 775.

M. Bander: "Theories of quark confinement", Physics Reports 75 (1981) 206.

C. Rebbi : "Lattice Gauge Theories and Monte Carlo simulations", World Scientific Publishing Company (1983).

M. Creutz, L. Jacobs et C. Rebbi : "Monte Carlo computations in Lattice Gauge Theories", Physics Reports 95 (1983) 202.

Ecole d'Eté de Physique des Particules, Gif-sur-Yvette 1986, IN2P3 (Paris)

Recent results

"Lattice gauge theory", Proceedings of the Wuppertal Meeting, Nov. 5-7 (1985), Plenum Press (N.Y.)

"Lattice gauge theory 1986" : Proceedings of the Brookhaven International Symposium, Sept. 15-19 (1986). Plenum Press (N.Y.)

"Lattice gauge theory 1987", Nucl. Phys. B (Proc. Suppl.) 4 (1988).

"Lattice gauge theory 1988", Nucl. Phys. B (Proc. Suppl.), to appear

ANTIPROTON ANNIHILATION IN QUANTUM CHROMODYNAMICS*

STANLEY J. BRODSKY

Stanford Linear Accelerator Center,
Stanford University, Stanford, California 94309, USA

1. INTRODUCTION

Quantum chromodynamics[1] has been extensively tested in high-momentum transfer inclusive reactions, where factorization theorems, asymptotic freedom and jet algorithms provide semiquantitative perturbative predictions. Tests of the confining nonperturbative aspects of the theory are either quite qualitative or at best indirect. In fact QCD is a theory of relatively low mass scales ($\bar{\Lambda}_{MS} \sim 200 \pm 100$ MeV, $(k\perp^2)^{1/2} \sim 300$ MeV, etc.) and its most critical tests of the theory as a viable theory of strong and nuclear interactions must involve relatively low energies and momentum transfer.

Anti-proton annihilation has a number of important advantages as a probe of QCD in the low energy domain. Exclusive reaction in which *complete* annihilation of the valance quarks occur ($\bar{p}p \to \ell\bar{\ell}$, $\gamma\gamma$, $\phi\phi$, etc.) necessarily involve impact distances $b_\perp$ smaller than $1/M_p = 5\ \mathrm{fm}^{-1}$ since baryon number is exchanged in the t-channel. There are a number of exclusive and inclusive $\bar{p}$ reactions in the intermediate momentum transfer domain which provide useful constraints on hadron wavefunctions or test novel features of QCD involving both perturbative and nonperturbative dynamics. In several cases ($\bar{p}p \to \bar{\ell}\ell$, $\bar{p}p \to J/\psi$, $\bar{p}p \to \gamma\gamma$), complete leading twist (leading power law) predictions are available. These reactions not only probe the subprocesses $\bar{q}\bar{q}\bar{q}\ qqq \to \gamma\gamma$, etc., but they also are sensitive to the normalization and shape of the proton distribution amplitude $\phi_p(x_i, x_2, x_3; Q)$, the basic measure of the proton's three-quark valance wavefunction. Additionally, one can explore such processes in terms of quasielastic reactions inside of nuclear targets, e.g., $\bar{p}A \to (J/\psi)\ (A-1)$, and study an extraordinary feature of QCD: "color transparency." There is another class of exclusive reactions in QCD involving light nuclei, such as $\bar{p}d \to \gamma n$ and $\bar{p}d \to \pi^- p$ which can probe quark and gluon degrees of freedom of the nucleus at surprisingly low energy. These will be discussed in sec. 11.

Inclusive reactions involving antiprotons have the advantage that the parton distributions are well understood. In these lectures, I will particularly focus on lepton pair production $\bar{p}A \to \ell\bar{\ell}X$ as a means to understand specific nuclear features in QCD, including collision broadening, breakdown of the QCD "target length condition." Thus studies of low to moderate energy antiproton reactions with laboratory energies under 10 GeV could give further insights into the full structure of QCD.

* Work supported by the Department of Energy contract DE–AC03–76SF00515.

Antiproton–Nucleon and Antiproton–Nucleus Interactions
Edited by F. Bradamante *et al.*
Plenum Press, New York, 1990

2. QCD TESTS AND HADRON LIGHT-CONE WAVEFUNCTIONS

QCD has two essential properties which make calculations of processes at short distance or high-momentum transfer tractable and systematic. The critical feature is asymptotic freedom: the effective coupling constant $\alpha_s(Q^2)$ which controls the interactions of quarks and gluons at momentum transfer Q^2 vanishes logarithmically at large Q^2. Complementary to asymptotic freedom is the existence of factorization theorems for both exclusive and inclusive processes at large momentum transfer. In the case of exclusive processes (in which the kinematics of all the final state hadrons are fixed at large invariant mass), the hadronic amplitude can be represented as the product of a hard-scattering amplitude for the constituent quarks convoluted with a distribution amplitude for each incoming or outgoing hadron.[2-6] The distribution amplitude contains all of the bound-state dynamics and specifies the momentum distribution of the quarks in the hadron.[2] The hard-scattering amplitude can be calculated perturbatively as a function of $\alpha_s(Q^2)$. The analysis can be applied to form factors, exclusive photon-photon reactions, photoproduction, fixed-angle scattering, etc. In the case of the simplest processes, $\gamma\gamma \to M\overline{M}$ and the meson form factors, rigorous all-order proofs can be given. As we shall see, many of these predictions are directly applicable to antiproton-initiated reactions.

The predictions of perturbative QCD have been strikingly confirmed in inclusive e^+e^- and $\gamma\gamma$ collisions, deep inelastic lepton reactions, massive lepton pair production, and the whole array of large p_T jet and photon reactions. Measurements of exclusive processes at high-momentum transfer, especially form factors and two-body photon-photon reactions have led to detailed checks on the scaling behavior of the theory. Recent results[7] for $\gamma\gamma \to M\overline{M}$ are shown in fig. 1. In general, the experimental results on the scaling behavior of exclusive and inclusive reactions appear consistent with short-distance subprocesses based on the elementary scattering of spin 1/2 quarks and spin 1 gluons, the fundamental degrees of freedom of QCD.

The key to understanding hadronization and hadron matrix elements is the hadron wavefunction itself. A convenient description of hadron wavefunctions is given by the set of n-body momentum space amplitudes, $\psi_n(x_i, k_{\perp_i}, \lambda_i)$, $i = 1, 2, \ldots n$, defined on the free quark and gluon Fock basis at equal "light-cone time" $\tau = t + z/c$ in the physical "light-cone" gauge $A^+ \equiv A^0 + A^3 = 0$. (Here $x_i = k_i^+/p^+$, $\sum x_i = 1$, is the light-cone momentum fraction of quark or gluon i in the n — particle Fock state; $k_{\perp_i}$, with $\sum k_{\perp_i} = 0$, is its transverse momentum relative to the total momentum p^μ; and λ_i is its helicity.) The quark and gluon structure functions $G_{q/H}(x, Q)$ and $G_{g/H}(x, Q)$ which control hard inclusive reactions and the hadron distribution amplitudes $\phi_H(x, Q)$ which control hard exclusive reactions are simply related to these wavefunctions:

$$G_{q/H}(x, Q) = \sum_n \int^{Q^2} \Pi d^2 k_{\perp_i} \int \Pi dx_i \; |\psi_n(x_i, k_{\perp_i})|^2 \delta(x_q - x) \quad ,$$

and

$$\phi_H(x_i, Q) = \int^{Q^2} \Pi d^2 k_{\perp_i} \; \psi_{valence}(x_i, k_{\perp_i}) \quad .$$

Thus an important tool is the use of light-cone quantization to construct a consistent relativistic Fock state basis for the hadrons and their observables in terms of quark and gluon quanta. The distribution amplitudes and the structure functions are defined directly in terms of these light-cone wavefunctions.[2] The form factor of a hadron can be computed exactly in terms of a convolution of initial and final light-cone Fock state wavefunctions.[8]

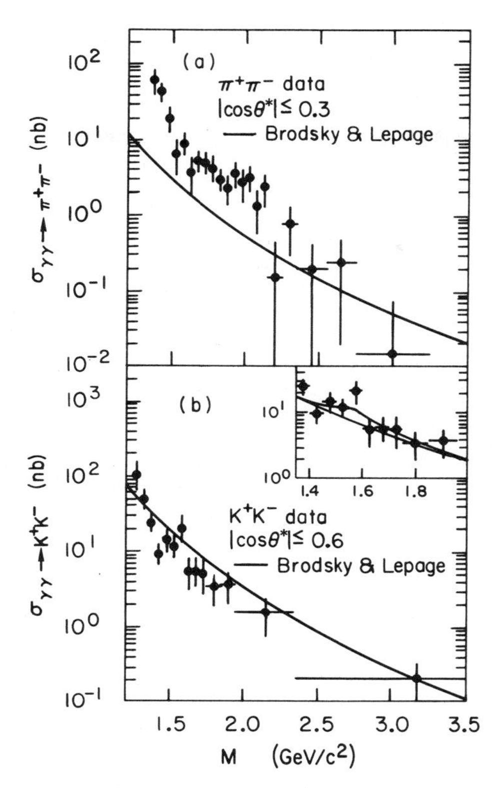

Fig. 1. Comparison of $\gamma\gamma \to \pi^+\pi^-$ and $\gamma\gamma \to K^+K^-$ meson pair production data with the parameter free perturbative QCD prediction of ref 2. The data are from ref 7.

The key to understanding hadronization and hadron matrix elements is the hadron wavefunction itself. A convenient description of hadron wavefunctions is given by the set of n-body momentum space amplitudes, $\psi_n(x_i, k_{\perp_i}, \lambda_i)$, $i = 1, 2, ...n$, defined on the free quark and gluon Fock basis at equal "light-cone time" $\tau = t + z/c$ in the physical "light-cone" gauge $A^+ \equiv A^0 + A^3 = 0$. (Here $x_i = k_i^+/p^+$, $\sum x_i = 1$, In the case of inclusive reactions all of the hadron Fock states generally participate; the necessity for higher-particle Fock states in the proton is apparent from its large gluon momentum fraction and the recent results from the EMC collaboration[9] suggesting that, on the average, little of the proton's helicity is carried by the light quarks.[10] In the case of high-momentum transfer Q exclusive reactions perturbative QCD predicts that only the lowest particle number (valence) Fock state contributes to leading order in $1/Q$. The essential gauge-invariant input is the distribution amplitude[11] $\phi_H(x, Q)$. Its dependence in log Q is controlled by evolution equations derivable from perturbation theory[11] or the operator product expansion.[12] A more detailed discussion of the light-cone Fock state wavefunctions and their relation to observables is given in ref. 13.

The phenomenology of hadron wavefunctions in QCD is now just beginning. Constraints on the baryon and meson distribution amplitudes have been recently obtained using QCD sum rules and lattice gauge theory. The results are expressed in terms of gauge-invariant moments $< x_j^m >= \int \Pi dx_i \; x_j^m \; \phi(x_i, \mu)$ of the hadron's distribution amplitude.

A particularly important challenge relevant to antiproton exclusive processes is the construction of baryon distribution amplitudes. A three-dimensional "snapshot" of the proton's *uud* wavefunction at equal light-cone time as deduced from QCD sum rules at $\mu \sim 1$ GeV by Chernyak et al.[14] is shown in fig. 2. The moments of the proton distribution amplitude computed by Chernyak et al., have now been confirmed in an independent analysis by Sachrajda and King.[15]

The moments of distribution amplitudes can also be computed using lattice gauge theory.[16] In the case of the pion distribution amplitudes, there is good agreement of the lattice gauge theory computations of Martinelli and Sachrajda[17] with the QCD sum rule results. This check has strengthened confidence in the reliability of the QCD sum rule method, although the shape of the meson distribution amplitudes are unexpectedly structured: the pion distribution amplitude is broad and has a dip at $x = 1/2$. In the case of the proton, the QCD sum rule prediction suggests that the u quark with helicity parallel to the proton helicity carries nearly 2/3 of the momentum in the three-quark valence Fock state. In fact, the QCD sum rule distributions, combined with the perturbative QCD factorization predictions, account well for the scaling, normalization of the pion form factor and also the branching ratio for $J/\psi \to p\overline{p}$. In addition, Maina has found that the data for large angle Compton scattering $\gamma p \to \gamma p$ are also well described.[18] However, a very recent lattice calculation of the lowest two moments by Martinelli and Sachrajda[18] does not show skewing of the average fraction of momentum of the valence quarks in the proton.

These initial results are interesting — suggesting a highly structured oscillating momentum-space valence wavefunctions. The results from both the lattice calculations and QCD sum rules demonstrate that the light quarks are highly relativistic, at least in mesons. This gives further indication that while nonrelativistic potential models are useful for enumerating the spectrum of hadrons (because they express the relevant degrees of freedom), they may not be reliable in predicting wave function structure.

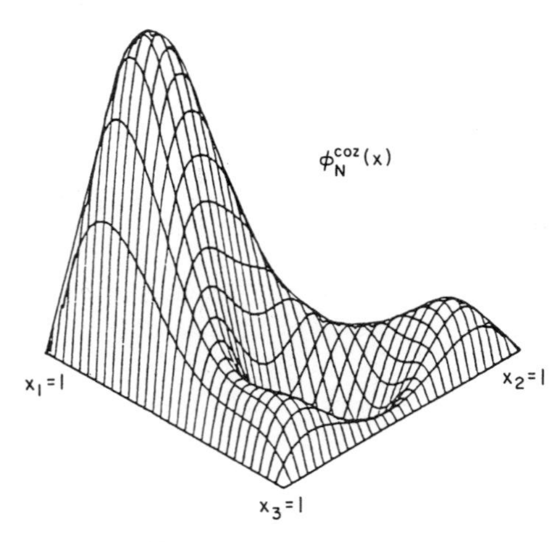

Fig. 2. The proton distribution amplitude $\phi_p(x_i, \mu)$ determined at the scale $\mu \sim 1$ GeV from QCD sum rules by Chernyak, Ogloblin and Zhitnitski.

The sum rule model form for the nucleon distribution amplitude together with the QCD factorization formulae, predicts the correct sign and magnitude as well as scaling behavior of the proton and neutron form factors.[19] (See fig. 3.)

A new nonperturbative method "discretized light-cone quantization," (DLCQ)[20] has been developed which has the potential for providing detailed information on all the hadron's Fock light-cone components. The basic idea is to diagonalize the QCD Hamiltonian on the light-cone Fock states, using a computationally-convenient

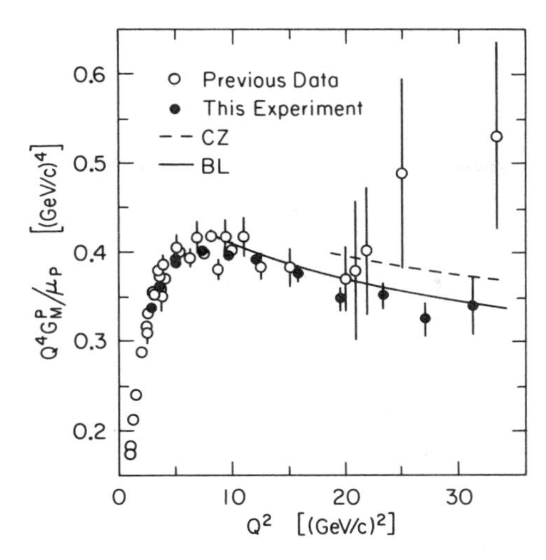

Fig. 3. Comparison of the scaling behavior of the proton magnetic form factor with the theoretical predictions of refs. 2 and 4. The CZ predictions[4] are normalized in sign and magnitude. The data are from ref. 19.

discrete momentum space basis. The eigenvalues M^2 of H_{LC} provide the spectrum of the theory; the eigenvectors yield the Fock state wavefunctions $\psi_n(x_i, k_{\perp_i}, \lambda_i)$. So far the method has been successfully applied to gauge theories and Yukawa theory (scalar gluons) in one-space and one-time dimension. New results for the spectrum and wavefunctions for QCD[1+1] are presented in sec. 15.

3. THE TIME-LIKE PROTON FORM FACTOR

As an introduction to the application of QCD to exclusive antiproton reactions, I will first review the leading-twist perturbative QCD predictions for $\bar{p}p \to \bar{\ell}\ell$, i.e., the time-like proton form factors $F_1(s)$ and $F_2(s)$. According to the QCD factorization analysis for exclusive processes, the main dynamics is contained [to leading order in $1/s, s = (\bar{p}+p)^2$] in the $\bar{q}\bar{q}\bar{q}\ qqq \to \gamma^* \to \bar{\ell}\ell$ amplitude. To leading order in $\alpha_s(Q^2)$ the latter can be computed from minimally connected PQCD tree graphs containing two off-shell gluons and two off-shell quark lines. (See fig. 4.) To leading order in $1/s$, it is sufficient to compute the subprocess amplitude taking the incident antiquarks and quarks collinear with their respective incident hadron direction, $p_i^\mu = x_i p^\mu$. One easily finds $T_H \sim \alpha_s^2(Q^2)/(Q^2)^2 f(x_i, y_j)$, where f is a rational function of the momentum fractions. By definition, higher-loop corrections to T_H are computed such that in intermediate states all quark propagators are noncollinear, and thus off-shell by order Q. This generates corrections of higher order in $\alpha_s(Q^2)$, with no additional logarithms. The contributions where partons in the intermediate state are collinear yields the evolution of the proton and antiproton distribution amplitudes.

After computing the hard-scattering amplitude T_H to the desired order in $\alpha_s(Q^2)$, one then convolutes the amplitude with the distribution amplitudes of the $\bar{p}$ and p. The distribution amplitude $\phi_p(x_i, s)$, $(\sum_{i=1,2,3} x_i = 1)$ is a nonperturbative wavefunction, but its logarithm dependence on s is computable from first principles (e.g., renormalization group plus operator product expansion) since this is controlled by the large momentum tail of the wavefunction:

$$\frac{\partial}{\partial \ell n\ s}\phi(x_i, s) = \int_0^1 dy_i\ V(y_i, x_i, \alpha_s(s))\phi(y_i, s) \quad .$$

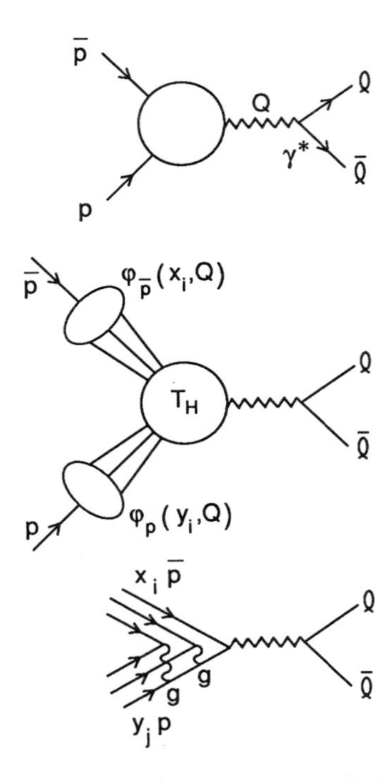

Fig. 4. Calculation of electromagnetic $\bar{p}p$ annihilation in PQCD.

This equation is of the form of an evolution equation where $\ell n\ s$ plays the role of the "time." The solution has the form

$$\phi(x_i, s) = \sum_n C_n(x_i) \left(\ell n \frac{s}{\Lambda^2}^{\gamma_n}\right)^{\gamma_n} \quad ,$$

where the $C_n(x_i)$ are a known set of orthonormal polynomials. The γ_n are computable fractional powers, anomalous dimensions characteristic of the interpolating local operators for three quarks with the proton quantum numbers.

Thus the $\bar{p}p \to \bar{\ell}\ell$ time-like Fock Dirac factor (helicity conserving) has the form

$$F_1(Q^2) = \left[\frac{\alpha_s(Q^2)}{Q^2}\right]^2 \sum_{n,m} b_{nm} \left(\frac{\ell n\ s}{\Lambda^2}\right)^{-\gamma_n - \gamma_m} \quad .$$

The coefficients $b_{n,m}$ reflect the nonperturbative input and are determined by the initial data for $\phi(x_i, s_o)$ in the evolution equation. The power law scaling, $F_1(s) \sim 1/s^2$, and $d\sigma/dt$ ($\bar{p}p \to \bar{\ell}\ell$) $\sim 1/s^6$ are consistent with the quark counting rules $F(Q^2) \sim (1/Q^2)^{n-1}$, $(d\sigma/dt)_{AB\to CD} \sim f(\theta_{cm})/s^{N_{tot}-2}$ where n is the number of fields in the bound state, and N_{tot} is the total number of incident and outgoing fields in $A + B \to C + D$; e.g., ($n = 3$, $\quad N_{tot} = 8 \quad$ for $\quad \bar{p}p \to \ell\bar{\ell}$).

It is important to note that the leading power-law behavior originates in the minimum three-particle Fock state of the $\bar{p}$ and p, at least in physical gauge, such as $A^+ = 0$. Higher Fock states give contributions higher order in $1/s$. For $\bar{p}p \to \ell\bar{\ell}$ this means that initial-state interaction such as one gluon exchange are dynamically suppressed. (See fig. 5.) Soft-gluon exchange is suppressed since the incident p or $\bar{p}$ color neutral wavefunction in the three-parton state with impact operation $b_\perp \sim 0(1/\sqrt{s}$. Hard-gluon exchange is suppressed by powers of $[\alpha s(s)]$.

The absence of a soft initial-state interaction in $\bar{p}p \to \bar{\ell}\ell$ is a remarkable consequence of gauge theory and is quite contrary to normal treatments of initial interactions

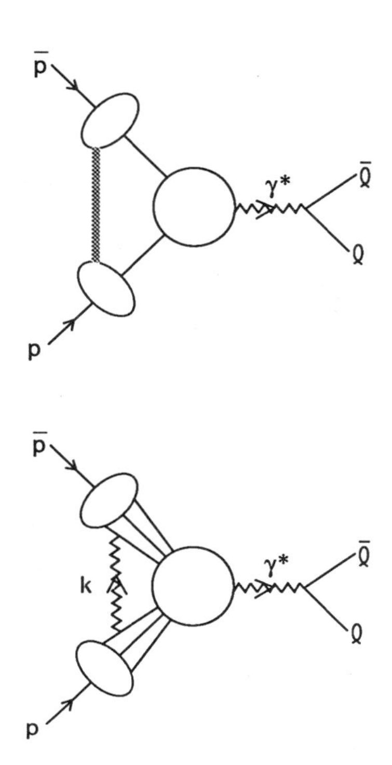

Fig. 5. Analysis of initial-state interactions in PQCD.

based on Glauber theory. This novel effect has experimental consequences: it can be studied in quasielastic $\bar{p}A \rightarrow \bar{\ell}\ell\ (A-1)$ reaction. Here we have in mind reactions in which there are no extra hadrons produced and the produced leptons are coplanar with the beam. (The nucleus $(A-1)$ can be left excited). Since PQCD predicts the absence of initial-state elastic and inelastic interactions, the number of such events could be strictly additive in the number Z of protons in the nucleus, every proton in the nucleus is equally available for short-distance annihilation. In traditional Glauber theory only the surface protons can participate because of the strong absorption of the $\bar{p}$ as it traverses the nucleus.

The above description is the ideal result for large s. QCD predicts that additivity is approached monotonically with increasing energy, corresponding to two effects: a) the effective transverse size of the $\bar{p}$ wavefunction is $b_\perp \sim 1/\sqrt{s}$, and b) the formation time for the $\bar{p}$ is sufficiently long, such that the Fock state stays small during transit of the nucleus.

The above example is an important example of the PQCD effect called "color transparency;" similar behavior is expected for all hard annihilation processing $\bar{p}p \rightarrow \gamma M$, $\bar{p}p \rightarrow J/\psi$, etc. In the case of exclusive high P_T processes in which hadrons are produced in the final state, each the final state hadron is produced with a small color singlet wavefunction; thus one predicts negligible attenuation of these hadrons in the quasielastic nucleus reaction.

A test of color transparency in quasielastic $pp \rightarrow pp$ scattering has recently been performed at BNL. This is discussed in more detail in secs. 4 and 9. In the next section I will discuss the general problem of hadronization in the nuclear environment for both exclusive and inclusive reactions. More detailed discussions of exclusive reactions in QCD are then given in secs. 5–13.

4. QCD HADRONIZATION IN NUCLEI

The least-understood process in QCD is *hadronization* — the mechanism which converts quark and gluon quanta to color-singlet integrally-charged hadrons. One way to study hadronization is to perturb the environment by introducing a nuclear medium surrounding the hard-scattering short distance reaction. This is obviously impractical in the theoretically simplest processes — e^+e^- or $\gamma\gamma$ annihilation. However, for large momentum transfer reactions occurring in a nuclear target, such as deep inelastic lepton scattering or massive lepton pair production, the nuclear medium provides a nontrivial perturbation to jet evolution through the influence of initial- and/or final-state interactions. In the case of large momentum transfer quasiexclusive reactions, one can use a nuclear target to filter and influence the evolution and structure of the hadron wavefunctions themselves. The physics of such nuclear reactions is surprisingly interesting and subtle — involving concepts and novel effects quite orthogonal to usual expectations.

In the case of inclusive reactions, the essential test of QCD involving $\bar{p}$ reactions is the Drell-Yan process $\bar{p}A \to \ell^+\ell^-X$ and $\bar{p}A \to \gamma\gamma X$. One of the remarkable consequences of QCD factorization for inclusive reactions at large p_T is the absence of inelastic initial- or final-state interactions of the high-energy particles in a nuclear target. Since structure functions measured in deep inelastic lepton scattering are essentially additive (up to the EMC deviations), factorization implies that the $q\bar{q} \to \mu^+\mu^-$ subprocesses in Drell–Yan reactions occurs with equal probability on each nucleon throughout the nucleus. At first sight this seems surprising since one expects energy loss from inelastic initial-state interactions.

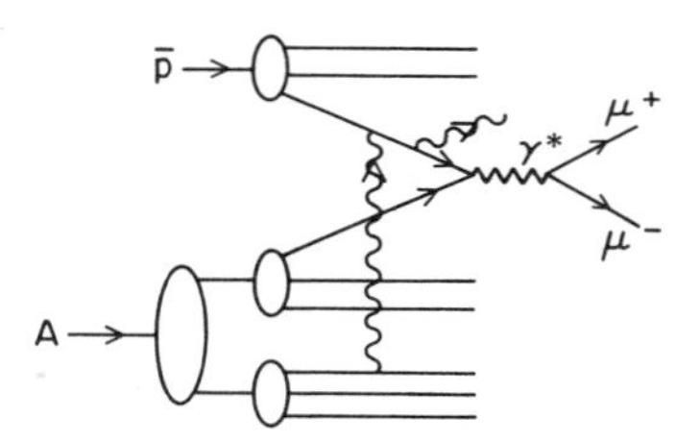

Fig. 6. Induced radiation from the propagation of an antiquark through a nuclear target in massive lepton production. Such inelastic interactions are coherently suppressed at parton energies large compared to a scale proportional to the length of the target.

In fact, inelastic reactions such as hard gluon bremsstrahlung induced in the nucleus which could potentially decrease the incident parton energy (illustrated in fig. 6) are suppressed by coherence if the quark energy (in the laboratory frame) is large compared to the target length: $E_q > \mu^2 \, L_A$. Here μ^2 is the difference of mass squared between the incident quark and the quark-gluon pair produced in the initial or final state collision. This phenomenon has its origin in studies of QED processes by Landau and Pomeranchuk. The QCD analysis is given by Bodwin, Lepage and myself.[26] The result can be derived by showing that the hard inelastic radiation emitted from differing scattering centers destructively interferes provided the target length condition is maintained. The destructive interference occurs when the momentum transfer μ^2/E_q due to the induced radiation is smaller than the inverse of the separation between two scattering centers in the nucleus. Soft radiation and elastic collisions, however, are still allowed, so one predicts collision broadening of the initial parton transverse momentum. Recent measurements of the Drell-Yan process $\pi A \to \mu^+\mu^-X$ by the NA–10 group[25] at the CERN–SPS confirm that the cross section for muon pairs at large transverse momentum is increased in a tungsten target relative to a deuteron target. (See fig. 7). Since the total cross section for lepton-pair production scales linearly with A (aside from relatively small EMC–effect corrections), there must be a corresponding decrease of the ratio of the differential cross section at low values of the di-lepton transverse momentum. This is also apparent in the data. Further measurements of low-energy $\bar{p}$ Drell–Yan reactions are needed to understand the limits of validity of QCD factorization and to explore the re-emergence of traditional Glauber inelastic scattering at low-antiquark energies. We discuss this further in sec. 14.

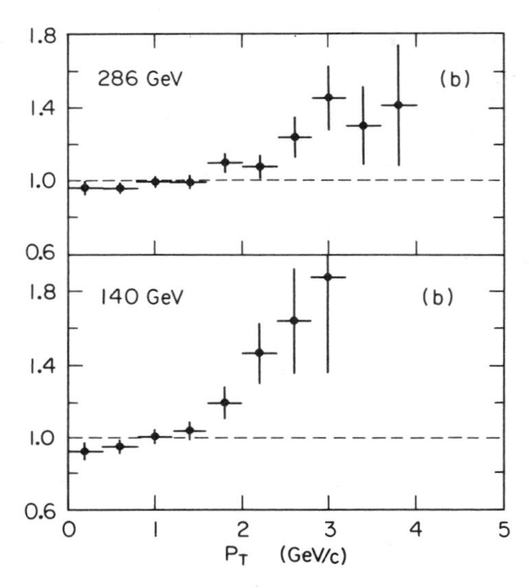

Fig. 7. The ratio $\sigma(\pi^- W \to \mu^+\mu^- X)/\sigma(\pi^- D \to \mu^+\mu^- X)$ as a function of the pair transverse momentum (from ref. 25).

The nucleus thus plays two complimentary roles in quantum chromodynamics:

1. A nuclear target can be used as a control medium or background field to modify or probe quark and gluon subprocesses. Some novel examples are *color transparency*, the predicted transparency of the nucleus to hadrons participating in high-momentum transfer exclusive reactions, and *formation zone phenomena*, the absence of hard, collinear, target-induced radiation by a quark or gluon interacting in a high-momentum transfer inclusive reaction if its energy is large compared to a scale proportional to the length of the target. (Soft radiation and elastic initial-state interactions in the nucleus still occur.) *Coalescence* with co-moving spectators is discussed as a mechanism which can lead to increased open charm hadroproduction, but which also suppresses forward charmonium production (relative to lepton pairs) in heavy ion collisions. There are also novel features of nuclear diffractive amplitudes — high energy hadronic or electromagnetic reactions which leave the entire nucleus intact and give nonadditive contributions to the nuclear structure function at low x_{Bj}.

2. Conversely, the nucleus can be studied as a QCD structure. At short distances nuclear wavefunctions and nuclear interactions necessarily involve *hidden color*, degrees of freedom orthogonal to the channels described by the usual nucleon or isobar degrees of freedom. At asymptotic momentum transfer, the deuteron form factor and distribution amplitude are rigorously calculable. One can also derive new types of testable scaling laws for exclusive nuclear amplitudes in terms of the reduced amplitude formalism. We discuss this topic in detail in sec. 11.

5. PERTURBATIVE QCD ANALYSIS OF EXCLUSIVE REACTIONS

Perturbative QCD predictions for exclusive processes such as $p\bar{p}$ annihilation into two photons at high-momentum transfer and high invariant pair mass can provide severe tests of the theory.[27] The simplest, but still very important example,[28] of the QCD analysis of an exclusive reaction is the calculation of the Q^2-dependence of the process $\gamma^*\gamma \to M$ where M is a pseudoscalar meson such as the η. The invariant amplitude contains only one form factor: $M_{\mu\nu} = \epsilon_{\mu\nu\sigma\tau} p_\eta^\sigma q^\tau F_{\gamma\eta}(Q^2)$.

It is easy to see from power counting at large Q^2 that the dominant amplitude (in light-cone gauge) gives $F_{\gamma\eta}(Q^2) \sim 1/Q^2$ and arises from diagrams (see fig. 8) which have the minimum path carrying Q^2; i.e., diagrams in which there is only a single quark propagator between the two photons. The coefficient of $1/Q^2$ involves only the two-particle $q\bar{q}$ Fock component of the meson wavefunction. More precisely the wavefunction is the distribution amplitude $\phi(x, Q)$, defined below, which evolves logarithmically on Q. Higher particle number Fock states give higher power-law falloff contributions to the exclusive amplitude.

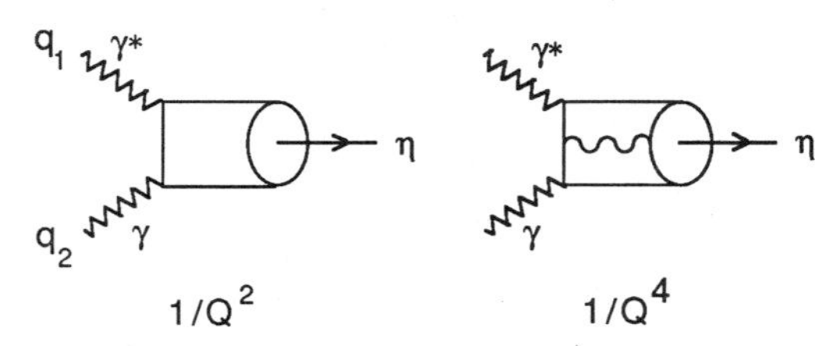

Fig. 8. Calculation of the $\gamma - \eta$ transition form factor in QCD from the valence $q\bar{q}$ and $q\bar{q}g$ Fock states.

The TPC/$\gamma\gamma$ data[29] shown in fig. 9 are in striking agreement with the predicted QCD power: a fit to the data gives $F_{\gamma\eta}(Q^2) \sim (1/Q^2)^n$ with $n = 1.05 \pm 0.15$. Data for the η' from Pluto and the TPC/$\gamma\gamma$ experiments give similar results, consistent with scale-free behavior of the QCD quark propagator and the point coupling to the quark current for both the real and virtual photons. In the case of deep inelastic lepton scattering, the observation of Bjorken scaling tests these properties when both photons are virtual.

The QCD power law prediction, $F_{\gamma\eta}(Q^2) \sim 1/Q^2$, is consistent with dimensional counting[30] and also emerges from current algebra arguments (when both photons are very virtual).[31] On the other hand, the $1/Q^2$ falloff is also expected in vector meson dominance models. The QCD and VDM predictions can be readily discriminated by studying $\gamma^*\gamma^* \to \eta$. In VDM one expects a product of form factors; in QCD the falloff of the amplitude is still $1/Q^2$ where Q^2 is a linear combination of Q_1^2 and Q_2^2. It is clearly very important to test this important feature of QCD.

The analysis of $\gamma^*\gamma \to \eta$ given here is the prototype of the general QCD analysis of exclusive amplitudes at high-momentum transfer:[32] at large p_T the power behavior of the amplitude is controlled by the minimum tree diagram connecting the valence quarks in the initial and final state — this is the hard scattering amplitude T_H which shrinks to a local operator at asymptotic momentum transfer — effectively the quarks interact when they are all at relative impact separation $b_T \sim 1/p_T$. One then convolutes T_H with the distribution amplitudes $\phi(x_i, Q)$ of the hadrons — analogs of the "wavefunction at the origin" in nonrelativistic quantum mechanics — to construct the hadronic amplitude. (See also sec. 3) This convolution is the basis of the factorization

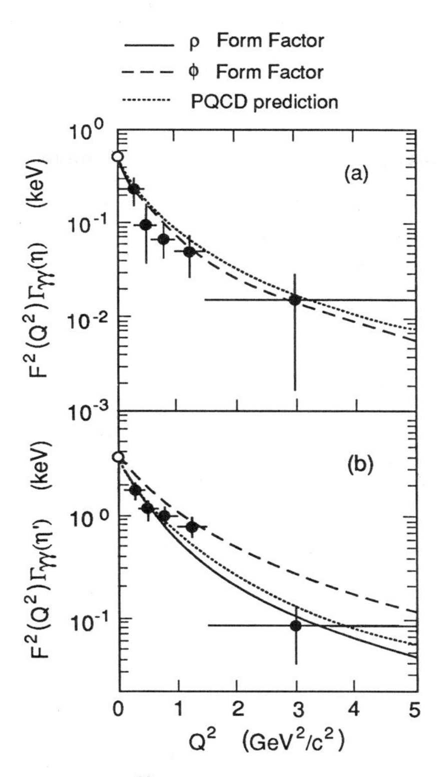

Fig. 9. Comparison of TPC/$\gamma\gamma$ data[29] for the $\gamma-\eta$ and $\gamma-\eta'$ transition form factors with the QCD leading twist prediction of ref. 27. The vector meson dominance predictions are also shown.

theorem for QCD exclusive reactions: to leading order in $1/p_T$, the nonperturbative dynamics associated with the hadronic bound states is isolated in universal, process-independent distribution amplitudes.[32] In cases such as $\gamma\gamma$ annihilation into meson pairs and meson form factors, the analysis is completely rigorous in the sense that it can be carried out systematically to all orders in perturbation theory.

A striking feature of the QCD description of exclusive processes is "color transparency:"[33] The only part of the hadronic wavefunction that scatters at large momentum transfer is its valence Fock state where the quarks are at small relative impact separation. Such a fluctuation has a small color-dipole moment and thus has negligible interactions with other hadrons. Since such a state stays small over a distance proportional to its energy, this implies that quasielastic hadron-nucleon scattering at large momentum transfer as illustrated in fig. 10 can occur additively on all of the nucleons in a nucleus with minimal attenuation due to elastic or inelastic final state interactions in the nucleus, i.e., the nucleus becomes "transparent." By contrast, in conventional Glauber scattering, one predicts strong, nearly energy-independent initial- and final-state attenuation.

A recent experiment[34] at BNL measuring quasielastic $pp \to pp$ scattering at $\theta_{\rm cm} = 90°$ in various nuclei appears to confirm the color transparency prediction — at least for p_{lab} up to 10 GeV/c. (See fig. 11.) Descriptions of elastic scattering which involve soft hadronic wavefunctions cannot account for the data. However, at higher energies, $p_{lab} \sim 12$ GeV/c, normal attenuation is observed in the BNL experiment. This is the same kinematical region $E_{\rm cm} \sim 5$ GeV where the large spin correlation in A_{NN} are observed.[35] Both features may be signaling new s-channel physics associated with the

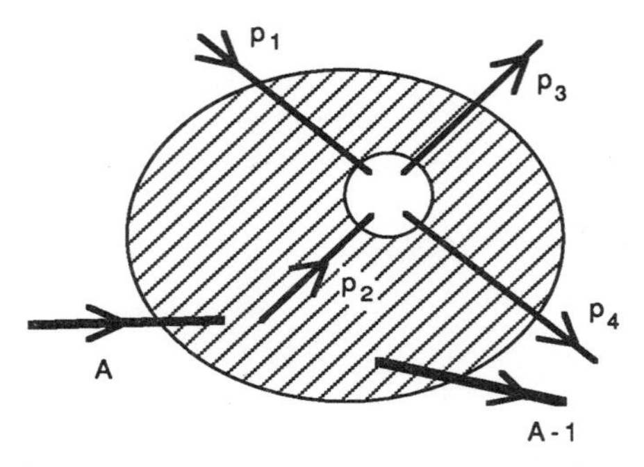

Fig. 10. Quasielastic pp scattering inside a nuclear target. Normally one expects such processes to be attenuated by elastic and inelastic interactions of the incident proton and the final-state interaction of the scattered proton. Perturbative QCD predicts minimal attenuation; i.e., "color transparency," at large momentum transfer.

onset of charmed hadron production[36] or interference with Landshoff pinch singularity diagrams.[37] Much more testing of the color transparency phenomena is required, particularly in quasielastic lepton-proton scattering, Compton scattering, antiproton-proton scattering, etc.

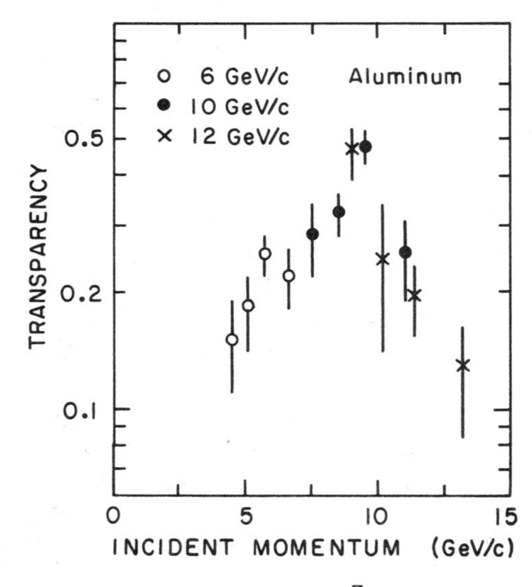

Fig. 11. Measurements of the transparency ratio $T = \frac{Z_{eff}}{Z} = \frac{d\sigma}{dt}[pA \to p(A-1)]/\frac{d\sigma}{dt}[pA \to pp]$ *near 90° on Aluminum (from ref. 34). Conventional theory predicts that T should be small and roughly constant in energy. Perturbative QCD[33] predicts a monotonic rise to* $T = 1$.

As I have discussed in the introduction, the essential nonperturbative input for exclusive reactions at high-momentum transfer is the hadron "distribution amplitude" $\phi(x, Q)$ which describe the longitudinal momentum distribution of the quarks in the valence, lowest particle-number Fock state.[28] Hadron wavefunctions can be conveniently defined as coefficients on a Fock basis at fixed $\tau = t + z/c$ in the light-cone gauge. Then

$$\phi(x,Q) = \int d^2k_\perp \theta(Q^2 - k_\perp^2)\psi_V(x,k_\perp) \ ;$$

i.e., $\phi(x,Q)$ is the probability amplitude to find the quark and antiquark in the meson (or three quarks in a baryon) collinear up to the transverse momentum scale Q. Here $x = (k^0 + k^z)/(p^0 + p^z)$. More generally, the distribution amplitude can be defined as a gauge-invariant matrix-element product of quark fields evaluated between the QCD vacuum and the hadron state. At large Q^2 one can use an operator product expansion or an evolution equation to determine $\phi(x,Q)$ from an initial value $\phi(x,Q_0)$ determined by nonperturbative input. The distribution amplitude contains all of the bound-state dynamics and specifies the momentum distribution of the quarks in the hadron. The hard-scattering amplitude can be calculated perturbatively as a function of $\alpha_s(Q^2)$. The analysis can be applied to form factors, exclusive photon-photon reactions, photoproduction, fixed-angle scattering, etc.

Exclusive two-body processes $\gamma\gamma \to H\overline{H}$ at large $s = W^2_{\gamma\gamma} = (q_1 + q_2)^2$ and fixed $\theta^{\gamma\gamma}_{\rm cm}$ provide a particularly important laboratory for testing QCD, since the large momentum-transfer behavior, helicity structure, and often even the absolute normalization can be rigorously predicted.[27,38] The angular dependence of some of the $\gamma\gamma \to H\overline{H}$ cross sections reflects the shape of the hadron distribution amplitudes $\phi_H(x_i,Q)$. The $\gamma_\lambda\gamma_{\lambda'} \to H\overline{H}$ amplitude can be written as a factorized form

$$\mathcal{M}_{\lambda\lambda'}(W_{\gamma\gamma},\theta_{\rm cm}) = \int_0^1 [dy_i]\, \phi^*_H(x_i,Q)\, \phi^*_{\overline{H}}(y_i,Q)\, T_{\lambda\lambda'}(x,y;W_{\gamma\gamma},\theta_{\rm cm}) \quad ,$$

where $T_{\lambda\lambda'}$ is the hard scattering helicity amplitude. To leading order $T \propto \alpha(\alpha_S/W^2_{\gamma\gamma})^{1,2}$ and $d\sigma/dt \sim W^{-4,-6}_{\gamma\gamma} f(\theta_{\rm cm})$ for meson and baryon pairs, respectively.

Lowest order predictions for pseudoscalar and vector-meson pairs for each helicity amplitude are given in ref. 27. In each case the helicities of the hadron pairs are equal and opposite to leading order in $1/W^2$. The normalization and angular dependence of the leading order predictions for $\gamma\gamma$ annihilation into charged meson pairs are almost model independent; i.e., they are insensitive to the precise form of the meson distribution amplitude. If the meson distribution amplitudes is symmetric in x and $(1-x)$, then the same quantity $\int_0^1 dx\ [\phi_\pi(x,Q)/(1-x)]$ controls the x-integration for both $F_\pi(Q^2)$ and to high accuracy $M(\gamma\gamma \to \pi^+\pi^-)$. Thus for charged pion pairs Lepage and I found the relation:

$$\frac{\frac{d\sigma}{dt}(\gamma\gamma \to \pi^+\pi^-)}{\frac{d\sigma}{dt}(\gamma\gamma \to \mu^+\mu^-)} \cong \frac{4|F_\pi(s)|^2}{1-\cos^4\theta_{\rm cm}} \quad .$$

Note that in the case of charged kaon pairs, the asymmetry of the distribution amplitude may give a small correction to this relation.

The scaling behavior, angular behavior and normalization of the $\gamma\gamma$ exclusive pair production reactions are nontrivial predictions of QCD. Recent Mark II meson pair data and PEP4/PEP9 data for separated $\pi^+\pi^-$ and K^+K^- production in the range $1.6 < W_{\gamma\gamma} < 3.2$ GeV near 90° are in satisfactory agreement with the normalization and energy dependence predicted by QCD. (See fig. 1.) In the case of $\pi^0\pi^0$ production, the $\cos\theta_{\rm cm}$ dependence of the cross section can be inverted to determine the x-dependence of the pion distribution amplitude. The one-loop corrections to the hard-scattering amplitude for meson pairs have been calculated by Nizic.[38] The QCD predictions for mesons containing admixtures of the $|gg\rangle$ Fock state is given by Atkinson, Sucher and Tsokos.[38]

The perturbative QCD analysis has been extended to baryon-pair production in comprehensive analyses by Farrar et al.[39] and by Gunion et al.[38] Predictions are given for the "sideways" Compton process $\gamma\gamma \to p\bar{p}$, $\Delta\overline{\Delta}$ pair production, and the entire decuplet set of baryon pair states. The arduous calculation of 280 $\gamma\gamma \rightarrow qqq\overline{qqq}$ diagrams in T_H required for calculating $\gamma\gamma \rightarrow B\overline{B}$ is greatly simplified by using two-component spinor techniques. The doubly charged Δ pair is predicted to have a fairly small normalization. Experimentally, such resonance pairs may be difficult to identify under the continuum background.

The normalization and angular distribution of the QCD predictions for proton-antiproton production shown in fig. 12 depend in detail on the form of the nucleon distribution amplitude, and thus provide severe tests of the model form derived by Chernyak, Ogloblin and Zhitnitsky from QCD sum rules.[40]

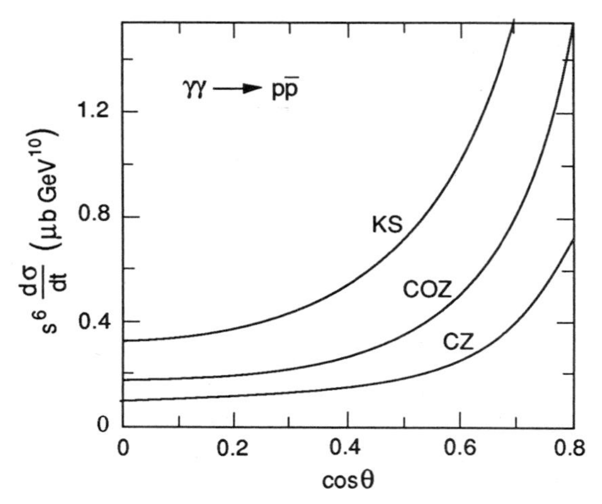

Fig. 12. Perturbative QCD predictions by Farrar and Zhang for the $\cos(\theta_{cm})$ dependence of the $\gamma\gamma \rightarrow p\bar{p}$ cross section assuming the King–Sachrajda (KS), Chernyak, Ogloblin and Zhitnitsky (COZ), and original Chernyak and Zhitnitsky (CZ) forms for the proton distribution amplitude, $\phi_p(x_i, Q)$.

An important check of the QCD predictions can be obtained by combining data from $\gamma\gamma \rightarrow p\bar{p}$ and the annihilation reaction, $p\bar{p} \rightarrow \gamma\gamma$, with large angle Compton scattering $\gamma p \rightarrow \gamma p$.[41]

This comparison checks in detail the angular dependence and crossing behavior expected from the theory. Furthermore, in $p\bar{p}$ collisions one can study time-like photon production into e^+e^- and examine the virtual photon mass dependence of the Compton amplitude. Predictions for the q^2 dependence of the $p\bar{p} \rightarrow \gamma\gamma^*$ amplitude can be obtained by crossing the results of Gunion and Millers.[38]

The region of applicability of the leading power-law predictions for $\gamma\gamma \rightarrow p\bar{p}$ requires that one be beyond resonance or threshold effects. It presumably is set by the scale where $Q^4 G_M(Q^2)$ is roughly constant, i.e., $Q^2 > 3$ GeV^2. Present measurements may thus be too close to threshold for meaningful tests.[42] It should be noted that unlike the case for charged meson-pair production, the QCD predictions for baryons are sensitive to the form of the running coupling constant and the endpoint behavior of the wavefunctions.

The QCD predictions for $\gamma\gamma \rightarrow H\overline{H}$ can be extended to the case of one or two virtual photons, for measurements in which one or both electrons are tagged. Because of the direct coupling of the photons to the quarks, the Q_1^2 and Q_2^2 dependence of the $\gamma\gamma \rightarrow H\overline{H}$ amplitude for transversely polarized photons is minimal at W^2 large and fixed θ_{cm}, since the off-shell quark and gluon propagators in T_H already transfer hard momenta;

i.e., the 2γ coupling is effectively local for Q_1^2, $Q_2^2 \ll p_T^2$. The $\gamma^*\gamma^* \to \bar{B}B$ and $M\overline{M}$ amplitudes for off-shell photons have been calculated by Millers and Gunion.[38] New results on charged $\pi\rho$ pair production were also presented to this meeting by Kessler and Tamazouzt. In each case, the predictions show strong sensitivity to the form of the respective baryon and meson distribution amplitudes.

We also note that photon-photon collisions provide a way to measure the running coupling constant in an exclusive channel, independent of the form of hadronic distribution amplitudes. The photon-meson transition form factors $F_{\gamma\to M}(Q^2)$, $M = \pi^0, \eta^0$, f, etc., are measurable in tagged $e\gamma \to e'M$ reactions. QCD predicts

$$\alpha_s(Q^2) = \frac{1}{4\pi} \frac{F_\pi(Q^2)}{Q^2|F_{\pi\gamma}(Q^2)|^2} \quad ,$$

where to leading order the pion distribution amplitude enters both numerator and denominator in the same manner.

6. APPLICABILITY OF PERTURBATIVE QCD TO EXCLUSIVE PROCESSES

Isgur and Llewellyn Smith[43] have recently raised a number of questions concerning the application of perturbative QCD to exclusive reactions in the momentum transfer range presently accessible to experiment. The issues involved are very important for understanding the basis of virtually all perturbative QCD predictions. In the following I will review and discuss the main points at issue:

(1) Isgur and Llewellyn Smith, and also Radyshkin,[44] argue that the normalization of the PQCD amplitude is of order $(\alpha_s/\pi)^n(\lambda^2/Q^2)^n$ where λ is a typical hadronic scale. If this were the correct estimate, the perturbative contributions would be too small to compete with the rapidly-falling "soft" nonperturbative contributions until very large momentum transfers Q.

In fact, the PQCD prediction for the pion form factor at large Q^2 is nominally of order $16\pi\alpha_s f_\pi^2$, a factor of order $16\pi^2$ times larger than the above estimate. The actual coefficient of the leading twist, leading power law term depends on the integral $\int_0^1 dx \, \frac{\phi_\pi(x,Q)}{(1-x)}$, and is thus only moderately sensitive to the shape of the meson distribution amplitude in the endpoint region.

The normalization and sign of the leading power law terms predicted by PQCD are in agreement with the measurements of the meson and baryon form factors as well as large invariant mass exclusive photon-photon meson pair production cross sections if one uses the hadron distribution amplitudes predicted by Chernyak et al.[40] and Sachrajda and King[15] from QCD sum rules. As discussed in sec. 2, the recent lattice gauge theory analysis of the moments of the meson distribution amplitude by Martinelli and Sachrajda[17] give results consistent with those of Chernyak and Zhitnitsky.

It might also be noted that in QED, the "soft" contributions to the positronium form factor from Coulomb photon exchange are the *same* order in α as the "hard" contributions from transverse photon exchange. There are no extra powers of α in the hard amplitude! Once the electrons are relativistic, i.e., for $Q^2 \sim M^2$, the hard, perturbative contribution dominates.[45]

(2) Isgur and Llewellyn Smith argue that the momentum transfer flowing through the gluon propagator in the hard scattering amplitude for an exclusive reaction is typically too small to trust the perturbative expansion. This seems to be of particular concern for the skewed, highly relativistic distribution amplitudes obtained from the QCD sum rule analysis of Chernyak et al. since the integration region where x is large tends to be emphasized. In the case of the hard-scattering T_H amplitude for the pion form

factor (illustrated in fig. 13), the struck quark is off-shell at order $(1-x)Q^2$ whereas the momentum transfer of the exchanged gluon is of order $(1-x)(1-y)Q^2$, which can be considerably smaller.

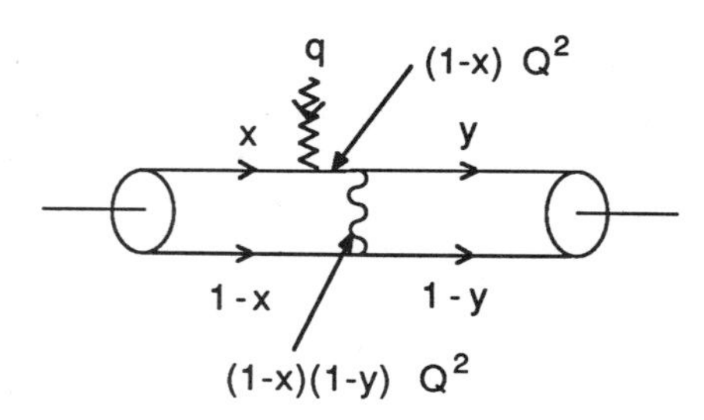

Fig. 13. Leading twist contribution to the meson form factor in QCD.

In fact, as shown by Lepage and myself,[28] the momentum transfer scale where one can analyze amplitudes perturbatively in QCD is controlled by the virtuality of the quark propagator not that of the exchanged gluon. [The range of the gluon virtuality is of course important in setting the scale of the effective coupling constant $\alpha_s(\bar{Q}^2)$.] If the struck quark is sufficiently off-shell, $|k_q^2| > \Lambda^2_{\rm QCD}$, one can easily show that multiple soft gluon exchange contributions are suppressed by powers of Q^2 relative to one-gluon exchange. The same considerations apply to the analysis of the evolution of deep inelastic structure functions: the critical scale is the off-shell virtuality of the quark propagators — not the minimum virtuality of the gluons. Even though the radiated gluons have low virtuality, one can compute the form of QCD evolution with elementary vector gluon couplings provided that the struck quark is sufficiently off-shell. Similarly, in computations of quark jet evolution, the perturbative gluon coupling dominates even though the gluon can be radiated near its mass shell. Requiring the gluon to have a minimum virtual mass corresponds to multiple jet production.

How can one reconcile the PQCD analysis with the concept that at low momentum transfer the interaction between quarks is nonperturbative? The concept of a nonperturbative potential (and estimates of scales involving the gluonium mass) can only be applicable to situations in which quarks are close to their mass shells and scatter at low relative velocity so that there is sufficient time to interact strongly. However, in the high momentum transfer form factor and deep inelastic scattering reactions, the struck quark is relatively far off its mass shell and interacts at high momentum relative to the spectator quarks. Thus its interactions may be computed perturbatively.

The above observations form the basis of the application of renormalization group equations and the operator product expansion to these reactions, and allows one to calculate the leading power behavior and the QCD logarithmic evolution of exclusive amplitudes for the pion form factor and $\gamma\gamma$ annihilation into meson pairs to all orders in perturbation theory.

The predictions[32] for the leading twist term in exclusive QCD hadronic amplitudes are thus unambiguous. Higher twist corrections to the quark and gluon propagator due to mass terms and intrinsic transverse momenta of a few hundred MeV give nominal corrections of higher order in $1/Q^2$. These finite mass corrections combine with the leading twist results to give a smooth approach to small Q^2. The PQCD scaling laws thus become valid at relatively low momentum transfer, the few GeV scale, consistent with what is observed in experiment, as in the results shown in figs. 1 and 9.[46]

(3) Independent of the underlying theory, the form factor of a hadron can be computed from the overlap of light-cone wavefunctions, summed over Fock states, as shown by Drell and Yan.[47] *This is the starting point for all relativistic calculations including the PQCD analysis. Isgur and Llewellyn Smith, and also Radyshkin, argue that one can obtain reasonable agreement with the form factor data by parameterizing the three-point vertex amplitude using various models for the bound state wavefunctions.*

However, phenomenological agreement with a parameterization of the vertex amplitude is not in contradiction with the PQCD analysis unless one can show that the QCD wavefunction with gluon exchange can be excluded in favor of purely nonperturbative forms. The analyses[48] of Dziembowski and Mankiewicz (which is consistent with QCD sum rules), Carlson and Gross, and Jacob and Kisslinger show that strictly soft wavefunctions, consistent with rotational invariance in the rest frame, and normalized correctly, cannot account for the pion or proton form factors in the power-law scaling regime.

Perhaps the most compelling evidence for the validity of the PQCD approach to exclusive processes is the observation[34] of color transparency in pp quasielastic scattering in nuclei, as discussed in sec. 4. The BNL data exclude models in which the scattering is dominated by soft wavefunctions.

The perturbative QCD predictions for the leading twist power-law contributions are generally consistent with data for exclusive processes when the momentum transfer exceeds several GeV.[49] It is difficult to understand the claim that these data are explained by higher twist or soft nonperturbative contributions since such effects necessarily fall at least one power of Q^2 faster than the dimensional counting prediction.

7. SUMMARY OF QCD PREDICTIONS FOR $\bar{\text{p}}$p EXCLUSIVE PROCESSES

Dimensional counting rules[50] give a direct connection between the degree of hadron compositeness and the power-law fall of exclusive scattering amplitudes at fixed center of mass angle: $M \sim Q^{4-n}F(\theta_{\rm cm})$ where n is the minimum number of initial and final state quanta. This rule gives the QCD prediction for the nominal power law scaling, modulo corrections from the logarithmic behavior of α_s, the distribution amplitude and small power-law corrections from Sudakov–suppressed Landshoff multiple scattering contributions. For $\bar{p}p$ one predicts

$$\frac{d\sigma}{d\Omega}\,(\bar{p}p \to \gamma\gamma) \simeq \frac{\alpha^2}{(p_T^2)^5}\, f^{\gamma\gamma}(\cos\theta, \ell n p_T)$$

$$\frac{d\sigma}{d\Omega}\,(\bar{p}p \to \gamma M) \simeq \frac{\alpha^2}{(p_T^2)^6}\, f^{\gamma M}(\cos\theta, \ell n p_T)$$

$$\frac{d\sigma}{d\Omega}\,(p\bar{p} \to M\overline{M}) \simeq \frac{1}{(p_T^2)^7}\, f^{M\overline{M}}(\cos\theta, \ell n p_T)$$

$$\frac{d\sigma}{d\Omega}\,(p\bar{p} \to B\overline{B}) \simeq \frac{1}{(p_T^2)^9}\, f^{B\overline{B}}(\cos\theta, \ell n p_T) \quad .$$

The angular dependence reflects the structure of the hard-scattering perturbative T_H amplitude, which in turn follows from the flavor pattern of the contributing duality diagrams. For example, a minimally-connected diagram such as that illustrated in fig. 14 is approximately characterized[51] as

$$T_H \sim \frac{1}{t^2}\,\frac{1}{s}\,\frac{1}{u} \quad .$$

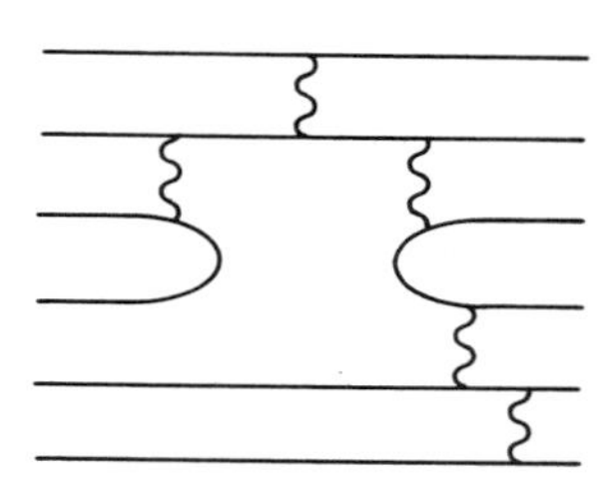

Fig. 14. A perturbative contribution to the hard-scattering amplitude in nucleon-nucleon collisions.

We emphasize that comparisons between channels related by crossing of the Mandelstam variables places a severe constraint on the angular dependence and analytic form of the underlying QCD exclusive amplitude. For example, it is possible to measure and compare

$$\bar{p}p \to \gamma\gamma \ : \ \gamma p \to \gamma p \ : \ \gamma\gamma \to \bar{p}p$$

$$\bar{p}p \to \gamma\pi^0 \ : \ \gamma p \to \pi^0 p \ : \ \pi^0 p \to \gamma p \ .$$

SLAC measurements[52] of the $\gamma p \to \pi^+ n$ cross section at $\theta_{CM} = \pi/2$ are consistent with the normalization and scaling (see fig. 15) $d\sigma/dt\ (\gamma p \to \pi^+ n) \simeq [1\,\text{nb}/(s/10\ \text{GeV})^7]\ f(t/s)$. We thus expect similar normalization and scaling for $d\sigma/dt\ (\bar{p}p \to \gamma\pi^0)$; all angle measurements up to $s \lesssim 15\ \text{GeV}^2$ appear possible given a high luminosity $\bar{p}$ beam.

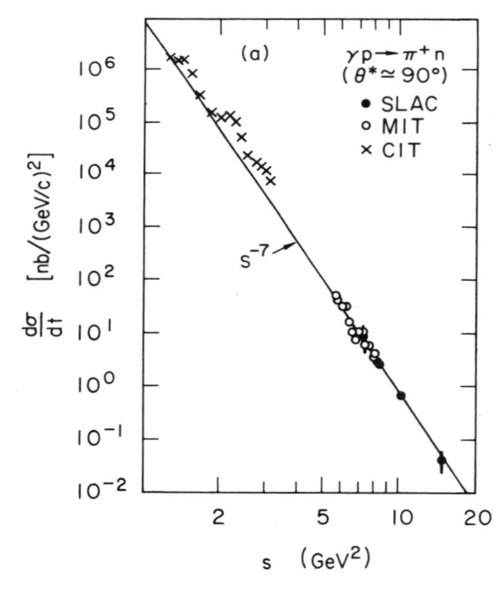

Fig. 15. Comparison of photoproduction data with the dimensional counting power-law prediction. The data are summarized in ref. 52.

Extensive measurements of the $pp \to pp$ cross section have been made at ANL, BNL and other laboratories.[51] The fixed-angle data on a log-log plot (see fig. 16) appears consistent with the nominal $s^{-10} f(\theta_{CM})$ dimensional counting production. However, as emphasized by Hendry,[53] the $s^{10} d\sigma/dt$ cross section exhibits oscillatory behavior with

p_T. Even more serious is the fact that polarization measurements[54] show significant spin-spin correlations (A_{NN}), and the single spin asymmetry (A_N) is not consistent with predictions based on hadron helicity conservation (see sec. 6) which is expected to be valid for the leading power behavior.[55] Recent discussions of these effects have been given by Farrar[56] and Lipkin.[57] I discuss a new explanation of all of these effects in sec. 9. Clearly, $\bar{p}p \to \bar{p}p$ data in the large-angle, large-energy regime will also be helpful in clarifying these fundamental issues.

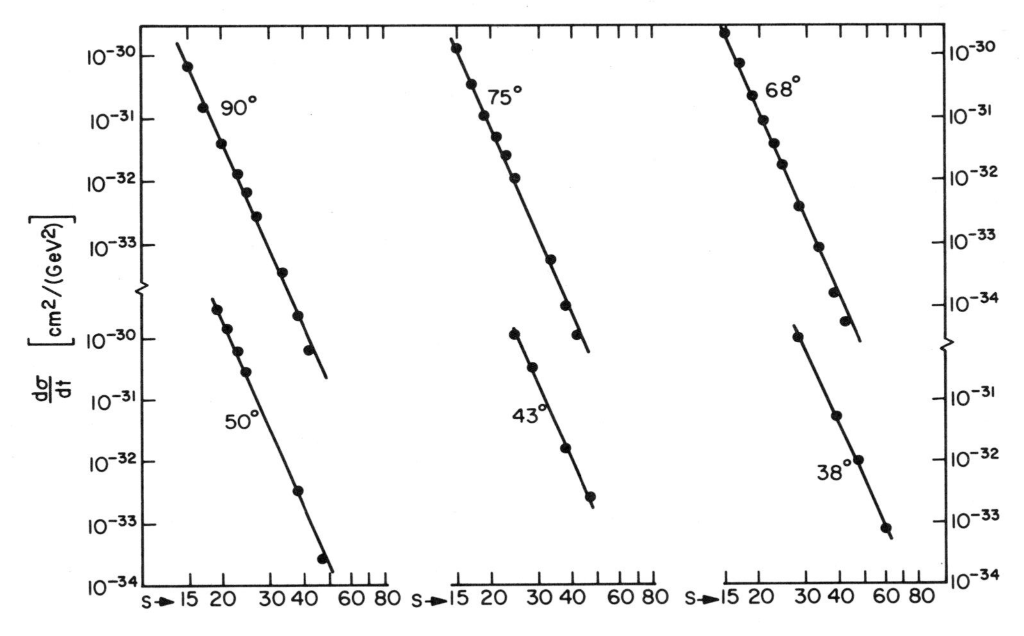

Fig. 16. Test of fixed θ_{CM} scaling for elastic pp scattering. The best fit gives the power $N = 9.7 \pm 0.5$ compared to the dimensional counting prediction N=10. Small deviations are not readily apparent on this log-log plot. The compilation is from Landshoff and Polkinghorne.

The simplest exclusive channels accessible to a $\bar{p}p$ facility are $\bar{p}p \to e^+e^-$, $\mu^+\mu^-$, $\tau^+\tau^-$ which to leading order in α provides a direct measurement of the Dirac and Pauli time-like proton form factor. The θ_{CM} angular dependence can be used to separate F_2 and F_1 and check the basic prediction,[2] $F_2(s)/F_1(s) \sim M^2/s$.

As discussed in sec. 3, perturbative QCD predicts asymptotic scaling of the form[2] $s^2 F_1(s) \sim f(\ell n\, s)$. A high-luminosity $\bar{p}$ facility could push time-like measurements of both form factors well beyond those available from e^+e^- storage rings. Since the normalization is similar to that of $p\bar{p} \to \gamma\gamma$, one should be able to measure the proton form factors out to center-of-mass energy squared as large as $s \sim 10\ \mathrm{GeV}^2$.

An important example of an exclusive process in QCD is the process $p\bar{p} \to \gamma\gamma$ as illustrated in fig. 17. To leading order in $1/p_T^2$,

$$\mathcal{M}_{p\bar{p}\to\gamma\gamma}(p_T^2, \theta_{CM}) = \int_0^1 [dx] \int_0^1 [dy] \phi_{\bar{p}}(x, p_T) T_H(qqq + \bar{q}\bar{q}\bar{q} \to \gamma\gamma) \phi_p(y, p_T) \quad ,$$

where $\phi_p(x, p_T)$ is the antiproton distribution amplitude and $T_H \sim \alpha_s^2(p_T^2)/(p_T^2)$ gives the scaling behavior of the minimally connected tree-graph amplitude for the two-photon annihilation of three quarks and three antiquarks collinear with the initial hadron directions. (See fig. 18.) QCD thus predicts

$$\frac{d\sigma}{d\Omega_{CM}}(p\bar{p}\to\gamma\gamma)\simeq\frac{\alpha_s^4(p_T^2)}{(p_T^2)^5}\,f(p_T,\theta_{CM},\ell n p_T^2)\ .$$

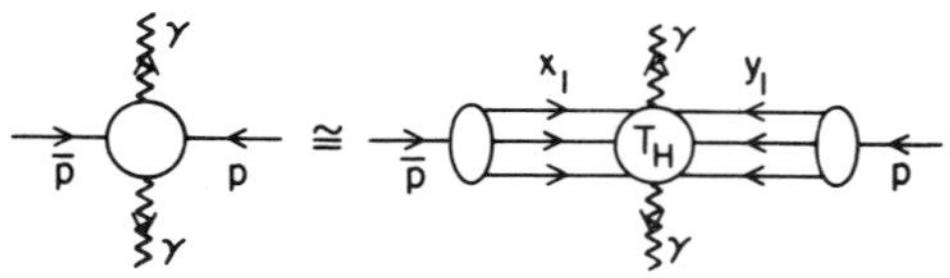

Fig. 17. *Application of QCD factorization to $\bar{p}p$ annihilation into photons.*

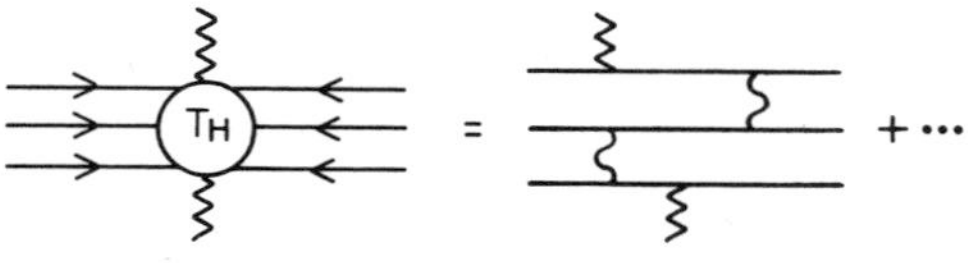

Fig. 18. *Example of a lowest-order perturbative contribution to T_H for the process $\bar{p}p\to\gamma\gamma$.*

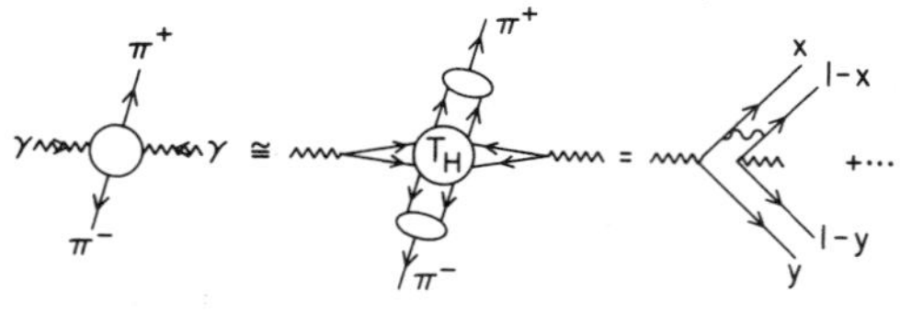

Fig. 19. *Application of QCD to two-photon production of meson pairs.*[58]

The complete calculations of the tree-graph structure (see figs. 19–21) of both $\gamma\gamma\to M\overline{M}$ and $\gamma\gamma\to B\overline{B}$ amplitudes has now been completed. One can use crossing to compute T_H for $p\bar{p}\to\gamma\gamma$ to leading order in $\alpha_s(p_T^2)$ from the calculations reported by Farrar, Maina and Neri[59] and Gunion and Millers.[60] Examples of the predicted angular distributions are shown in figs. 22 and 23. The region of applicability of the leading power-law results is presumed to be set by the scale where $Q^4G_M(Q^2)$ is roughly constant, i.e., $Q^2>3$ GeV^2. (See fig. 3.) Present two-photon collision measurements[61] are at energies too close to the $\bar{p}p$ threshold to meaningfully test the predictions.

As discussed in sec. 2, a model form for the proton distribution amplitude has been proposed by Chernyak and Zhitnitskii[4] based on QCD sum rules which leads to normalization and sign consistent with the measured proton form factor. (See fig. 2.) The CZ sum rule analysis has been recently corrected and modified by King and Sachrajda[62] but the final results are not known at this time. The CZ proton distribution

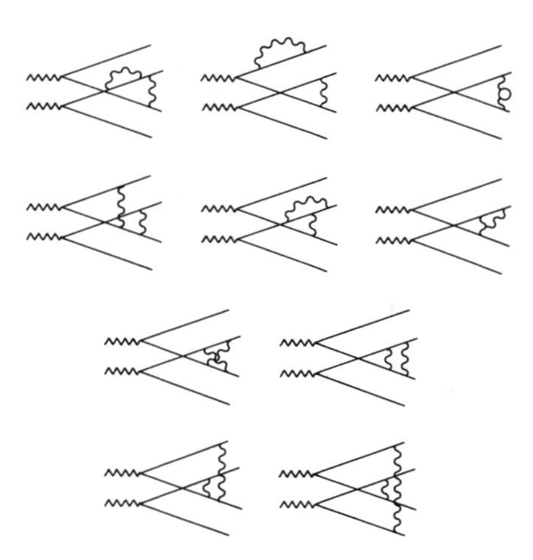

Fig. 20. Next-to-leading perturbative contribution to T_H for the process $\gamma\gamma \longrightarrow M\bar{M}$. The calculation has been done by Nizic.[58]

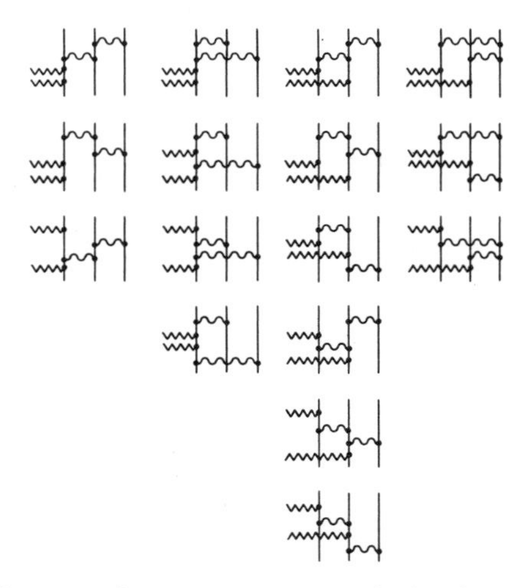

Fig. 21. Leading diagrams for $\gamma + \gamma \longrightarrow \bar{p} + p$ calculated in refs. 59 and 60.

amplitude yields predictions for $\gamma\gamma \to p\bar{p}$ in rough agreement with the experimental normalization, although the production energy is too low for a clear test. It should be noted that unlike meson pair production[58] the QCD predictions for baryons are highly sensitive to the form of the running coupling constant and the endpoint behavior of the wavefunctions.

The $\gamma^*\gamma^* \to \bar{B}B$ and $M\overline{M}$ amplitudes for off-shell photons have also been calculated by Millers and Gunion.[63] The results show important sensitivity to the form of the respective baryon and meson distribution amplitudes.The consequences of $|gg\rangle$ mixing in singlet mesons in $\gamma\gamma$ processes is discussed in ref. 64. It is possible that data from $p\bar{p}$ collisions at energies up to 10 GeV could greatly clarify the question of whether the perturbative QCD predictions are reliable at moderate momentum trans-

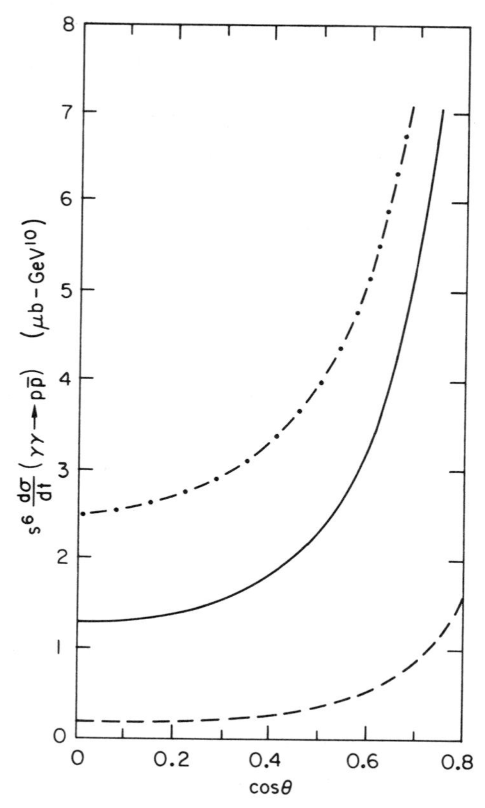

Fig. 22. QCD prediction for the scaling and angular distribution for $\gamma + \gamma \to \bar{p} + p$ *calculated by Farrar et al.*[59] *The dashed-dot curve corresponds to* $4\Lambda^2/s = 0.0016$ *and a maximum running coupling constant* $\alpha_s^{max} = 0.8$. *The solid curve corresponds to* $4\Lambda^2/s = 0.016$ *and a maximum running coupling constant* $\alpha_s^{max} = 0.5$. *The dashed curve corresponds to a fixed* $\alpha_s = 0.3$. *The results are very sensitive to the endpoint behavior of the proton distribution amplitude. The CZ form is assumed.*

fer.[65,66] As emphasized in sec. 4, an important check of the QCD predictions can be obtained by combining data from $p\bar{p} \to \gamma\gamma$, $\gamma\gamma \to p\bar{p}$ with large angle Compton scattering $\gamma p \to \gamma p$. This comparison checks in detail the angular dependence and crossing behavior expected from the theory. Furthermore, in $p\bar{p}$ collisions one can even study time-like photon production into e^+e^- and examine the virtual photon mass dependence of the Compton amplitude. Predictions for the q^2 dependence of the $p\bar{p} \to \gamma\gamma^*$ amplitude can be obtained by crossing the results of Gunion and Millers.[63]

8. HELICITY SELECTION RULE AND EXCLUSIVE CHARMONIUM DECAYS

One of the simplest predictions of perturbative QCD for exclusive processes including $p\bar{p}$ reactions is hadron helicity conservation: to leading order in $1/Q$, the total helicity of hadrons in the initial state must equal the total helicity of hadrons in the final state. This selection rule is independent of any photon or lepton spin appearing in the process. The result follows from (a) neglecting quark mass terms, (b) the vector coupling of gauge particles, and (c) the dominance of valence Fock states with zero angular momentum projection.[67]

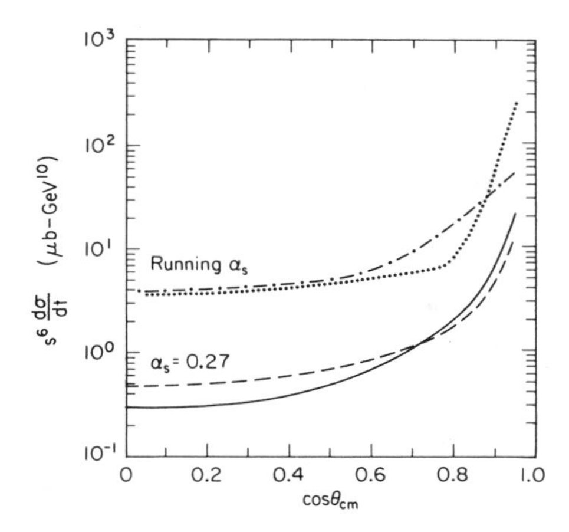

Fig. 23. QCD prediction for the scaling and angular distribution for $\gamma + \gamma \rightarrow \bar{p} + p$ calculated by Gunion, Sparks and Millers.[60] *CZ distribution amplitudes*[4] *are assumed. The solid and running curves are for real photon annihilation. The dashed and dot-dashed curves correspond to one photon space-like, with $Q_b^2/s = 0.1$.*

Hadron helicity conservation may be relevant to an interesting puzzle concerning the exclusive decays J/ψ and $\psi' \rightarrow \rho\pi, K^*\overline{K}$ and possibly other Vector-Pseudoscalar (VP) combinations. One expects $J/\psi(\psi')$ to decay to hadrons via three gluons or, occasionally, via a single direct photon. In either case the decay proceeds via $|\Psi(0)|^2$, where $\Psi(0)$ is the wave function at the origin in the nonrelativistic quark model for $c\bar{c}$. Thus it is reasonable to expect on the basis of perturbative QCD, that for any final hadronic state h:

$$Q_h \equiv \frac{B(\psi' \rightarrow h)}{B(J/\psi \rightarrow h)} \cong \frac{B(\psi' \rightarrow e^+e^-)}{B(J/\psi \rightarrow e^+e^-)} = 0.135 \pm 0.023 \quad .$$

Usually this is true, as is well documented in ref. 68 for $p\bar{p}\pi^0$, $2\pi^+2\pi^-\pi^0$, $\pi^+\pi^-\omega$, and $3\pi^+3\pi^-\pi^0$, hadronic channels. The startling exceptions occur for $\rho\pi$ and $K^*\overline{K}$ where the present experimental limits[68] are $Q_{\rho\pi} < 0.0063$ and $Q_{K^*\overline{K}} < 0.0027$. Recently San Fu Tuan, Peter Lepage and I[69] have proposed an explanation of the puzzle by assuming (a) the general validity of the perturbative QCD theorem that total hadron helicity is conserved in high-momentum transfer exclusive processes, but supplemented by (b) violation of the QCD theorem when the J/ψ decay to hadrons via three hard gluons is modulated by the gluons forming an intermediate gluonium state $\mathcal{O}$ before transition to hadrons. In essence the model of Hou and Soni[70] takes over in this latter stage.

Since the vector state $\mathcal{O}$ has to be produced with helicity $\lambda = \pm 1$, the VP decays should be suppressed by a factor $1/s$ in the rate. The ψ' seems to respect this rule. The J/ψ does *not*, and that is the mystery. Put in more quantitative terms, we expect on the basis of perturbative QCD[67] $Q_{\rho\pi} \equiv [B(\psi' \rightarrow \rho\pi)/B(J/\psi \rightarrow \rho\pi)] \sim [M_{J/\psi}/M_{\psi'}]^6$ assuming quark helicity is conserved in strong interactions. This includes a form factor suppression proportional to $[M_{J/\psi}/M_{\psi'}]^4$. This suppression is not nearly large enough, though, to account for the data.

One can question the validity of the QCD helicity conservation theorem at the charmonium mass scale. Helicity conservation has received important confirmation in $J/\psi \to p\overline{p}$ where the angular distribution is known experimentally to follow $[1 + \cos^2\theta]$ rather than $\sin^2\theta$ for helicity flip. The ψ' decays clearly respect hadron helicity conservation. It is difficult to understand how the J/ψ could violate this rule since the J/ψ and ψ' masses are so close. Corrections from quark mass terms, soft gluon corrections and finite energy corrections would not be expected to lead to large J/ψ differences. It is hard to imagine anything other than a resonant or interference effect that could account for such dramatic energy dependence.

A relevant violation of the QCD theorem which does have significance to this problem, is the recognition that the theorem is built on the underlying assumption of short-range "point-like" interactions among the constituents throughout. For instance, $J/\psi(c\overline{c}) \to 3g$ has a short-range $\cong 1/m_c$ associated with the short time scale of interaction. If, however, subsequently the three gluons were to resonate forming a gluonium state σ which has large transverse size $\cong 1/M_H$ covering an extended (long) time period, then the theorem is invalid. Note that even if the gluonium state $\mathcal{O}$ has large mass, close to $M_{J/\psi}$, its size could still be the standard hadronic scale of 1 fm, just as the case for the D-meson and B-mesons.

We have thus proposed, following Hou and Soni, that the enhancement of $J/\psi \to K^*\overline{K}$ and $J/\psi \to \rho\pi$ decay modes is caused by a quantum mechanical mixing of the J/ψ with a $J^{PC} = 1^{--}$ vector gluonium state $\mathcal{O}$ which causes the breakdown of the QCD helicity theorem. The decay width for $J/\psi \to \rho\pi(K^*\overline{K})$ via the sequence $J/\psi \to \mathcal{O} \to \rho\pi(K^*\overline{K})$ must be substantially larger than the decay width for the (non-pole) continuum process $J/\psi \to$ 3 gluons $\to \rho\pi(K^*\overline{K})$. In the other channels (such as $p\overline{p}, p\overline{p}\pi^0, 2\pi^+2\pi^-\pi^0$, etc.), the branching ratios of the $\mathcal{O}$ must be so small that the continuum contribution governed by the QCD theorem dominates over that of the $\mathcal{O}$ pole. For the case of the ψ' the contribution of the $\mathcal{O}$ pole must always be inappreciable in comparison with the continuum process where the QCD theorem holds. The experimental limits on $Q_{\rho\pi}$ and $Q_{K^*\overline{K}}$ are now substantially more stringent than when Hou and Soni made their estimates of M_V, $\Gamma_{\mathcal{O}\to\rho\pi}$ and $\Gamma_{\mathcal{O}\to K^*\overline{K}}$ in 1982.

It is interesting, indeed, that the existence of such a gluonium state $\mathcal{O}$ was first postulated by Freund and Nambu[71] based on OZI dynamics soon after the discovery of the J/ψ and ψ' mesons. In fact, Freund and Nambu predicted that the $\mathcal{O}$ would decay copiously precisely into $\rho\pi$ and $K^*\overline{K}$ with severe suppression of decays into other modes like e^+e^- as required for the solution of the puzzle.

Final states h which can proceed only through the intermediate gluonium state satisfy the ratio:

$$Q_h = \frac{B(\psi' \to e^+e^-)}{B(J/\psi \to e^+e^-)} \frac{(M_{J/\psi} - M_{\mathcal{O}})^2 + \frac{1}{4}\Gamma_0^2}{(M_{\psi'} - M_{\mathcal{O}})^2 + \frac{1}{4}\Gamma_0^2} .$$

We have assumed that the coupling of the J/ψ and ψ' to the gluonium state scales as the e^+e^- coupling. The value of Q_h is small if the $\mathcal{O}$ is close in mass to the J/ψ. Thus we require $(M_{J/\psi} - M_{\mathcal{O}})^2 + \frac{1}{4}\Gamma_{\mathcal{O}}^2 \lesssim 2.6\, Q_h$ GeV^2. The experimental limit for $Q_{K^*\overline{K}}$ then implies $[(M_{J/\psi} - M_{\mathcal{O}})^2 + \frac{1}{4}\Gamma_{\mathcal{O}}^2]^{1/2} \lesssim 80$ MeV. This implies $| M_{J/\psi} - M_{\mathcal{O}} | < 80$ MeV and $\Gamma_{\mathcal{O}} < 160$ MeV. Typical allowed values are $M_{\mathcal{O}} = 3.0$ GeV , $\Gamma_{\mathcal{O}} = 140$ MeV or $M_{\mathcal{O}} = 3.15$ GeV , $\Gamma_{\mathcal{O}} = 140$ MeV. Notice that the gluonium state could be either lighter or heavier than the J/ψ. The branching ratio of the $\mathcal{O}$ into a given channel must exceed that of the J/ψ.

It is not necessarily obvious that a $J^{PC} = 1^{--}$ gluonium state with these parameters would necessarily have been found in experiments to date. One must remember that though $\mathcal{O} \to \rho\pi$ and $\mathcal{O} \to K^*\overline{K}$ are important modes of decay, at a mass of order 3.1 GeV many other modes (all be it less important) are available. Hence, a total width $\Gamma_{\mathcal{O}} \cong 100$ to 150 MeV is quite conceivable. Because of the proximity of $M_{\mathcal{O}}$ to $M_{J/\psi}$, the most important signatures for an $\mathcal{O}$ search via exclusive modes $J/\psi \rightarrow K^*\overline{K}h$, $J/\psi \rightarrow \rho\pi h$; $h = \pi\pi, \eta, \eta'$, are no longer available by phase-space considerations. However, the search could still be carried out using $\psi' \rightarrow K^*\overline{K}h$, $\psi' \rightarrow \rho\pi h$; with $h = \pi\pi$, and η. Another way to search for $\mathcal{O}$ in particular, and the three-gluon bound states in general, is via the inclusive reaction $\psi' \rightarrow (\pi\pi) + X$, where the $\pi\pi$ pair is an isosinglet. The three-gluon bound states such as $\mathcal{O}$ should show up as peaks in the missing mass (i.e., mass of X) distribution.

Perhaps the most direct way to search for the $\mathcal{O}$ is to scan $\overline{p}p$ or e^+e^- annihilation at $\sqrt{s}$ within $\sim$ 100 MeV of the J/ψ, triggering on vector/pseudoscalar decays such as $\pi\rho$ or $\overline{K}K^*$.

The fact that the $\rho\pi$ and $K^*\overline{K}$ channels are strongly suppressed in ψ' decays but not in J/ψ decays clearly implies dynamics beyond the standard charmonium analysis. As we have shown, the hypothesis of a three-gluon state $\mathcal{O}$ with mass within $\cong$ 100 MeV of the J/ψ mass provides a natural, perhaps even compelling, explanation of this anomaly. If this description is correct, then the ψ' and J/ψ hadronic decays are not only confirming hadron helicity conservation (at the ψ' momentum scale) but are also providing a signal for bound gluonic matter in QCD. An alternative model by Chaichian Tornquist, based on nonperturbative exponential vertex functions, has also been proposed.

The production of heavy quark resonances $p\bar{p} \rightarrow \psi, \chi, \eta_c$, etc. can be analyzed in a systematic way in QCD using the exclusive amplitude formalism of ref. 2. Since quark helicity is conserved in the basic subprocesses to leading order, and the distribution amplitude is the azimuthal angle symmetric $L_z = 0$ projection of the valence hadron Fock wavefunction, total hadron helicity is conserved for $A + B \rightarrow C + D$:[55] $\lambda_A + \lambda_B = \lambda_C + \lambda_D$. The result is predicted to hold to all orders in $\alpha_s(Q^2)$. Thus an essential feature of the perturbative QCD is the prediction of hadron helicity conservation up to kinematical and dynamical corrections of order m/Q and $\langle\psi\bar{\psi}\rangle^{1/3}/Q$ where Q is the momentum transfer or heavy mass scale, m is the light quark mass, and $\langle\psi\bar{\psi}\rangle$ is a measure of nonperturbative effects due to chiral symmetry breaking of the QCD vacuum. Applying this prediction to $p\bar{p}$ annihilation, one predicts $\lambda_p + \lambda_{\bar{p}} = 0$, i.e., $S_z = J_z = \pm 1$ is the leading amplitude for heavy resonance production.[55] Thus the ψ is expected to be produced with $J_z = \pm 1$, whereas the χ and η_c cross sections should be suppressed, at least to leading power in the heavy quark mass. The analogous tests in e^+e^- annihilation appear to be verified for ψ' decays but not the ψ. Hou and Soni[72] have suggested this effect may be due to the ψ mixing with $J = 1$ gluonium states. Antiproton-proton production of narrow resonances should be able to clarify these basic QCD issues.

9. SPIN CORRELATIONS, QCD COLOR TRANSPARENCY, AND HEAVY QUARK THRESHOLDS IN pp SCATTERING

One of the most serious challenges to quantum chromodynamics is the behavior of the spin-spin correlation asymmetry $A_{NN} = \frac{[d\sigma(\uparrow\uparrow) - d\sigma(\uparrow\downarrow)]}{[d\sigma(\uparrow\uparrow) + d\sigma(\uparrow\downarrow)]}$ measured in large momentum transfer pp elastic scattering. (See fig. 24.) At $p_{lab} = 11.75$ GeV/c and $\theta_{\rm cm} = \pi/2$, A_{NN} rises to $\simeq 60\%$, corresponding to four times more probability for protons to scatter with their incident spins both normal to the scattering plane and parallel, rather than normal and opposite.

The polarized cross section shows a striking energy and angular dependence not expected from the slowly-changing perturbative QCD predictions.[74] However, the unpolarized data is in first approximation consistent with the fixed angle scaling law $s^{10} d\sigma/dt(pp \rightarrow pp) = f(\theta_{CM})$ expected from the perturbative analysis. (See fig. 16.)

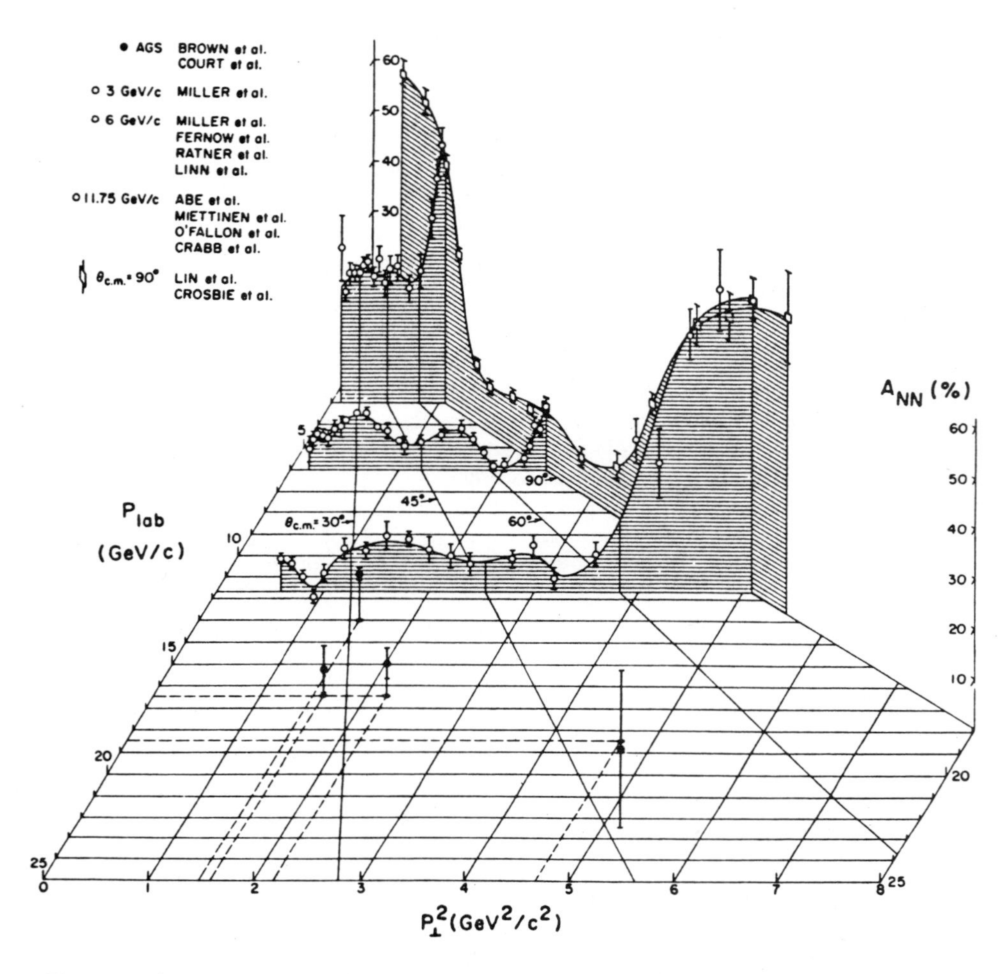

Fig. 24. The spin-spin correlation A_{NN} for elastic pp scattering with beam and target protons polarized normal to the scattering plane.[73] *(From ref. 73.) $A_{NN} = 60\%$ implies that it is four times more probable for the protons to scatter with spins parallel rather than antiparallel.*

The onset of new structure[75] at $s \simeq 23$ GeV^2 is a sign of new degrees of freedom in the two-baryon system. In this section, I will discuss an explanation by Guy de Teramond and myself[36] for (1) the observed spin correlations, (2) the deviations from fixed-angle scaling laws, and (3) the anomalous energy dependence of absorptive corrections to quasielastic pp scattering in nuclear targets, in terms of a simple model based on two $J = L = S = 1$ broad resonances (or threshold enhancements) interfering with a perturbative QCD quark-interchange background amplitude. The structures in the $pp \to pp$ amplitude may be associated with the onset of strange and charmed thresholds. If this view is correct, large angle pp elastic scattering would have been virtually featureless for $p_{lab} \geq 5$ GeV/c, had it not been for the onset of heavy flavor production. As a further illustration of the threshold effect, we also show the effect in A_{NN} due to a narrow $^3F_3$ pp resonance at $\sqrt{s} = 2.17$ GeV ($p_{lab} = 1.26$ GeV/c) associated with the $p\Delta$ threshold.

The perturbative QCD analysis[76] of exclusive amplitudes assumes that large momentum transfer exclusive scattering reactions are controlled by short distance quark-gluon subprocesses, and that corrections from quark masses and intrinsic transverse momenta can be ignored. The main predictions are fixed-angle scaling laws[77] (with small corrections due to evolution of the distribution amplitudes, the running coupling constant, and pinch singularities), hadron helicity conservation,[78] and the novel phenomenon, "color transparency."

As discussed in secs. 2 and 5, the power-law scaling quark-counting predictions for form factors, two-body elastic hadron-hadron scattering,[79] and exclusive two-photon reactions are generally consistent with experiment at transverse momenta beyond a few GeV. (See figs. 3, 9, 15 and 16). In leading order in $1/p_T$, only the lowest particle-number "valence" Fock state wavefunction with all the quarks within an impact distance $b_\perp \leq 1/p_T$ contributes to the high momentum transfer scattering amplitude in QCD. Such a Fock state component has a small color dipole moment and thus interacts only weakly with hadronic or nuclear matter.[33] This minimally interacting proton configuration can retain its small size as it propagates in the nucleus over a distance which grows with energy. Thus, unlike traditional Glauber theory, QCD predicts that large momentum transfer quasielastic reactions occurring in a nucleus suffer minimal initial and final state attenuation; i.e., one expects a volume rather than surface dependence in the nuclear number. This is the QCD "color transparency" prediction.

As discussed in the introduction, a test of color transparency in large momentum transfer quasielastic pp scattering at $\theta_{\rm cm} \simeq \pi/2$ has recently been carried out at BNL using several nuclear targets (C, Al, Pb).[34] The attenuation at $p_{lab} = 10$ GeV/c in the various nuclear targets was observed to be in fact much less than that predicted by traditional Glauber theory. (See fig. 11.) This appears to support the color transparency prediction. However at $p_{lab} = 12$ GeV/c, normal attenuation was observed, in contradiction to the expectation from perturbative QCD that the transparency effect should become even more apparent! Our observation is that one can explain this surprising result if the scattering at $p_{lab} = 12$ GeV/c ($\sqrt{s} = 4.93$ GeV), is dominated by an s-channel B=2 resonance (or resonance-like structure) with mass near 5 GeV, since unlike a hard-scattering reaction, a resonance couples to the fully-interacting large-scale structure of the proton. If the resonance has spin $S = 1$, this can also explain the large spin correlation A_{NN} measured nearly at the same momentum, $p_{lab} = 11.75$ GeV/c. Conversely, in the momentum range $p_{lab} = 5$ to 10 GeV/c we predict that the perturbative hard-scattering amplitude is dominant at large angles. The experimental observation of diminished attenuation at $p_{lab} = 10$ GeV/c thus provides support for the QCD description of exclusive reactions and color transparency.

What could cause a resonance at $\sqrt{s} = 5$ GeV, more than 3 GeV beyond the pp threshold? We can think of several possibilities: (a) a multigluonic excitation such as $|qqqqqqggg\rangle$, (b) a "hidden color" color singlet $|qqqqqq\rangle$ excitation,[80] or (c) a "hidden flavor" $|qqqqqqQ\overline{Q}\rangle$ excitation, which is the most interesting possibility, since it is so predictive. As in QED, where final state interactions give large enhancement factors for attractive channels in which $Z\alpha/v_{rel}$ is large, one expects resonances or threshold enhancements in QCD in color-singlet channels at heavy quark production thresholds since all the produced quarks have similar velocities.[81] One thus can expect resonant behavior at $M^* = 2.55$ GeV and $M^* = 5.08$ GeV, corresponding to the threshold values for open strangeness: $pp \to \Lambda K^+ p$, and open charm: $pp \to \Lambda_c D^0 p$, respectively. In any case, the structure at 5 GeV is highly inelastic: we find that its branching ratio to the proton-proton channel is $B^{pp} \simeq 1.5\%$.

We now proceed to a description of the model. We have purposely attempted not to overcomplicate the phenomenology; in particular, we have used the simplest Breit–Wigner parameterization of the resonances, and we have not attempted to optimize the parameters of the model to obtain a best fit. It is possible that what we identify as a single resonance is actually a cluster of resonances.

The background component of the model is the perturbative QCD amplitude. Although complete calculations are not yet available, many features of the QCD predictions are understood, including the approximate s^{-4} scaling of the $pp \to pp$ amplitude at fixed θ_{cm} and the dominance of those amplitudes that conserve hadron helicity.[78] Furthermore, recent data comparing different exclusive two-body scattering channels from BNL[79] show that quark interchange amplitudes[82] dominate quark annihilation or gluon exchange contributions. Assuming the usual symmetries, there are five independent pp helicity amplitudes: $\phi_1 = M(++,++)$, $\phi_2 = M(--,++)$, $\phi_3 = M(+-,+-)$, $\phi_4 = M(-+,+-)$, $\phi_5 = M(++,+-)$. The helicity amplitudes for quark interchange have a definite relationship.[74] For definiteness, we will assume the following form

$$\phi_1(\text{PQCD}) = 2\phi_3(\text{PQCD}) = -2\phi_4(\text{PQCD})$$

$$= 4\pi C F(t)F(u)[\frac{t - m_d^2}{u - m_d^2} + (u \leftrightarrow t)]e^{i\delta} \quad .$$

The hadron helicity nonconserving amplitudes, $\phi_2(\text{PQCD})$ and $\phi_5(\text{PQCD})$ are zero. This form is consistent with the nominal power-law dependence predicted by perturbative QCD[76] and also gives a good representation of the angular distribution over a broad range of energies.[83] Here $F(t)$ is the helicity conserving proton form factor, which for simplicity, we take as the standard dipole form, $F(t) = (1 - t/m_d^2)^{-2}$, with $m_d^2 = 0.71 \text{ GeV}^2$. As shown in ref. 74, the PQCD-quark-interchange structure alone predicts $A_{NN} \simeq 1/3$, nearly independent of energy and angle.

Because of the rapid fixed-angle s^{-4} falloff of the perturbative QCD amplitude, even a very weakly-coupled resonance can have a sizeable effect at large momentum transfer. The large empirical values for A_{NN} suggest a resonant $pp \to pp$ amplitude with $J = L = S = 1$ since this gives $A_{NN} = 1$ (in absence of background) and a smooth angular distribution. Because of the Pauli principle, an $S = 1$ di-proton resonances must have odd parity and thus odd orbital angular momentum. We parameterize the two non-zero helicity amplitudes for a $J = L = S = 1$ resonance in Breit–Wigner form:

$$\phi_3(\text{resonance}) = 12\pi \frac{\sqrt{s}}{p_{\text{cm}}} d^1_{1,1}(\theta_{\text{cm}}) \frac{\frac{1}{2}\Gamma^{pp}(s)}{M^* - E_{\text{cm}} - \frac{i}{2}\Gamma} \quad ,$$

$$\phi_4(\text{resonance}) = -12\pi \frac{\sqrt{s}}{p_{\text{cm}}} d^1_{-1,1}(\theta_{\text{cm}}) \frac{\frac{1}{2}\Gamma^{pp}(s)}{M^* - E_{\text{cm}} - \frac{i}{2}\Gamma} \quad .$$

(The ${}^3F_3$ resonance amplitudes have the same form with $d^3_{\pm1,1}$ replacing $d^1_{\pm1,1}$.) Since we are far from threshold, threshold factors in the pp channel can be treated as constants. As in the case of a narrow resonance like the Z^0, we expect that the partial width into nucleon pairs is proportional to the square of the time-like proton form factor: $\Gamma^{pp}(s)/\Gamma = B^{pp}|F(s)|^2/|F(M^{*2})|^2$, corresponding to the formation of two protons at this invariant energy. The resonant amplitudes then die away by one inverse power of $(E_{\text{cm}} - M^*)$ relative to the dominant PQCD amplitudes. (In this sense, they are higher twist contributions relative to the leading twist perturbative QCD amplitudes.) The model is thus very simple: each pp helicity amplitude ϕ_i is the coherent sum of PQCD plus resonance components: $\phi = \phi(\text{PQCD}) + \Sigma\phi(\text{resonance})$. Because of pinch singularities and higher-order corrections, the hard QCD amplitudes are expected to have a nontrivial phase;[84] we have thus allowed for a constant phase δ in $\phi(\text{PQCD})$. Because of the absence of the ϕ_5 helicity-flip amplitude, the model predicts zero single spin asymmetry A_N. This is consistent with the large angle data at $p_{lab} = 11.75$ GeV/c.[85]

At low transverse momentum, $p_T \leq 1.5$ GeV, the power-law fall-off of $\phi(\text{PQCD})$ in s disagrees with the more slowly falling large-angle data, and we have little guidance from basic theory. Our interest in this low-energy region is to illustrate the effects of resonances and threshold effects on A_{NN}. In order to keep the model tractable, we have simply extended the background quark interchange and the resonance amplitudes at low energies using the same forms as above but replacing the dipole form factor by a phenomenological form $F(t) \propto e^{-1/2\beta\sqrt{|t|}}$. We have also included a kinematic factor of $\sqrt{s}/2p_{\rm cm}$ in the background amplitude. The value $\beta = 0.85$ GeV^{-1} then gives a good fit to $d\sigma/dt$ at $\theta_{\rm cm} = \pi/2$ for $p_{lab} \leq 5.5$ GeV/c.[86] The normalizations are chosen to maintain continuity of the amplitudes.

The predictions of the model and comparison with experiment are shown in figs. 25–30. The following parameters are chosen: $C = 2.9 \times 10^3$, $\delta = -1$ for the normalization and phase of ϕ(PQCD). The mass, width and pp branching ratio for the three resonances are $M_d^* = 2.17$ GeV, $\Gamma_d = 0.04$ GeV, $B_d^{pp} = 1$; $M_s^* = 2.55$ GeV, $\Gamma_s = 1.6$ GeV, $B_s^{pp} = 0.65$; and $M_c^* = 5.08$ GeV, $\Gamma_c = 1.0$ GeV, $B_c^{pp} = 0.0155$, respectively. As shown in figs. 25 and 26, the deviations from the simple scaling predicted by the PQCD amplitudes are readily accounted for by the resonance structures. The cusp which appears in fig. 26 marks the change in regime below $p_{lab} = 5.5$ GeV/c where PQCD becomes inapplicable. It is interesting to note that in this energy region normal attenuation of quasielastic pp scattering is observed.[34] The angular distribution (normalized to the data at $\theta_{\rm cm} = \pi/2$) is predicted to broaden relative to the steeper perturbative QCD form, when the resonance dominates. As shown in fig. 27 this is consistent with experiment, comparing data at $p_{lab} = 7.1$ and 12.1 GeV/c.

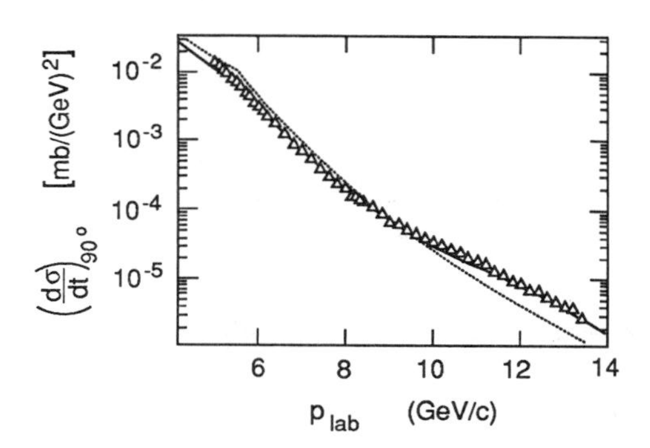

Fig. 25. Prediction (solid curve) for $d\sigma/dt(pp \to pp)$ at $\theta_{\rm cm} = \pi/2$ compared with the data of Akerlof et al.[86] The dotted line is the background PQCD prediction.

The most striking test of the model is its prediction for the spin correlation A_{NN} shown in fig. 28. The rise of A_{NN} to $\simeq 60\%$ at $p_{lab} = 11.75$ GeV/c is correctly reproduced by the high energy J=1 resonance interfering with ϕ(PQCD). The narrow peak which appears in the data of fig. 28 corresponds to the onset of the $pp \to p\Delta(1232)$ channel which can be interpreted as a $uuuuddq\bar{q}$ resonant state. Because of spin-color statistics one expects in this case a higher orbital momentum state, such as a $pp\ {}^3F_3$ resonance. The model is also consistent with the recent high-energy data point for A_{NN} at $p_{lab} = 18.5$ GeV/c and $p_T^2 = 4.7$ GeV^2 (see fig. 29). The data show a dramatic decrease of A_{NN} to zero or negative values. This is explained in our model by the destructive interference effects above the resonance region. The same effect accounts for the depression of A_{NN} for $p_{lab} \approx 6$ GeV/c shown in fig. 28. The comparison of the angular dependence of A_{NN} with data at $p_{lab} = 11.75$ GeV/c is shown in fig. 30. The agreement with the data[87] for the longitudinal spin correlation A_{LL} at the same p_{lab} is somewhat worse.

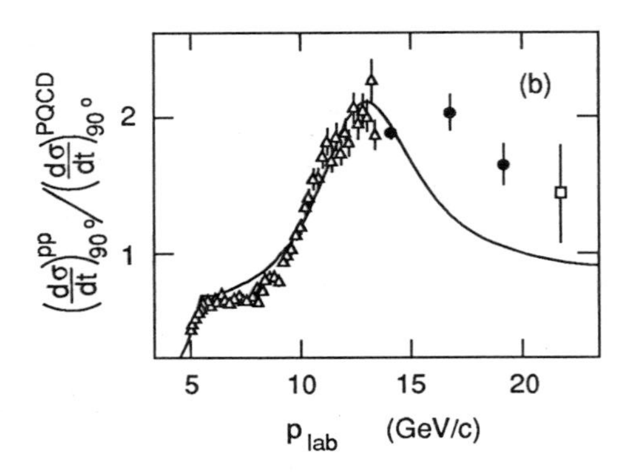

Fig. 26. Ratio of $d\sigma/dt(pp \rightarrow pp)$ at $\theta_{\rm cm} = \pi/2$ to the PQCD prediction. The data[86] *are from Akerlof et al. (open triangles), Allaby et al. (solid dots) and Cocconi et al. (open square). The cusp at $p_{lab} = 5.5$ GeV/c indicates the change of regime from PQCD.*

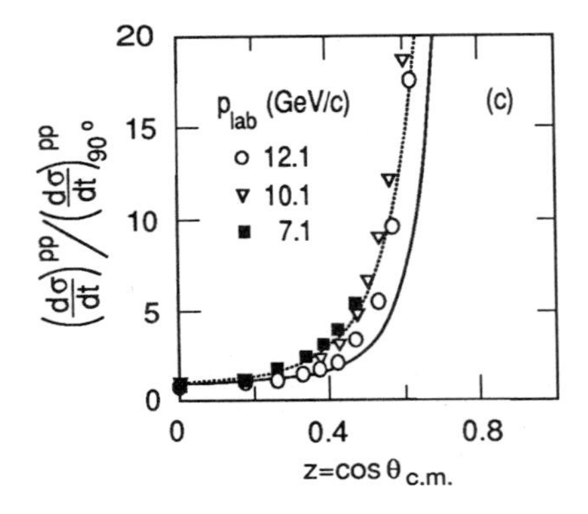

Fig. 27. The $pp \rightarrow pp$ angular distribution normalized at $\theta_{\rm cm} = \pi/2$. The data are from the compilation given in Sivers et al., ref. 79. The solid and dotted lines are predictions for $p_{lab} = 12.1$ and 7.1 GeV/c, respectively, showing the broadening near resonance.

Thus far we have not attempted a global fit to all the *pp* elastic scattering data, but rather have tried to show that many features can be naturally explained with only a few ingredients: a perturbative QCD background plus resonant amplitudes associated with rapid changes of the inelastic *pp* cross section. The model provides a good description of the s and t dependence of the differential cross section, including its "oscillatory" dependence[88] in s at fixed $\theta_{\rm cm}$, and the broadening of the angular distribution near the resonances. Most important, it gives a consistent explanation for the striking behavior of both the spin-spin correlations and the anomalous energy dependence of the attenuation of quasielastic *pp* scattering in nuclei. We predict that color transparency should reappear at higher energies ($p_{lab} \geq 16$ GeV/c), and also at smaller angles ($\theta_{\rm cm} \approx 60°$) at $p_{lab} = 12$ GeV/c where the perturbative QCD amplitude dominates. If the J=1 resonance structures in A_{NN} are indeed associated with heavy quark degrees of freedom, then the model predicts inelastic *pp* cross sections of the order of 1 mb and 1μb for the production of strange and charmed hadrons near their respective thresholds.[89] Thus a crucial test of the heavy quark hypothesis for explaining A_{NN}, rather than hidden color or gluonic excitations, is the observation of significant charm hadron production at $p_{lab} \geq 12$ GeV/c.

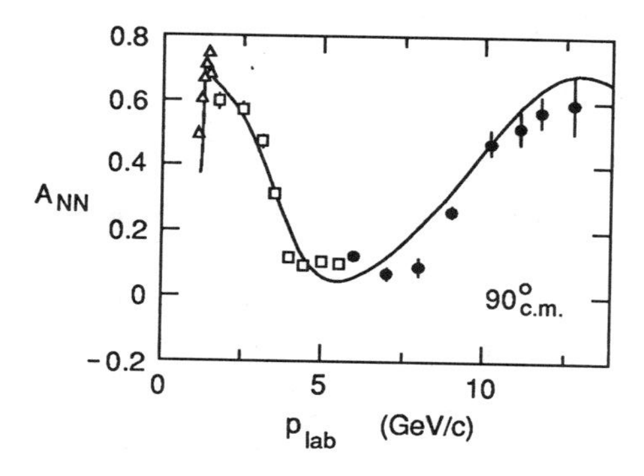

Fig. 28. A_{NN} *as a function of* p_{lab} *at* $\theta_{cm} = \pi/2$. *The data*[1] *are from Crosbie et al. (solid dots), Lin et al. (open squares) and Bhatia et al. (open triangles). The peak at* $p_{lab} = 1.26$ *GeV/c corresponds to the* $p\Delta$ *threshold. The data are well reproduced by the interference of the broad resonant structures at the strange* ($p_{lab} = 2.35$ *GeV/c) and charm* ($p_{lab} = 12.8$ *GeV/c) thresholds, interfering with a PQCD background. The value of* A_{NN} *from PQCD alone is 1/3.*

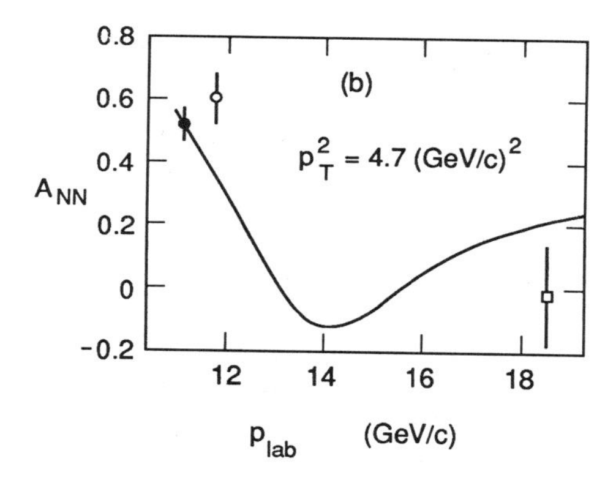

Fig. 29. A_{NN} *at fixed* $p_T^2 = 4.7$ *GeV/c)*^{2}. *The data point*[1] *at* $p_{lab} = 18.5$ *GeV/c is from Court et al.*

10. HEAVY HADRON PAIR PRODUCTION IN $p\bar{p}$ EXCLUSIVE REACTIONS

One of the few areas of high-energy phenomenology which is apparently not well-understood in perturbative QCD is the production of charmed hadrons. The simple fusion subprocesses $q\bar{q} \rightarrow Q\bar{Q}$ and $gg \rightarrow Q\bar{Q}$ are expected to dominate heavy quark inclusive reactions at least for very large M_Q.[90] However, in the case of charm production cross sections, the predictions for the energy and x_L dependence appear to contradict experiment.[91] It is possible that there are significant nonperturbative contributions to charm production such as the intrinsic heavy quark contributions[92] associated with loop interactions in the hadronic wavefunction, strong binding effects at low relative velocity,[93] and other nonperturbative or higher twist effects. A review of some of these issues is given in ref. 91.

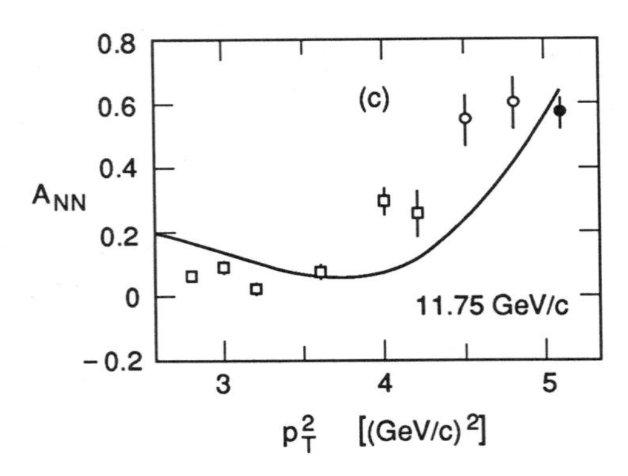

Fig. 30. A_{NN} as a function of transverse momentum. The data[1] *are from Crabb et al. (open circles) and O'Fallon et al. (open squares). Diffractive contributions should be included for $p_T^2 \leq 3$ GeV^2.*

Here we want to address the equally provocative question of heavy flavor production in exclusive $p\bar{p}$ reactions, e.g., $\bar{p}p \to \overline{\Lambda}_{\bar{Q}}\Lambda_Q$ where $Q = s, c, b$. The following arguments are heuristic, but they may give a guide to the expected scaling laws and features of these reactions.

Consider the diagram of fig. 31. for the $\bar{p}p \to \overline{\Lambda}_{\overline{Q}}\Lambda_Q$ matrix element. If the Λ's are produced in the forward direction with $p_T^2 \lesssim \mu^2 \sim (300 \text{ MeV})^2$ then there is maximal kinematic overlap for the light quarks between the initial and final light wavefunctions. The hard subprocess cross section $\bar{u}u \to c\bar{c}$ would normally give cross sections of order

$$\frac{d\sigma}{d\Omega} \sim \frac{\alpha_s^2(s) f(\Omega)}{s} \sim \frac{\alpha_s^2(4m_Q^2)}{4M_Q^2} f(\Omega) \quad ,$$

but the alignment restriction $p_T^2 < \mu^2$ gives an extra $\mu^2/4m_Q^2$ suppression in the angular integral. Therefore we expect the scaling law

$$\sigma(\bar{p}p \to \overline{\Lambda}_{\overline{Q}}\Lambda_Q) \sim \mu^2 \frac{\alpha_s^2(4M_Q^2)}{m_Q^4} F\left(\frac{4m_Q^2}{s}\right) \quad ,$$

i.e., $\overline{\Lambda}_{\bar{s}}\Lambda_s : \overline{\Lambda}_{\bar{c}}\Lambda_c : \overline{\Lambda}_{\bar{b}}\Lambda_b = 1 : 10^{-2}$ to $10^{-3} : 10^{-4}$ to 10^{-4} for $s \gg 4m_Q^2$ similar rates have been recently estimated in the "diquark" model of Kroll and collaborators. Thus it may not be hopeless to actually measure exclusive pairs of heavy charmed baryons in $\bar{p}p$ collisions. The above analysis can be readily extended to other heavy flavor baryon and meson pair exclusive cross sections. The issues are important for clarifying the OZI rule in QCD and the connection between exclusive and inclusive production mechanisms.

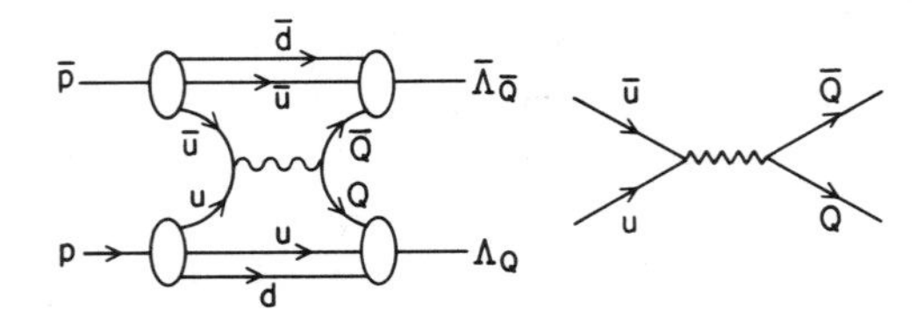

Fig. 31. Perturbative diagram for exclusive production of heavy baryon pairs in $\bar{p}p$ annihilation.

Mass corrections to QCD hard scattering amplitudes for a number of heavy quark production amplitudes have been computed. Exclusive pair production of heavy hadrons $|Q_1\overline{Q}_2\rangle$, $|Q_1Q_2Q_3\rangle$ consisting of higher generation quarks ($Q_i = t, b, c$ and possibly s) can be reliably predicted[94] within the framework of perturbative QCD, since the required wave function input is essentially determined from nonrelativistic considerations. The results can be applied to e^+e^- annihilation, $\gamma\gamma$ annihilation, and W and Z decay into higher generation pairs. The normalization, angular dependence and helicity structure can be predicted away from threshold, allowing a detailed study of the basic elements of heavy quark hadronization. A particularly striking feature of the QCD predictions is the existence of a zero in the form factor and e^+e^- annihilation cross section for zero-helicity hadron pair production close to a specific time-like value $q^2/4M_H^2 = m_h/2m_\ell$ where m_h and m_ℓ are the heavier and lighter quark masses, respectively. This zero reflects the destructive interference between the spin-dependent and spin-independent (Coulomb exchange) couplings of the gluon in QCD. In fact, all pseudoscalar meson form factors are predicted in QCD to reverse sign from space-like to time-like asymptotic momentum transfer because of their essentially monopole form. For $m_h > 2m_\ell$ the form factor zero occurs in the physical region. An interesting question is whether this type of numerator zero structure applies to the gluonic diagram amplitudes appropriate to $\bar{p}p$ reactions.

11. EXCLUSIVE NUCLEAR REACTIONS — REDUCED AMPLITUDES

An ultimate goal of QCD phenomenology is to describe the nuclear force and the structure of nuclei in terms of quark and gluon degrees of freedom. Explicit signals of QCD in nuclei have been elusive, in part because of the fact that an effective Lagrangian containing meson and nucleon degrees of freedom must be in some sense equivalent to QCD if one is limited to low-energy probes. On the other hand, an effective local field theory of nucleon and meson fields cannot correctly describe the observed off-shell falloff of form factors, vertex amplitudes, Z-graph diagrams, etc. because hadron compositeness is not taken into account.

The distinction between the QCD and other treatments of nuclear amplitudes is particularly clear in the reaction $\gamma d \to np$; i.e., photodisintegration of the deuteron at fixed center of mass angle. Using dimensional counting, the leading power-law prediction from QCD is simply $\frac{d\sigma}{dt}(\gamma d \to np) \sim \frac{1}{s^{11}} f(\theta_{\rm cm})$. Of course in the nuclear amplitude, the virtual momenta are partitioned among many quarks and gluons, so that finite mass corrections will be significant at low to medium energies. Nevertheless, there is an elegant way to test the basic QCD dynamics in these reactions taking into account much of the finite-mass, higher-twist corrections by using the "reduced amplitude" formalism.[95] The basic observation is that for vanishing nuclear binding energy $\epsilon_d \to 0$, the deuteron can be regarded as two nucleons sharing the deuteron four-momentum (see fig. 32). The momentum ℓ is limited by the binding and can thus be neglected. Thus the photodisintegration amplitude contains the probability amplitude (i.e., nucleon form factors) for the proton and neutron to each remain intact after absorbing momentum transfers $p_p - 1/2p_d$ and $p_n - 1/2p_d$, respectively. After the form factors are removed, the remaining "reduced" amplitude should scale as $f(\theta_{\rm cm}/p_T)$. The single inverse power of transverse momentum p_T is the slowest conceivable in any theory, but it is the unique power predicted by PQCD.

A test of the prediction that $f(\theta_{\rm cm})$ is energy dependent at high-momentum transfer is compared with experiment in fig. 33. It is particularly striking to see the QCD prediction verified at incident photon lab energies as low as 1 GeV. A comparison with a standard nuclear physics model with exchange currents is also shown for comparison in fig. 33(a). The fact that this prediction falls less fast than the data suggests that meson and nucleon compositeness are not taken to into account correctly.

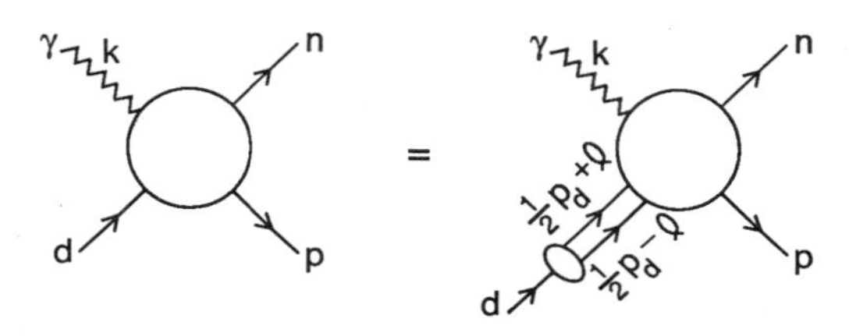

Fig. 32. Construction of the reduced nuclear amplitude for two-body inelastic deuteron reactions.[95]

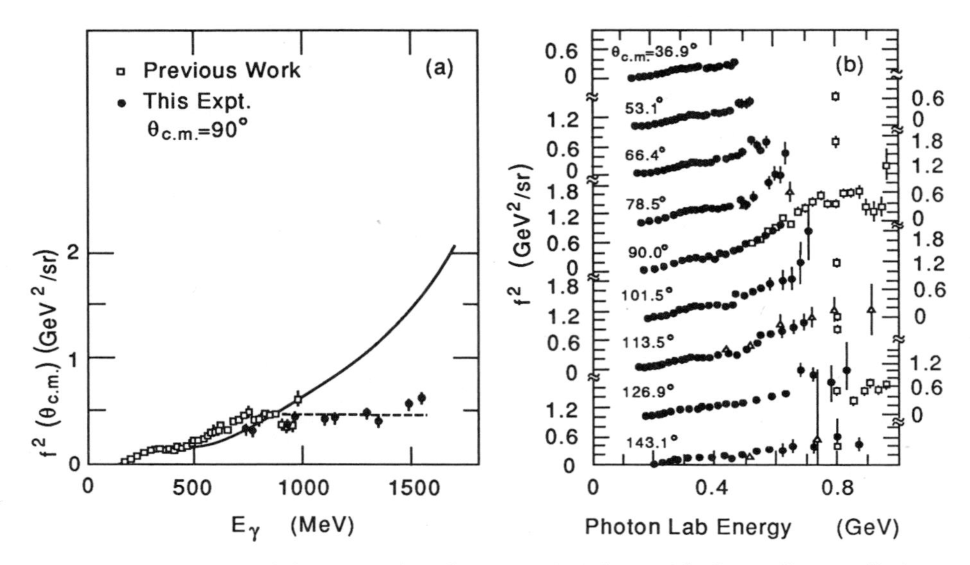

Fig. 33. Comparison of deuteron photodisintegration data with the scaling prediction which requires $f^2(\theta_{cm})$ to be at most logarithmically dependent on energy at large momentum transfer. The data in (a) are from the recent experiment of ref. 46. The nuclear physics prediction shown in (a) is from ref. 96. The data in (b) are from ref. 97.

There are a number of related tests of QCD and reduced amplitudes which require $\bar{p}$ beams.[98] including $\bar{p}d \to \gamma n$ and $\bar{p}d \to \pi^- p$ in the fixed $\theta_{\rm cm}$ region. These reactions are particularly interesting tests of QCD in nuclei. Dimensional counting rules predict the asymptotic behavior $\frac{d\sigma}{dt}\,(\bar{p}d \to \pi^- p) \sim \frac{1}{(p_T^2)^{12}}\, f(\theta_{\rm cm})$ since there are 14 initial and final quanta involved. One cannot expect the onset of such scaling laws until p_T is well into the multi-GeV regime since each hard propagator must carry significant momentum transfer. However, again one notes that the $\bar{p}d \to \pi^- p$ amplitude contains a factor representing the probability amplitude (i.e., form factor) for the proton to remain intact after absorbing momentum transfer squared $\hat{t} = (p - 1/2pd)^2$ and the $\bar{N}N$ time-like form factor at $\hat{s} = (\bar{p} + 1/2p_d)^2$. Thus $\mathcal{M}_{\bar{p}d\to\pi^- p} \sim F_{1N}(\hat{t})\; F_{1N}(\hat{s})\, \mathcal{M}_r$, where $\mathcal{M}_r$ has the same QCD scaling properties as quark meson scattering. We thus predict

$$\frac{\frac{d\sigma}{d\Omega}\,(\bar{p}d \to \pi^- p)}{F_{1N}^2(\hat{t})\, F_{1N}^2(\hat{s})} \sim \frac{f(\Omega)}{p_T^2}\ .$$

The analogous analysis of the deuteron form factor as defined in

$$\frac{d\sigma}{dt}(\ell d \to \ell d) = \left.\frac{d\sigma}{dt}\right|_{point} (F_d(Q^2))^2$$

yields a scaling law for the reduced form factor (see fig. 34):

$$f_d(Q^2) \equiv \frac{F_d(Q^2)}{F_{1N}\left(\frac{Q^2}{4}\right) F_{1N}\left(\frac{Q^2}{4}\right)} \sim \frac{1}{Q^2} \quad ,$$

i.e., the same scaling law as a meson form factor. As shown in fig. 35, this scaling is consistent with experiment for $Q = p_T \gtrsim 1$ GeV. We have also seen the evidence[46] for reduced amplitude scaling for $\gamma d \to pn$ at large angles and $p_T \gtrsim 1$ GeV. (see fig. 33). We thus expect similar precocious scaling behavior to hold for $\bar{p}d \to \pi^- p$ and other $\bar{p}d$ exclusive reduced amplitudes.

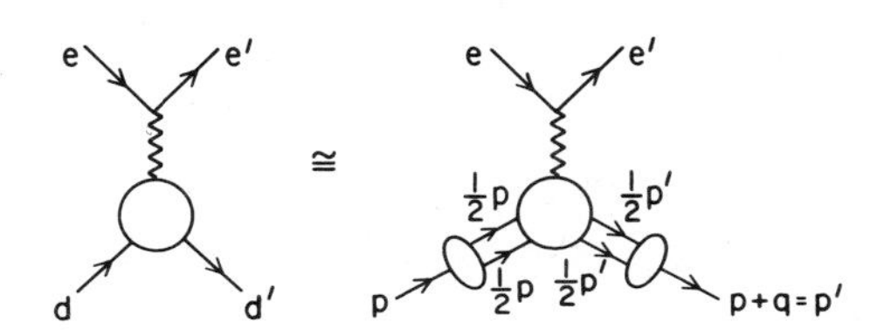

Fig. 34. Application of the reduced amplitude formalism to the deuteron form factor at large momentum transfer.

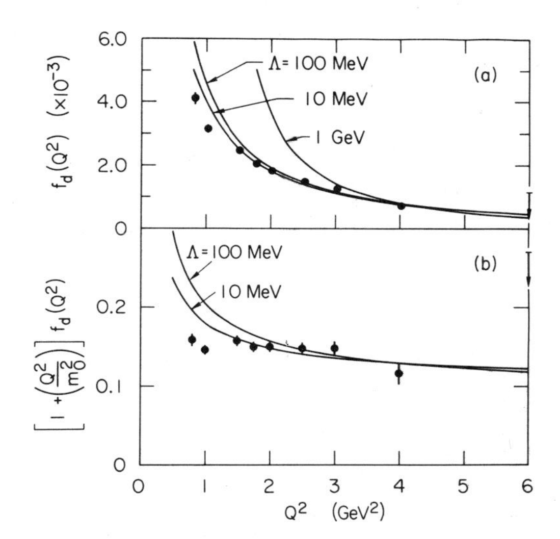

Fig. 35. Scaling of the deuteron reduced form factor. The data are summarized in ref. 95.

12. QUASI-EXCLUSIVE NUCLEAR PROCESSES

As discussed in sec. 2, one of the most novel features of QCD is "color transparency" which predicts a small absorption cross section for hadrons in specific kinematic configurations.[100] This concept can be tested in quasiexclusive antiproton-nuclear reactions. For large p_T one predicts

$$\frac{d\sigma}{dt\,dy}\left(\bar{p}A \to \pi^+\pi^- + (A-1)\right) \simeq \sum_{p\epsilon A} G_{p/A}(y)\,\frac{d\sigma}{dt}\left(\bar{p}p \to \pi^+\pi^-\right) \quad ,$$

where $G_{p/A}(y)$ is the probability distribution to find the proton in the nucleus with light-cone momentum fraction $y = (p^0 + p^z)/(p^0_A + p^z_A)$, and

$$\frac{d\sigma}{dt}(\bar{p}p \to \pi^+\pi^-) \simeq \left(\frac{1}{p_T^2}\right)^8 f(\cos\theta_{\rm cm}) \ .$$

The distribution $G_{p/A}(y)$ can be measured in $eA \to ep(A-1)$ quasiexclusive reactions. A remarkable feature of the above prediction is that there are no corrections required from initial-state absorption of the $\bar{p}$ as it traverses the nucleus, nor final-state interactions of the outgoing pions. The point is that the only part of hadron wavefunctions which is involved in the large p_T reaction is $\psi_H(b_\perp \sim \mathcal{O}(1/p_T))$. i.e., the amplitude where all the valence quarks are at small relative impact parameter. These configurations correspond to small color singlet states which, because of color cancellations, have negligible hadronic interactions in the target. Measurements of these reactions thus test a fundamental feature of the Fock state description of large p_T exclusive reactions.

Another interesting feature which can be probed in such reactions is the behavior of $G_{p/A}(y)$ for y well away from the Fermi distribution peak at $y \sim m_N/M_A$. For $y \to 1$ spectator counting rules[101] predict $G_{p/A}(y) \sim (1-y)^{2N_s-1} = (1-y)^{6A-7}$ where $N_s = 3(A-1)$ is the number of quark spectators required to "stop" ($y_i \to 0$) as $y \to 1$. This simple formula has been quite successful in accounting for distributions measured in the forward fragmentation of nuclei at the BEVALAC.[102]

13. COLOR TRANSPARENCY AND QUASIEXCLUSIVE J/ψ PRODUCTION IN p̄A COLLISIONS

Novel features of QCD, including color transparency, can be studied by measuring quasiexclusive J/ψ production by anti-protons in a nuclear target. We are particularly interested in the quasiexclusive annihilation process $\bar{p}A \to J/\psi(A-1)$ where the nucleus is left in a ground or excited state, but extra hadrons are not created. (See fig. 36.) The cross section involves a convolution of the $\bar{p}p \to J/\psi$ subprocess cross section with the distribution $G_{p/A}(y)$ where $y = (p^0+p^3)/(p^0_A+p^3_A)$ is the boost-invariant light-cone fraction for protons in the nucleus. This distribution can be determined from quasiexclusive lepton-nucleon scattering $\ell A \to \ell p(A-1)$.

In first approximation $\bar{p}p \to J/\psi$ involves $qqq+\bar{q}\bar{q}\bar{q}$ annihilation into three charmed quarks. The transverse momentum integrations are controlled by the charm mass scale and thus only the Fock state of the incident antiproton which contains three antiquarks at small impact separation can annihilate. Since this state has a relatively small color dipole moment it should have a longer than usual mean-free path in nuclear matter; i.e., "color transparency." Thus unlike traditional expectations, QCD predicts that the $\bar{p}p$ annihilation into charmonium is not restricted to the front surface of the nucleus. The exact nuclear dependence also depends on the formation time for the physical $\bar{p}$ to couple to the small $\bar{q}\bar{q}\bar{q}$ configuration, $\tau_F \propto E_p$. It may be possible to study the effect of finite formation time by varying the beam energy, E_p, and using the Fermi-motion of the nucleon to stay at the J/ψ resonance.

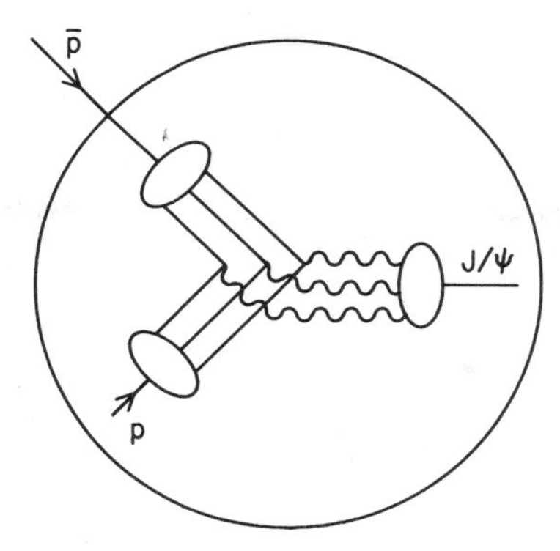

Fig. 36. Schematic representation of quasielastic charmonium production in $\bar{p}A$ reactions.

Since the J/ψ is produced at nonrelativistic velocities in this low energy experiment, it is formed inside the nucleus. The A-dependence of the quasiexclusive reaction can thus be used to determine the J/ψ-nucleon cross section at low energies. For a normal hadronic reaction $\bar{p}A \to HX$, we expect $A_{\text{eff}} \sim A^{1/3}$, corresponding to absorption in the initial and final state. In the case of $\bar{p}A \to J/\psi\ X$ we expect A_{eff} much closer to A^1 if color transparency is fully effective and $\sigma(J/\psi\ N)$ is small.

14. INCLUSIVE p̄ REACTIONS AND THE QCD CRITICAL LENGTH

The factorization structure of QCD implies that the structure functions and distribution amplitudes that control high momentum transfer reactions are process-independent. The proofs are highly nontrivial. In the case of inclusive massive lepton-pair production (The Drell–Yan process), the $\bar{p}p \to \ell\bar{\ell}X$ cross section to leading order in $1/Q^2$ takes the form (see fig. 37):

$$\frac{d\sigma}{dx_a dx_b d\Omega} = \frac{1}{3} \sum_{q\bar{q}} G_{\bar{q}/\bar{p}}(x_a, Q)\, G_{q/p}(x_b, Q)\, \frac{d\sigma}{d\Omega}\, (q\bar{q} \to \mu^+\mu^-)\ .$$

This factorization separates the long-distance (nonperturbative) dynamics contained in the universal-process independent structure functions $G_{q/p} = G_{\bar{q}/\bar{p}}$ from the short-distance perturbative physics contained in the subprocess $q\bar{q} \to \mu^+\mu^-$ cross section. Antiproton tests of this classic QCD prediction are crucial since the beam and target structure functions for the valence quark and antiquarks are measured directly in deep inelastic lepton scattering.

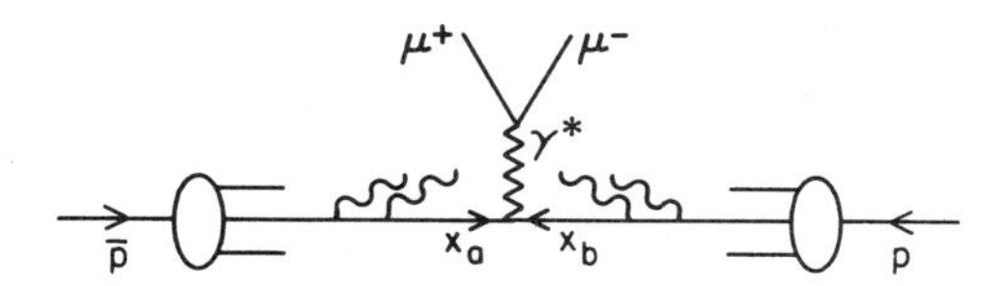

Fig. 37. Schematic representation of factorization of the Drell–Yan cross section in QCD.

Despite the simple form of the inclusive cross section, all-orders factorization for the Drell–Yan process has only just recently been analyzed to all orders in perturbation theory (by G. Bodwin[103] and J. Collins, D. Soper and G. Sterman.[104]) The most serious complications are due to the elastic and inelastic initial-state hadronic interactions which potentially could affect the color correlations, and momentum distribution of the annihilating q and $\bar{q}$. (See fig. 38.) Clearly such effects ruin factorization in a macroscopic target. In fact, as shown in ref. 26 a necessary condition to eliminate the initial state effects is that the incident parton energy must be large compared to a scale proportional to the length of the target. This translates into a necessary condition for factorization: $Q^2 = M_N L \mu^2 / x_b$. For a uranium target this implies that factorization can only be valid if the lepton pair mass is greater than a few GeV at large x_b. It is clearly interesting to study this phenomena experimentally, since it involves the transition between perturbative and soft dynamics and the propagation of antiquarks in nuclear matter. This important area of physics could be studied systematically using a low to medium energy $\bar{p}$ beam.

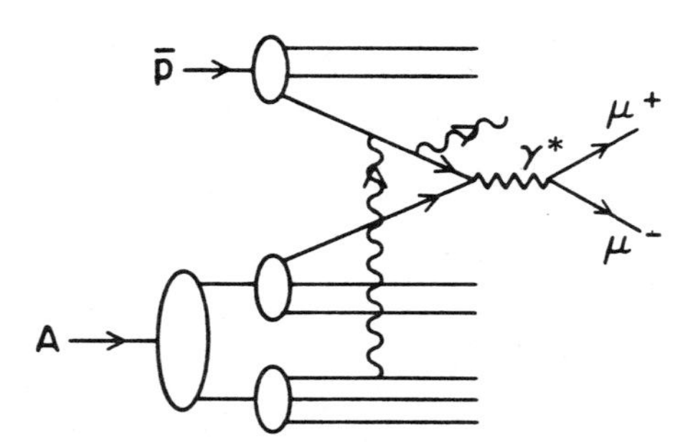

Fig. 38. Induced radiation from the interaction of the active antiquark with target spectators in the Drell–Yan process. The inelastic interactions are suppressed at parton energies which are large compared to a scale set by the length of the target.[26]

These results have striking implications for the interaction of the recoil quark jet in deep inelastic electron-nucleus scattering, and other inclusive reactions involving recoil jets. For the quark (and gluons) satisfying the length condition, there should be no extra radiation induced as the parton traverses the nucleus. However, low-energy gluons, emitted in the deep inelastic electron-quark collision, can suffer radiative losses, leading to cascading of soft particles in the nucleus. It is clearly very important to study this phenomena as a function of recoil quark energy and nuclear size. It should be emphasized that the absence of inelastic initial- or final-state collisions for high-energy partons does not preclude collision broadening due to elastic initial- or final-state interactions. The elastic corrections are unitary to leading order in $1/Q$ and do not effect the normalization of the deep inelastic cross section. Thus one predicts that the mean square transverse momentum of the recoil quark and its leading particles will increase as $A^{1/3}$.

The transverse momentum of the recoil quark reflects the intrinsic transverse momentum of the nucleon wave function. The EMC effect[106] implies that quarks in a nucleus have smaller average longitudinal momentum than in a nucleon.[105] Independent of the specific physical mechanism underlying the EMC effect,[105] the quarks in a nucleus would also be expected to have smaller transverse momentum. This effect can counteract to a certain extent the collision broadening of the outgoing jet.

Unlike the struck quark the remnant of the target system does not evolve with the probe momentum Q. However, the quantum numbers of the spectator system is $\overline{3}$ in color, so nonperturbative hadronization must occur. Since the transverse momentum of the leading particles in the spectator jet is not affected by the QCD radiative corrections, it more closely reflects the intrinsic transverse momentum of the hadron state.

It is also interesting to study the behavior of the transverse momentum of the quark and spectator jets as a function of x_{Bj}. For $x_{Bj} \sim 1$, the three-quark Fock state dominates the reaction. If the valence state has a smaller transverse size[23] than that of the nucleon, averaged over all of its Fock components, then one expects an increase of $\langle k_\perp^2 \rangle$ in that regime. Evidence for a significant increase of $\langle k_\perp^2 \rangle$ in the projectile fragmentation region at large quark momentum fractions has been reported by the SFM group[107] at the ISR for $pp \to$ di-jet $+X$ reactions.

It should be noted that the factorization proofs have not yet been extended to reactions such as $p\bar{p} \to$ Jet + Jet + X where the subprocess channel is not in a color singlet. In addition, due to infrared noncancellations, factorization is known to break down beyond leading twist.[108]

15. DISCRETIZED LIGHT–CONE QUANTIZATION

A central goal of QCD analysis is not only to obtain a complete description of the hadronic spectrum but also to evaluate their current matrix elements. Thus a key problem in the application of QCD to hadron and nuclear physics is how to determine the wavefunction of a relativistic multiparticle composite system. This is obviously a formidable task. Although composite systems in QCD can be represented formally in terms of the covariant Bethe–Salpeter formalism, calculations beyond ladder approximation appear intractable, and the ladder approximation itself is usually inadequate. For example, in order to derive the Dirac equation for the electron in a static Coulomb field from the Bethe–Salpeter equation for muonium with $m_\mu/m_e \to \infty$, one requires an infinite number of irreducible crossed-graph kernel contributions to the QED potential. Similarly, the matrix elements of currents and the wave function normalization also require, at least formally, the consideration of an infinite sum of irreducible kernels. The relative-time dependence of the Bethe Salpeter amplitudes for states with three or more constituent fields adds even more complexities.

A more intuitive procedure would be to extend the Schrödinger wavefunction description of bound states to the relativistic domain by developing a relativistic many-body Fock expansion for the hadronic state. Formally this can be done by quantizing QCD at equal time, and calculating matrix elements from the time-ordered expansion of the S-matrix. However, the calculation of each covariant Feynman diagram with n-vertices requires the calculation of $n!$ frame-dependent time-ordered amplitudes. Even worse, the calculation of the normalization of a bound state wave function (or the matrix element of a charge or current operator) requires the computation of contributions from all amplitudes involving particle production from the vacuum. (Note that even after normal ordering, the interaction Hamiltonian density for QED, $H_I = e : \overline{\psi}\gamma_\mu\psi A^\mu :$, contains contributions $b^\dagger d^\dagger a^\dagger$ which create particles from the perturbative vacuum.) For this reason, it is not possible to represent a relativistic field-theoretic bound system limited to a fixed number of constituents at a given time in a standard Hamiltonian framework since the interactions create new quanta from the vacuum.[109] Lorentz invariance is also difficult to incorporate in an equal time formalism.

Fortunately, there is a natural and consistent covariant framework, originally due to Dirac,[110] (quantization on the "light front ") for describing bound states in gauge theory analogous to the Fock state in nonrelativistic physics. This framework is the light-cone quantization formalism in which

$$|\pi\rangle = |q\bar{q}\rangle\,\psi^{\pi}_{q\bar{q}} + |q\bar{q}g\rangle\,\psi^{\pi}_{q\bar{q}g} + \cdots$$

$$|p\rangle = |qqq\rangle\,\psi^{p}_{qqq} + |qqqg\rangle\,\psi^{p}_{qqqg} + \cdots \quad .$$

Each wavefunction component ψ_n describes a state of fixed number of quark and gluon quanta evaluated in the interaction picture at equal light-cone "time" $\tau = t + z/c$. As discussed in the introduction, given the $\{\psi_n\}$, virtually any hadronic property can be computed, including anomalous moments, form factors, structure functions for inclusive processes, distribution amplitudes for exclusive processes, etc. As shown by Drell and Yan, space-like form factors are given by a simple overlap of the light-cone wavefunctions, summed over Fock states.[111] At high momentum transfer only the valence Fock-state enters, to leading order in $1/Q$.

As noted above, in an equal time formalism one must allow for fluctuations in which three or four particles appear with zero total three-momentum. In the light-cone formalism such fluctuations cannot appear since the total k^+ is conserved and each particle has to have positive k^+. Accordingly, the perturbative vacuum is an eigenstate of the total Hamiltonian on the light cone. Light-cone quantization and equal τ wave functions, rather than equal t wavefunctions, thus provides a sensible Fock state expansion. It also turns out to be convenient to use τ-ordered light-cone perturbation theory in place of covariant perturbation theory to analyze light cone dominated processes such as deep inelastic scattering and large momentum transfer exclusive reactions. Light-cone quantization and perturbation theory are developed in detail in ref. 23.

H. C. Pauli and I[20] have proposed a direct approach to solving QCD by attempting to diagonalize the light-cone Hamiltonian on a free particle discretized momentum Fock state basis. Since H_{LC}, P^+, $\vec{P}_\perp$, and the conserved charges all commute, H_{LC} is block diagonal. By choosing periodic (or antiperiodic) boundary conditions for the basis states along the negative light-cone $\psi(z^- = +L) = \pm\psi(z^- = -L)$, the Fock basis becomes restricted to finite dimensional representations. The eigenvalue problem thus reduces to the diagonalization of a finite Hermitian matrix. To see this, note that periodicity in z^- requires $P^+ = \frac{2\pi}{L}K$, $k_i^+ = \frac{2\pi}{L}\,n_i$, $\sum_{i=1}^{n} n_i = K$. The dimension of the representation corresponds to the number of partitions of the integer K as a sum of positive integers n. For a finite resolution K, the wavefunction is sampled at the discrete points

$$x_i = \frac{k_i^+}{P^+} = \frac{n_i}{K} = \left\{\frac{1}{K},\ \frac{2}{K},\ \ldots\frac{K-1}{K}\right\} \quad .$$

The continuum limit is clearly $K \to \infty$.

One can easily show that P^- scales as L. We thus define $P^- \equiv \frac{L}{2\pi}H$. The eigenstates with $P^2 = M^2$ at fixed P^+ and $\vec{P}_\perp = 0$ thus satisfy $H_{LC}\,|\Psi\rangle = KH\,|\Psi\rangle = M^2\,|\Psi\rangle$, independent of L (which corresponds to a Lorentz boost factor).

The basis of the DLCQ method is thus conceptually simple: one quantizes the independent fields at equal light-cone time τ and requires them to be periodic or antiperiodic in light-cone space with period $2L$. The commuting operators, the light-cone momentum $P^+ = \frac{2\pi}{L}K$ and the light cone energy $P^- = \frac{L}{2\pi}H$ are constructed explicitly in a Fock space representation and diagonalized simultaneously. The eigenvalues give the physical spectrum: the invariant mass squared $M^2 = P^\nu P_\nu$. The eigenfunctions give the wavefunctions at equal τ and allow one to compute the current matrix elements, structure functions, and distribution amplitudes required for physical processes. All of these quantities are manifestly independent of L, since $M^2 = P^+P^- = HK$. Lorentz-invariance is violated by periodicity, but re-established at the end of the calculation by going to the continuum limit: $L \to \infty$, $K \to \infty$ with P^+ finite. In the case of gauge theory, the use of the light-cone gauge $A^+ = 0$ eliminates negative metric states in both Abelian and non–Abelian theories.

Since continuum as well as single hadron color singlet hadronic wavefunctions are obtained by the diagonalization of H_{LC}, one can also calculate scattering amplitudes as well as decay rates from overlap matrix elements of the interaction Hamiltonian for the weak or electromagnetic interactions. An important point is that all higher Fock amplitudes including spectator gluons are kept in the light-cone quantization approach; such contributions cannot generally be neglected in decay amplitudes involving light quarks.

Eller, Pauli and I[112] have used DLCQ to obtain detailed results for the bound state and continuum spectrum and wavefunctions for QED in one-space and one-time dimension for arbitrary mass and coupling constant. I will give here only a brief discussion of the method. The commuting operators K, Q and $H = H_0 + V$ have the form

$$K = \sum n(b_n^\dagger b_n + d_n^\dagger d_n) + n(a_n^\dagger a_n)$$

$$Q = \sum (b_n^\dagger b_n - d_n^\dagger d_n)$$

$$H_0 = \sum \frac{m_\perp^2}{n}(b_n^\dagger b_n + d_n^\dagger d_n) + \frac{k_\perp^2}{n} a_n^\dagger a_n$$

$$V = \frac{g^2}{\pi} \sum_{n \neq m, k \neq \ell} b_k^\dagger b_\ell d_n^\dagger d_m \frac{\delta_{n+k,m+\ell}}{(n-m)^2} + \cdots \quad .$$

Only the one fermion antifermion (Abelian) interaction, corresponding to "instantaneous" gluon exchange, is displayed. The $Q = 0$ Fock state basis states are of the form $b_n^\dagger d_m^\dagger a_\ell^\dagger |0\rangle = |n; m; \ell\rangle$, $(n + m + \ell = K)$ where $|0\rangle$ is the perturbative vacuum. (Spin, color and transverse momentum for any number of dimensions are represented as extra internal variables.) We then solve $HK|\Psi\rangle = M^2|\Psi\rangle$ on the free particle basis $|\Psi\rangle = \sum_i C_i |i\rangle$. Note that the eigenvalues of H_{LC} give not only the bound state spectrum, but also all of the multiparticle scattering states with the same quantum numbers.

In the case of gauge theory in 3+1 dimensions, one also takes the $k_\perp^i = (2\pi/L_\perp)n_\perp^i$ as discrete variables on a finite cartesian basis. The theory is covariantly regulated if one restricts states by the condition

$$\sum_i \frac{k_{\perp i}^2 + m_i^2}{x_i} \leq \Lambda^2 \quad ,$$

where Λ is the ultraviolet cutoff. In effect, states with total light-cone kinetic energy beyond Λ^2 are cut off. In a renormalizable theory physical quantities are independent of physics beyond the ultraviolet regulator; the only dependence on Λ appears in the coupling constant and mass parameters of the Hamiltonian, consistent with the renormalization group.[113] The resolution parameters need to be taken sufficiently large such that the theory is controlled by the continuum regulator Λ, rather than the discrete scales of the momentum space basis.

The simplest application of DLCQ to local gauge theory is QED in one-space and one-time dimensions. Since $A^+ = 0$ is a physical gauge there are no photon degrees of freedom. Explicit forms for the matrix representation of H_{QED} are given in ref. 112. The basic interactions which occur in H_{LC}(QCD) are illustrated in fig. 39.

For the general case $m^2 \neq 0$, $(\mathrm{QED})_{1+1}$ can be solved by numerical diagonalization. The complete charge-zero spectrum (normalized to the ground state mass) for $K = 16$ is shown as a function of coupling constant in fig. 40. Since the physics can only depend

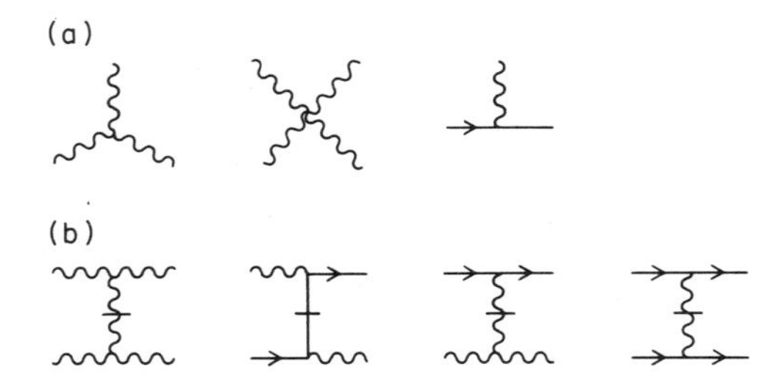

Fig. 39. Diagrams which appear in the interaction Hamiltonian for QCD on the light cone. The propagators with horizontal bars represent instantaneous gluon and quark exchange which arise from reduction of the dependent fields in $A^+ = 0$ gauge.

on the ratio m/g, it is convenient to introduce the parametrization $\lambda = \sqrt{\frac{1}{1+\pi(m/g)^2}}$ which maps the entire range of m and g onto the finite interval $0 \leq \lambda \leq 1$.

In the zero coupling limit the spectrum is that of the free theory. In the infinite coupling limit $\lambda = 1$ the theory is essentially equivalent to the limit of zero fermion mass. Schwinger has shown that massless $(\mathrm{QED})_{1+1}$ is equivalent to a free boson theory. In the light-cone formalism one can solve the $m = 0$ theory explicitly. One defines[114] bilinear operators in the fermion fields a_n and $a_n^\dagger$ which have normal boson commutation rules. Then for $Q = 0$

$$H = m^2 \sum_{n+1}^{\infty} \frac{1}{n} \left(b_n^\dagger b_n + d_n^\dagger d_n\right) + \frac{g^2}{\pi} \sum_{n=1}^{\infty} \frac{1}{n} a_n^\dagger a_n \ .$$

Thus for $m^2 = 0$ (or $g^2/\pi \gg m^2$), H_{QED} is equivalent to free boson theory with $m_b^2 = g^2/\pi$. The distinction between the theories in the limit of zero fermion mass is discussed by McCartor.[115]

Figure 41 shows the structure function for the ground state of $(\mathrm{QED})_{1+1}$ as a function of λ. In the weak binding limit $g \to 0$ or $(m \to \infty)$, the structure function becomes a delta function at equal partition of the constituent momentum, as expected. In the strong coupling limit $g \to \infty$ $(m \to 0)$ the structure function becomes flat. This is consistent with the interpretation of the Schwinger boson as a point-like composite of a fermion and antifermion. The contribution to higher Fock states to the lowest mass structure function is strikingly small; the probability of nonvalence states is less than 1% for any value of λ.

It is interesting that there is analytic agreement between the DLCQ results and the exact solutions of the Schwinger model for finite K, as well as in the continuum limit. This can be traced to the fact that the structure function of the Schwinger boson is flat and thus needs minimal resolution. In the case of the massive Schwinger model (QED_2), we established the existence of the continuum limit numerically; for sufficiently large resolution K the results become independent of K. The essential criteria for convergence is that the intrinsic dynamical structure of the wavefunctions is sufficiently resolved at the rational values $x = n/K$, $n = 1, 2, ..., K-1$ accessible at a given K.

In the large K limit, the eigenvalues agree quantitatively with the results of Bergknoff[114] and with those of a lattice gauge calculation by Crewther and Hamer.[116] This result is important in establishing the equivalence of different complementary nonperturbative methods. We also verified numerically that different Fock space representations yield the same physical results. In particular, we solved the QED_2

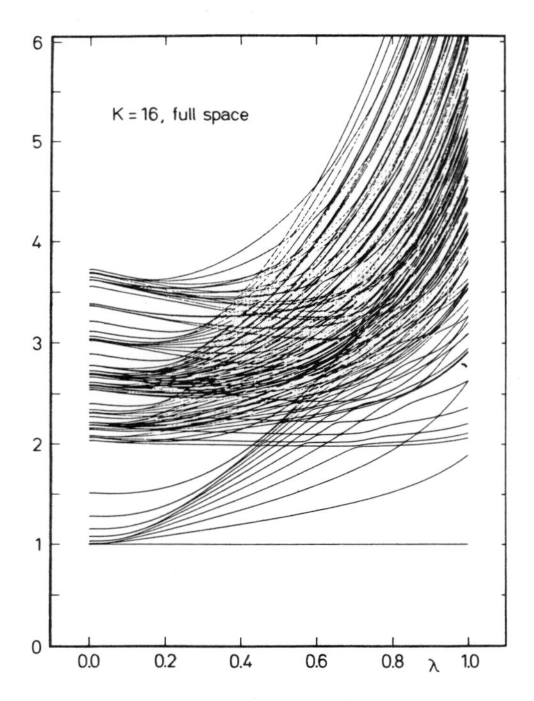

Fig. 40. Spectrum of QED in one-space and one-time dimension for harmonic resolution $K = 16$. The ratios M_i/M_1 are plotted as a function of the scaled coupling constant λ. The Schwinger limit is $\lambda = 1$. (From ref. 112.)

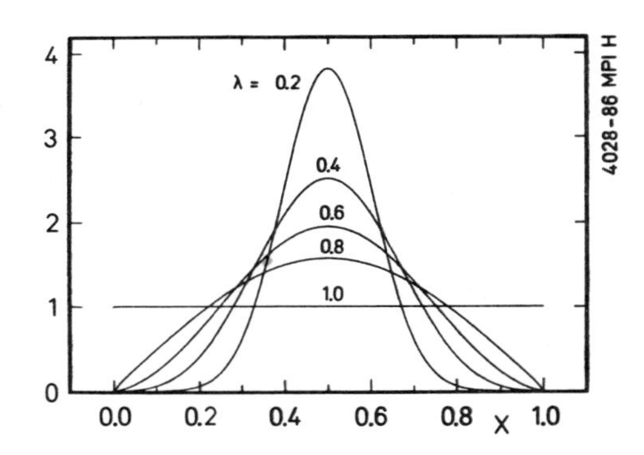

Fig. 41. The structure function of the lowest mass bound state for QED in 1+1 space-time dimensions, as calculated in the DLCQ formalism.[112]

spectrum in the space corresponding to the solutions of the free, massive Dirac equation $(i\gamma^\mu\partial_\mu + m_F)\psi = 0$ as well as of the massless equation $i\gamma^\mu\partial_\mu\psi = 0$. Convergence is slow in $1/K$ only at very large coupling λ near 1 .

Even for moderately large values of the resolution, DLCQ provides one with a qualitatively correct picture of the whole spectrum of eigenfunctions. This aspect becomes important for the development of scattering theory within the DLCQ approach. For example, we have found the rather surprising result that the lowest eigenfunction has very small probability (less than 1%) for $|2f; 2\bar{f}\rangle$ and higher particle Fock states (i.e., no 'sea quarks'). We have also obtained the spectrum of the Yukawa theory with spin-zero bosons, a theory with a more complicated Fock structure. Also Harindranath and Vary[117] have recently used a DLCQ approach to analyze ϕ^4 theory, a model with a nontrivial vacuum structure.

Recently Hornbostel[118] has used DLCQ to obtain the complete color-singlet spectrum of QCD in one space and one time dimension for $N_C = 2, 3, 4$. The hadronic spectra are obtained as a function of quark mass and QCD coupling constant. (See fig. 42.) Where they are available, the spectra agree with results obtained earlier; in particular, the lowest meson mass in SU(2) agrees within errors with lattice Hamiltonian results.[120] The meson mass at $N_C = 4$ is close to the value obtained in the large N_C limit. The method also provides the first results for the baryon spectrum in a non–Abelian gauge theory. The lowest baryon mass is shown in fig. 42 as a function of coupling constant. The ratio of meson to baryon mass as a function of N_C also agrees at strong coupling with results obtained by Frishman and Sonnenschein.[119] Precise values for the mass eigenvalue can be obtained by extrapolation to large K since the functional dependence in $1/K$ is understood.

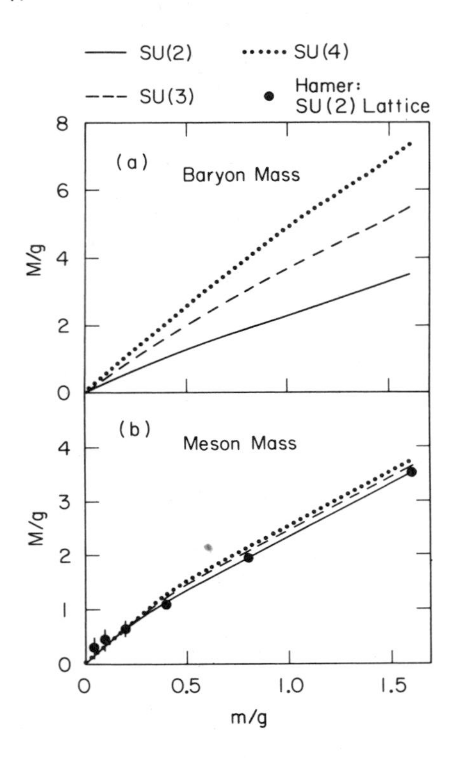

Fig. 42. The baryon and meson spectrum in QCD [1+1] computed in DLCQ for $N_C = 2, 3, 4$ *as a function of quark mass and coupling constant. (From ref. 118.)*

As emphasized above, when the light-cone Hamiltonian is diagonalized for a finite resolution K, one gets a complete set of eigenvalues corresponding to the total dimension of the Fock state basis. A representative example of the spectrum is shown in fig. 43 for baryon states ($B = 1$) as a function of the dimensionless variable $\lambda = 1/(1 + \pi m^2/g^2)$. Antiperiodic boundary conditions are used. Note that spectrum automatically includes continuum states with $B = 1$.

The structure functions for the lowest meson and baryon states in SU(3) at two different coupling strengths $m/g = 1.6$ and $m/g = 0.1$ are shown in figs. 44 and 45.

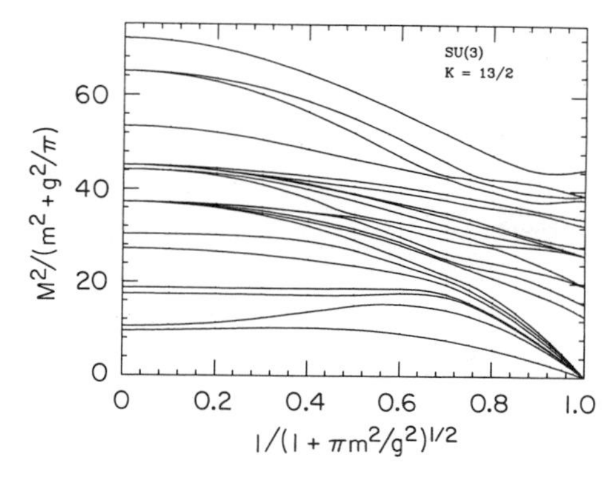

Fig. 43. Representative baryon spectrum for QCD in one-space and one-time dimension. (From ref. 118.)

Higher Fock states have a very small probability; representative contributions to the baryon structure functions are shown in figs. 46 and 47. For comparison, the valence wavefunction of a higher mass state which can be identified as a composite of meson pairs (analogous to a nucleus) is shown in fig. 48. The interactions of the quarks in the pair state produce Fermi motion beyond $x = 0.5$.

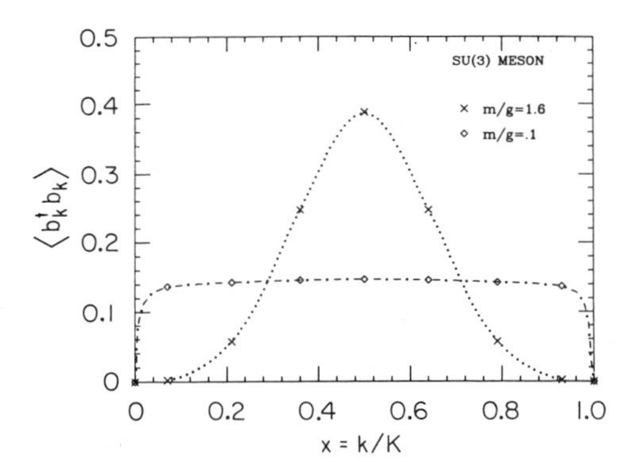

Fig. 44. The meson quark momentum distribution in QCD[1+1] computed using DLCQ. (From ref. 118.)

There are a number of important advantages of the DLCQ method which have emerged from this study of two-dimensional field theories. They are as follows:

1. The Fock space is denumerable and finite in particle number for any fixed resolution K. In the case of gauge theory in 3+1 dimensions, one expects that photon or gluon quanta with zero four-momentum decouple from neutral or color-singlet bound states, and thus need not be included in the Fock basis. The transverse momenta are additive and can be introduced on a cartesian grid. Hornbostel[118] has developed methods to implement the color degrees of freedom for the non–Abelian theories. Tang[121] is currently studying QED[3+1] in DLCQ as a function of the QED coupling constant.
2. Because we are using discrete momentum space representation, rather than a space-time lattice, there are no special difficulties with fermions: e.g., no fermion doubling, fermion determinants, or necessity for a quenched approximation. Furthermore, the discretized theory has basically the same ultraviolet structure as

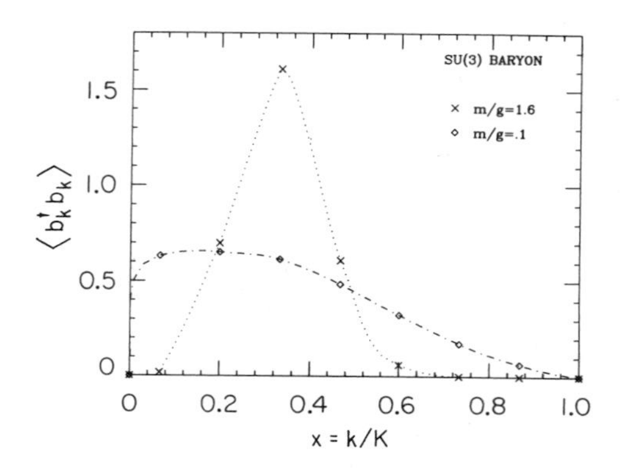

Fig. 45. The baryon quark momentum distribution in QCD[1+1] computed using DLCQ. (From ref. 118.)

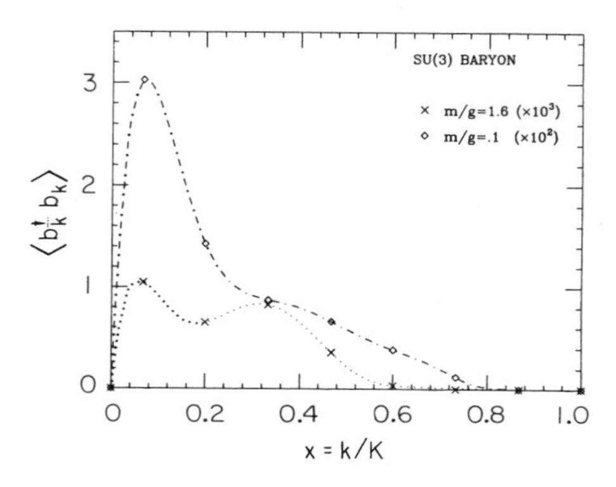

Fig. 46. Contribution to the baryon quark momentum distribution from $qqqq\bar{q}$ states for QCD[1+1]. (From ref. 118.)

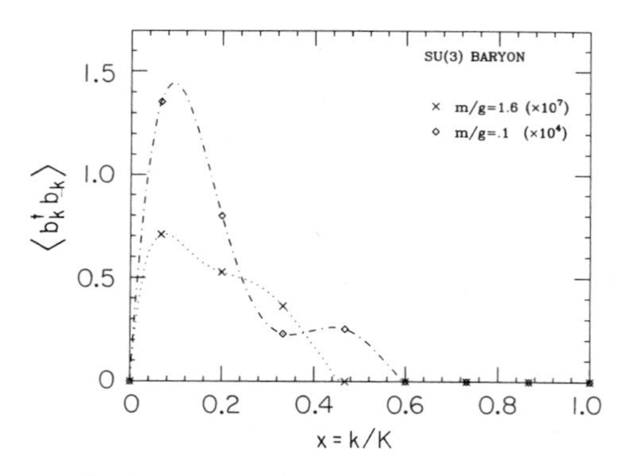

Fig. 47. Contribution to the baryon quark momentum distribution from $qqqqq\bar{q}\bar{q}$ states for QCD[1+1]. (From ref. 118.)

the continuum theory. It should be emphasized that unlike lattice calculations, there is no constraint or relationship between the physical size of the bound state and the length scale L.

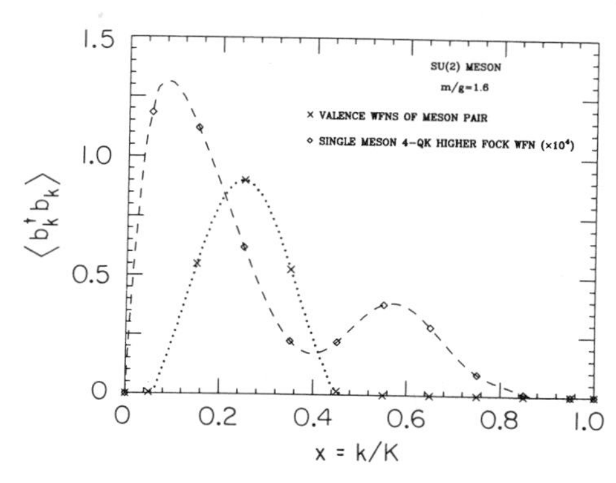

Fig. 48. Comparison of the meson quark distributions in the $qq\bar{q}\bar{q}$ Fock state with that of a continuum meson pair state. The structure in the former may be due to the fact that these four-particle wavefunctions are orthogonal. The analysis is for $N_C = 2$ in 1+1 dimensions. (From ref. 118.)

3. The DLCQ method has the remarkable feature of generating the complete spectrum of the theory; bound states and continuum states alike. These can be separated by tracing their minimum Fock state content down to small coupling constant since the continuum states have higher particle number content. In lattice gauge theory it appears intractable to obtain information on excited or scattering states or their correlations. The wavefunctions generated at equal light cone time have the immediate form required for relativistic scattering problems.
4. DLCQ is basically a relativistic many-body theory, including particle number creation and destruction, and is thus a basis for relativistic nuclear and atomic problems. In the nonrelativistic limit the theory is equivalent to the many-body Schrödinger theory.

The immediate goal is gauge theory in 3+1 dimensions. Already Klabucar and Pauli[122] have studied QCD[3+1] in the $q\bar{q}$ sector for strong coupling. In the Abelian case it will be interesting to analyze QED and the positronium spectrum in the large α limit. Whether the non–Abelian theory can be solved using DLCQ — considering its greater number of degrees of freedom and its complex equal-time vacuum is an open question. The studies for Abelian and non–Abelian gauge theory carried out so far in 1+1 dimensions give grounds for optimism.

16. SUMMARY

With the advent of new methods to attack nonperturbative QCD, such as sum rule constraints, implementation of effective Lagrangians such as the Skyrme model, extensions of lattice gauge theory, and promising methods to solve the light-cone Hamiltonian for its spectrum and Fock state solutions, a renaissance of interest is developing for understanding hadron and nuclear dynamics from first principles.

As I have outlined in these lectures, An experimental program with antiprotons of energies under 10 GeV can serve as an important test of QCD dynamics, especially for exclusive channels. Already there are extensive calculations available for $\bar{p}p \to \gamma\gamma$ for both real and virtual channels. Fixed angle scattering, meson-pair and lepton-pair final states also give sensitive tests of the theory. We have emphasized the possibility that the production of charmed hadrons in exclusive $\bar{p}p$ channels may have a non-negligible cross section. All of these channels bear on the question at what momentum scale perturbative factorization methods apply.

Inclusive measurements are usually studied at much higher energies than those potentially available at a SUPER–LEAR or AMPLE facility. Nevertheless, as discussed in these lectures, there are interesting novel effects involving the interface between perturbative and nonperturbative dynamics and quark propagation in hadronic matter — all of which can be explored at $\bar{p}$ energies below 10 GeV. In particular, $\bar{p}$ — nuclear collisions can play an important role in clarifying fundamental QCD issues such as color transparency, critical length phenomena, and the validity of the reduced nuclear amplitude phenomenology.

17. ACKNOWLEDGEMENTS

I wish to acknowledge the support of the Alexander von Humboldt Foundation and the hospitality of the Max Planck Institute for Nuclear Physics, Heidelberg. Parts of these lectures were also presented at the Third Lake Louise Winter Institute on QCD and the Fermilab Low Energy Facility Workshop. Some of the material presented here is based on collaborations with others, particularly G. de Teramond, J. F. Gunion, J. R. Hiller, K. Hornbostel, G. P. Lepage, A. H. Mueller, H. C. Pauli, D. E. Soper, A. Tang and S. F. Tuan.

REFERENCES

1. For reviews see, e.g., E. Reya, Phys. Rept. 69, 195 (1981) and A. H. Mueller, Lectures on perturbative QCD given at the Theoretical Advanced Study Institute, New Haven, 1985.
2. G. P. Lepage and S. J. Brodsky, Phys. Rev. D22, 2157 (1980); G. P. Lepage, S. J. Brodsky, Tao Huang and P. B. Mackenzie, CLNS–82/522, published in the *Proceedings of the Banff Summer Institute*, 1981.
3. A. H. Mueller, Phys. Rept. 73, 237 (1981).
4. V. L. Chernyak and I. R. Zhitnitskii, Phys. Rept. 112, 1783 (1984). Xiao-Duang Xiang, Wang Xin-Nian and Huang Tao, BIHEP–TH–84, 23 and 29, 1984.
5. A. V. Efremov and A. V. Radyushkin, Phys. Lett. 94B, 245 (1980).
6. S. J. Brodsky, Y. Frishman, G. P. Lepage and C. Sachrajda, Phys. Lett. 91B, 239 (1980).
7. J. Boyer et al., Phys. Rev. Lett. 56, 207 (1986); TPC/Two Gamma Collaboration (H. Aihara et al.), UCR–TPC–86–01, April 1986.
8. S. D. Drell and T. M. Yan, Phys. Rev. Lett. 24, 181 (1970).
9. J. Ashman et al., CERN preprint CERN–EP/87–230.
10. S. J. Brodsky, J. Ellis and M. Karliner, SLAC–PUB–4519 (1988).
11. G. P. Lepage and S. J. Brodsky, Phys. Rev. D22, 2157 (1980); G. P. Lepage, S. J. Brodsky, T. Huang and P. B. Mackenzie, CLNS–82/522, published in the *Proceedings of the Banff Summer Institute*, 1981.
12. S. J. Brodsky, Y. Frishman, G. P. Lepage and C. Sachrajda, Phys. Lett. 91B, 239 (1980).
13. S. J. Brodsky, SLAC–PUB–4342 and in the *Proceedings of the VIIIth Nuclear and Particle Physics Summer School*, Launceston, Australia, 1987.
14. V. L. Chernyak, A. A. Ogloblin and I. R. Zhitnitsky, Novosibirsk preprints 87–135,136, and references therein.

15. I. D. King and C. T. Sachrajda, Nucl. Phys. B279, 785 (1987).
16. S. Gottlieb and A. S. Kronfeld, Phys. Rev. D33 (1986) 227–233; CLNS–85/646, June 1985, 22 pp.
17. G. Martinelli and C. T. Sachrajda, CERN–TH 4909 (1987).
18. G. Martinelli and C. T. Sachrajda, SHEP 87/88–16, 1988.
19. R. G. Arnold et al., SLAC–PUB–3810, April 1986.
20. H. C. Pauli and S.J. Brodsky, Phys. Rev. D32, 1993 (1985); Phys. Rev. D32, 2001 (1985).
21. J. Ashman et al., CERN preprint CERN–EP/87–230.
22. S. J. Brodsky, J. Ellis and M. Karliner, SLAC–PUB–4519 (1988).
23. G. P. Lepage and S. J. Brodsky, Phys. Rev. D22, 2157 (1980); G. P. Lepage, S. J. Brodsky, T. Huang and P. B. Mackenzie, CLNS–82/522, published in the *Proceedings of the Banff Summer Institute*, 1981.
24. S. J. Brodsky, Y. Frishman, G. P. Lepage and C. Sachrajda, Phys. Lett. 91B, 239 (1980).
25. P. Bordalo et al., CERN EP/87–67 and 68 (1987).
26. S. J. Brodsky, G. T. Bodwin and G. P. Lepage, in the *Proceedings of the Volendam Multiparticle Dynamics Conference*, 1982, p. 841; *Proceedings of the Banff Summer Institute*, 1981, p. 513; and to be published. This effect is related to the formation zone principle of L. Landau and I. Pomeranchuk, Dok. Akademii Nauk SSSR 92, 535,735 (1953).
27. S. J. Brodsky and G. P. Lepage, Phys. Rev. D24, 1808 (1981).
28. G. P. Lepage and S. J. Brodsky, Phys. Rev. D22, 2157 (1980).
29. H. Aihara et al., Phys. Rev. Lett. 57, 51 (1986). The Mark II data for combined charged meson pair production are also in good agreement with the PQCD predictions. See J. Boyer et al., Phys. Rev. Lett. 56, 207 (1986).
30. S. J. Brodsky and G. R. Farrar, Phys. Rev. Lett. 31, 1153 (1973), Phys. Rev. D11, 1309 (1975); V. A. Matveev, R. M. Muradyan and A. V. Tavkhelidze, Lett. Nuovo Cim. 7, 719 (1973).
31. H. Suura, T. F. Walsh and B. L. Young, Lett. Nuovo Cimento 4, 505 (1972). See also, M. K. Chase, Nucl. Phys. B167, 125 (1980).
32. General QCD analyses of exclusive processes are given in: S. J. Brodsky and G. P. Lepage, SLAC–PUB–2294, presented at the Workshop on Current Topics in High Energy Physics, Cal Tech (Feb. 1979), S. J. Brodsky, in the *Proceedings of the La Jolla Institute Summer Workshop on QCD*, La Jolla (1978), G. P. Lepage and S. J. Brodsky, Phys. Lett. B87, 359 (1979), Phys. Rev. Lett. 43, 545, 1625(E) (1979), Phys. Rev. D22, 2157 (1980), A. V. Efremov and A. V. Radyushkin, Phys. Lett. B94, 245 (1980), V. L. Chernyak, V. G. Serbo and A. R. Zhitnitskii, Yad. Fiz. 31, 1069 (1980), S. J. Brodsky, Y. Frishman, G. P. Lepage and C. Sachrajda, Phys. Lett. 91B, 239 (1980), and A. Duncan and A. H. Mueller, Phys. Rev. D21, 1636 (1980). The QCD prediction for the pion form factor at asymptotic Q^2 was first obtained by V. L. Chernyak, A. R. Zhitnitskii and V. G. Serbo, JETP Lett. 26, 594 (1977), D. R. Jackson, Ph.D. Thesis, Cal Tech (1977), and G. Farrar and D. Jackson, Phys. Rev. Lett. 43, 246 (1979). See also, A. M. Polyakov, *Proceedings of the International Symposium on Lepton and Photon Interactions at High Energies*, Stanford (1975), and G. Parisi, Phys. Lett. 84B, 225 (1979).

33. A. H Mueller and S. J. Brodsky, talks presented at the Fermilab Workshop on A^{α} Physics, 1982; A. H. Mueller, *Proceedings of the XVII Recontre de Moriond* (1982); S. J. Brodsky, *Proceedings of the XIII International Symposium on Multiparticle Dynamics*, Volendam (1982). See also, G. Bertsch, A. S. Goldhaber and J. F. Gunion, Phys. Rev. Lett. 47, 297 (1981).

34. A. S. Carroll et al., BNL report (1988), to be published in Phys. Rev. Letters; S. Heppelmann et al., DPF meeting (Salt Lake City, 1987).

35. G. R. Court et al., Phys. Rev. Lett. 57, 507 (1986).

36. S. J. Brodsky and G. de Teramond, Phys. Rev. Lett. 60, 1924 (1988).

37. J. P. Ralston and B. Pire, Univ. of Kansas preprint (1988).

38. G. W. Atkinson, J. Sucher, and K. Tsokos, Phys. Lett. 137B, 407 (1984); G. R. Farrar, E. Maina and F. Neri, Nucl. Phys. B259, 702 (1985); E. Maina, Rutgers Ph.D. Thesis (1985); J. F. Gunion, D. Millers and K. Sparks, Phys. Rev. D33, 689 (1986); P. H. Damgaard, Nucl. Phys. B211, 435 (1983); B. Nezic, Ph.D. Thesis, Cornell University (1985); D. Millers and J. F. Gunion, Phys. Rev. D34, 2657 (1986).

39. G. R. Farrar, H. Zhang, A. A. Globlin and I. R. Zhitnitsky, Rutgers preprint RU–88–14; G. R. Farrar, E. Maina and F. Neri, Nucl. Phys. B259, 702 (1985), Err.-ibid. B263, 746 (1986).

40. V. L. Chernyak, A. A. Ogloblin and I. R. Zhitnitsky, Novosibirsk preprint INP–134 (1987); V. L. Chernyak and A. R. Zhitnitsky, Phys. Rept. 112, 1783 (1984). See also, Xiao-Duang Xiang, Wang Xin-Nian and Huang Tao, BIHEP–TH–84, 23 and 29, 1984, and M. J. Lavelle, ICTP–84–85–12; Nucl. Phys. B260, 323 (1985). The sensitivity of the proton form factor to the Chernyak et al. wavefunctions is investigated in C. R. Ji, A. Sills and R. Lombard-Nelsen, Phys. Rev. D36, 165 (1987).

41. E. Maina and G. R. Farrar, ref. 39.

42. A simple method for estimating hadron pair production cross sections near threshold in $\gamma\gamma$ collisions is given in S. J. Brodsky, G. Köpp and P. M. Zerwas, Phys. Rev. Lett. 58, 443 (1987).

43. N. Isgur and C. H. Llewellyn Smith, reports presented to this meeting and the *Third Conference on the Intersection between Particle and Nuclear Physics* (1988), and Phys. Rev. Lett. 52, 1080 (1984).

44. A. V. Radyushkin, *Proceedings of the Ninth European Conference on Few Body Problems in Physics*, Tbilisi (1984).

45. For an explicit calculation in hadrons containing only heavy quarks, see S. J. Brodsky and C. R. Ji, Phys. Rev. Lett. 55, 2257 (1985).

46. For a remarkable confirmation of the PQCD predictions for $\gamma d \to np$, see sec. 11, and J. Napolitano et al., ANL preprint PHY–5265–ME–88 (1988).

47. S. D. Drell and T. M. Yan, Phys. Rev. Lett. 24, 181 (1970); S. J. Brodsky and S. D. Drell, Phys. Rev. D22, 2236 (1980).

48. O. Jacob and L. S. Kisslinger, Phys. Rev. Lett. 56, 225 (1986); C. Carlson and F. Gross, Phys. Rev. Lett.; Z. Dziembowski and L. Mankiewicz, Phys. Rev. D37, 778, 2030 (1980).

49. The exceptions involve spin effects, which are sensitive to threshold and non-leading power law corrections.

50. S. J. Brodsky and G. R. Farrar, Phys. Rev. Lett. 31, 1153 (1973); Phys. Rev. D11, 1309 (1975).

51. J. F. Gunion, S. J. Brodsky and R. Blankenbecler, Phys. Rev. D8, 287 (1973); Phys. Lett. 39B, 649 (1972); D. Sivers, S. J. Brodsky and R. Blankenbecler, Phys. Rept. 23C, 1 (1976). Extensive references to fixed angle scattering are given in this review.

52. R. L. Anderson et al., Phys. Rev. Lett. 30, 627 (1973).

53. A. W. Hendry, Phys. Rev. D10, 2300 (1974).

54. G. R. Court et al., UM–HE–86–03, April 1986, 14 pp.

55. S. J. Brodsky and G. P. Lepage, Phys. Rev. D24, 2848 (1981).

56. G. R. Farrar, RU–85–46, 1986.

57. S. J. Brodsky, C. E. Carlson and H. J. Lipkin, Phys. Rev. D20 2278 (1979); H. J. Lipkin, (private communication).

58. S. J. Brodsky and G. P. Lepage, Phys. Rev. D24, 1808 (1981). The next-to-leading order evaluation of T_H for these processes is given by B. Nezic, Ph.D. Thesis, Cornell University (1985).

59. G. R. Farrar, E. Maina and F. Neri, Nucl. Phys. B259, 702 (1985), Err.-ibid. B263, 746 (1986).

60. J. F. Gunion, D. Millers and K. Sparks, Phys. Rev. D33, 689 (1986).

61. R. Brandelik et al., Phys. Lett. 108B, 67 (1982). See also the *Proceedings of the VIth International Workshop on Two Photon Reactions*, Lake Tahoe, CA (1984), edited by R. L. Lander.

62. I. D. King and C. T. Sachrajda, SHEP–85/86–15, April 1986, 36 pp.

63. D. Millers and J. F. Gunion, UCD–86–04, 1986;

64. G. W. Atkinson, N. Tsokos and J. Sucher, Phys. Lett. 137B, 407 (1984).

65. N. Isgur and C. H. Llewellyn Smith, Phys. Rev. Lett. 52, 1080 (1984).

66. O. C. Jacob and L. S. Kisslinger, Phys. Rev. Lett. 56, 225 (1986).

67. S. J. Brodsky and G. P. Lepage, Phys. Rev. D24, 2848 (1981).

68. M. E. B. Franklin, Ph.D Thesis (1982), SLAC–254, UC–34d; M. E. B. Franklin et al., Phys. Rev. Lett. 51, 963 (1983); G. Trilling, in *Proceedings of the Twenty-First International Conference on High Energy Physics*, Paris, July 26–31, 1982; E. Bloom, ibid.

69. S. J. Brodsky, G. P. Lepage and San Fu Tuan, Phys. Rev. Lett. 59, 621 (1987).

70. Wei-Shou Hou and A. Soni, Phys. Rev. Lett. 50, 569 (1983).

71. P. G. O. Freund and Y. Nambu, Phys. Rev. Lett. 34, 1645 (1975).

72. Wei-Shu Hou and A. Soni, Phys. Rev. Lett. 50, 569 (1983).

73. G. R. Court et al., Phys. Rev. Lett. 57, 507 (1986); T. S. Bhatia et al., Phys. Rev. Lett. 49, 1135 (1982); E. A. Crosbie et al., Phys. Rev. D23, 600 (1981); A. Lin et al., Phys. Lett. 74B, 273 (1978); D. G. Crabb et al., Phys. Rev. Lett. 41, 1257 (1978); J. R. O'Fallon et al., Phys. Rev. Lett. 39, 733 (1977); For a review, see A. D. Krisch, UM–HE–86–39 (1987).

74. S. J. Brodsky, C. E. Carlson and H. J. Lipkin, Phys. Rev. D20, 2278 (1979); G. R. Farrar, S. Gottlieb, D. Sivers and G. Thomas, Phys. Rev. D20, 202 (1979).

75. For other attempts to explain the spin correlation data, see C. Avilez, G. Cocho and M. Moreno, Phys. Rev. D24, 634 (1981); G. R. Farrar, Phys. Rev. Lett.

56, 1643 (1986), Err-ibid. 56, 2771 (1986); H. J. Lipkin, Nature 324, 14 (1986); S. M. Troshin and N. E. Tyurin, JETP Lett. 44, 149 (1986) [Pisma Zh. Eksp. Teor. Fiz. 44, 117 (1986)]; G. Preparata and J. Soffer, Phys. Lett. 180B, 281 (1986); S. V. Goloskokov, S. P. Kuleshov and O. V. Seljugin, *Proceedings of the VII International Symposium on High Energy Spin Physics*, Protvino (1986); C. Bourrely and J. Soffer, Phys. Rev. D35, 145 (1987).

76. G. P. Lepage and S. J. Brodsky, Phys. Rev. D22, 2157 (1980); S. J. Brodsky, Y. Frishman, G. P. Lepage and C. Sachrajda, Phys. Lett. 94B, 245 (1980); A. Duncan and A. H. Mueller, Phys. Lett. 90B, 159 (1980); A. V. Efremov and A. V. Radyushkin, Phys. Lett. 94B, 245 (1980); A. H. Mueller, Phys. Rept. 73, 237 (1981); V. L. Chernyak and A. R. Zhitnitskii, Phys. Rept. 112, 173 (1984).

77. S. J. Brodsky and G. R. Farrar, Phys. Rev. Lett. 31, 1153 (1973); V. Matveev, R. Muradyan and A. Tavkhelidze, Nuovo Cimento Lett. 7, 719 (1973).

78. S. J. Brodsky and G. P. Lepage, Phys. Rev. D24, 2848 (1981).

79. G. C. Blazey et al., Phys. Rev. Lett. 55, 1820 (1985); G. C. Blazey, Ph.D. Thesis, University of Minnesota (1987); B. R. Baller, Ph.D. Thesis, University of Minnesota (1987); D. S. Barton, et al., J. de Phys. 46, C2, Supp. 2 (1985). For a review, see D. Sivers, S. J. Brodsky and R. Blankenbecler, Phys. Rept. 23C, 1 (1976).

80. There are five different combinations of six quarks which yield a color singlet B=2 state. It is expected that these QCD degrees of freedom should be expressed as B=2 resonances. See, e.g., S. J. Brodsky and C. R. Ji, Phys. Rev. D34, 1460 (1986).

81. For other examples of threshold enhancements in QCD, see S. J. Brodsky, J. F. Gunion and D. E. Soper, Phys. Rev. D36, 2710 (1987); S. J. Brodsky, G. Kopp and P. M. Zerwas, Phys. Rev. Lett. 58, 443, (1987). Resonances are often associated with the onset of a new threshold. For a discussion, see D. Bugg, Presented at the IV LEAR Workshop, Villars-Sur-Ollon, Switzerland, September 6–13, 1987.

82. J. F. Gunion, R. Blankenbecler and S. J. Brodsky, Phys. Rev. D6, 2652 (1972).

83. With the above normalization, the unpolarized pp elastic cross section is $d\sigma/dt = \Sigma_{i=1,2,\ldots 5} \mid \phi_i^2 \mid /(128\pi s p_{\rm cm}^2)$.

84. J. P. Ralston and B. Pire, Phys. Rev. Lett. 57, 2330 (1986).

85. At low momentum transfers one expects the presence of both helicity-conserving and helicity nonconserving pomeron amplitudes. Preliminary calculations indicate that the data for A_N at $p_{lab} = 11.75$ GeV/c can be understood over the full angular range in these terms. The large value of $A_N = 24 \pm 8\%$ at $p_{lab} = 28$ GeV/c and $p_T^2 = 6.5$ GeV^2 remains an open problem. See P. R. Cameron et al., Phys. Rev. D32, 3070 (1985).

86. K. Abe et al., Phys. Rev. D12, 1 (1975), and references therein. The high energy data for $d\sigma/dt$ at $\theta_{\rm cm} = \pi/2$ are from C. W. Akerlof et al., Phys. Rev. 159, 1138 (1967); G. Cocconi et al., Phys. Rev. Lett. 11, 499 (1963); J. V. Allaby et al., Phys. Lett. 23, 389 (1966).

87. I. P. Auer et al., Phys. Rev. Lett. 52, 808 (1984). Comparison with the low energy data for A_{LL} at $\theta_{\rm cm} = \pi/2$ suggests that the resonant amplitude below $p_{lab} = 5.5$ GeV/c has more structure than the single resonance form adopted here. See I. P. Auer et al., Phys. Rev. Lett. 48, 1150 (1982).

88. A. W. Hendry, Phys. Rev. D10, 2300 (1974); N. Jahren and J. Hiller, University of Minnesota preprint, 1987.

89. The neutral strange inclusive pp cross section measured at $p_{lab} = 5.5$ GeV/c is 0.45 ± 0.04 mb; G. Alexander et al., Phys. Rev. 154, 1284 (1967).
90. See, e.g., J. R. Cudell, F. Halzen and K. Hikasa, MAD/PH/276 (1986), and references therein.
91. S. J. Brodsky, SLAC–PUB–3770, *Proceedings of the XVI International Symposium on Multiparticle Dynamics*, Tel Aviv (1985).
92. S. J. Brodsky, P. Hoyer, C. Peterson and N. Sakai, Phys. Lett. 93B, 451 (1980).
93. S. J. Brodsky, J. C. Collins, S. D. Ellis, J. F. Gunion and A. H. Mueller, Published in Snowmass Summer Study 1984, 227; S. J. Brodsky and J. F. Gunion, *Proceedings of the SLAC Summer Institute*, 1984, 603; S. J. Brodsky, H. E. Haber and J. F. Gunion, SSC/DPF Workshop 1984, 100.
94. S. J. Brodsky and C. R. Ji, Phys. Rev. Lett. 55, 2257 (1985).
95. S. J. Brodsky and B. T. Chertok, Phys. Rev. Lett. 37, 269 (1976), Phys. Rev. D114, 3003 (1976); S. J. Brodsky and J. R. Hiller, Phys. Rev. C28, 475 (1983).
96. T. S.-H. Lee, ANL preprint (1988).
97. H. Myers et al., Phys. Rev. 121, 630 (1961); R. Ching and C. Schaerf, Phys. Rev. 141, 1320 (1966); P. Dougan et al., Z. Phys. A 276, 55 (1976).
98. See, e.g., C. R. Ji and S. J. Brodsky, SLAC–PUB–3148, to be published in Phys. Rev. D; SLAC–PUB–3747, Lectures given at Stellenbosch Advanced Course in Theoretical Physics (Quarks and Leptons), Jan. 21–Feb 1, 1985; Phys. Rev. D33, 1951, 1406, 2653, (1986).
99. The data are compiled in Brodsky and Hiller, ref. 95.
100. A. H. Mueller, *Proceedings of the Moriond Conference*, 1982; S. J. Brodsky and B. T. Chertok, ref. 95; G. Bertsch, S. J. Brodsky, A. S. Goldhaber and J. F. Gunion, Phys. Rev. Lett. 47, 297 (1981).
101. R. Blankenbecler and S. J. Brodsky, Phys. Rev. D10, 2973 (1974).
102. I. A. Schmidt and R. Blankenbecler, Phys. Rev. D15, 3321 (1977).
103. G. T. Bodwin, Phys. Rev. D31, 2616 (1985); G. T. Bodwin, S. J. Brodsky and G. P Lepage, ANL–HEP–CP–85–32–mc, 1985, *Presented at the 20th Rencontre de Moriond*, Les Arcs, France, March 10–17, 1985;
104. J. C. Collins, D. E. Soper and G. Sterman, Phys. Lett. 134B 263 (1984).
105. For a recent review and further theoretical references, see E. L. Berger and F. Coester, ANL–HEP–PR–87–13 (1987).
106. J. J. Aubert et al., Phys. Lett. 123B, 275 (1983); For recent reviews, see E. L. Berger, ANL–HEP–PR–87–45, and E. L. Berger and F. Coester, ANL–HEP–PR–87–13 (to be published in Ann. Rev. of Nucl. Part. Sci.).
107. H. G. Fischer, presented at the Leipzig Conference, 1984.
108. C. E. Carneiro, M. Day, J. Frenkel, J. C. Taylor and M. T. Thomaz, Nucl. Phys. B183, 445 (1981); C. Sachrajda, *Proceedings of the Tahoe Multiparticle Dynamics Conference,* 1983, 415; W. W. Lindsay, D. A. Ross and C. Sachrajda, Nucl. Phys. B222, 189 (1983).
109. The spectrum of Yukawa theory is calculated in the usual Fock space by E. D. Brooks and S. C. Frautschi, Z. Phys. C23, 263 (1984).
110. P. A. M. Dirac, Rev. Mod. Phys. 21, 392 (1949). Further references to light-cone quantization are given in ref. 112.
111. See S. J. Brodsky and S. D. Drell, Phys. Rev. D22, 2236 (1980), and references therein.

112. T. Eller, H. C. Pauli and S. J. Brodsky, Phys. Rev. D35, 1493 (1987).

113. For a discussion of renormalization in light-cone perturbation theory, see S. J. Brodsky, R. Roskies and R. Suaya, Phys. Rev. D8, 4574 (1974), and also ref. 23.

114. H. Bergknoff, Nucl. Phys. B122, 215 (1977).

115. G. McCartor, Z. Phys. C36, 329 (1987), and to be published.

116. D. P. Crewther and C. J. Hamer, Nucl. Phys. B170, 353 (1980).

117. A. Harindranath and J. P. Vary, Phys. Rev. D36, 1141 (1987).

118. K. Hornbostel, to be published.

119. Y. Frishman and J. Sonnenschein, Nucl. Phys. B294, 801 (1987), and preprint WIS–87/65–PH.

120. C. J. Burden and C. J. Hamer, Phys. Rev. D37, 479 (1988), and references therein.

121. A. Tang, in preparation.

122. D. Klabucar and H. C. Pauli, MPI H–1988–V4 (1988). This paper gives a compendium of results on the quantization of gauge theories in DLCQ and early references.

BARYON-ANTIBARYON NUCLEAR INTERACTIONS

I.S. Shapiro

Lebedev Physical Institute
Leninsky Prospect 53
Moscow 117924, USSR

I BASIC PROPERTIES OF BARYON-ANTIBARYON NUCLEAR FORCES

Nuclear interaction and the low energy antiproton physics

It seems at the first glance that the main difference between low energy physics of Baryon-Baryon (BB) and Baryon-Antibaryon ($B\bar{B}$) is the presence of annihilation in the latter case. The distinction of BB and $B\bar{B}$ nuclear forces as well as the significance of this feature is not so obvious. But actually a set of unusual phenomena observed in experiments with low energy antiprotons is caused by the $B\bar{B}$ nuclear interaction. The annihilation process at low energies is very much affected by the nuclear forces. The physical ground of this fact is of the same nature as the influence of Coulomb attraction on the low-energy electron-positron system including the formation of bound states (positronium) and the enhancement of the e^-e^+ annihilation into photons cross section (the latter process can be considered in the frame of QED perturbation theory). To clear up the influence of $B\bar{B}$ nuclear interaction on the observable effects in low energy antiproton physics we have to overcome two difficulties: some uncertainties in our knowledge of nuclear forces and the non applicability of the perturbation approach to the treatment of the $B\bar{B}$ annihilation into mesons. This obstacle is the most serious one. At non-relativistic energies the nuclear scattering amplitude as well as its poles which correspond to the bound and resonant $B\bar{B}$ states caused by the attractive nuclear forces can be properly calculated from the Shrödinger equation with appropriate potential. In contrast, the $B\bar{B}$ annihilation even at rest is a relativistic quantum field process which today is hardly possible to study more or less quantitatively in any reliable way.

As the nuclear forces and the annihilation both are generated by the same strong interactions it appears a natural question how to extract the effects of the nuclear forces from the observable phenomena of $B\bar{B}$ collisions at low energies.

The solution of this problem exists and is based on the fact that the range R of the nuclear forces and characteristic space distances r_a realized in the annihilation processes are quite different: the first is much greater than the latter ($R >> r_a$). This inegality is a key point in the low energy antiproton physics. However, since it has not yet been fully accepted, even by some specialists familiar with this field, we will start our lectures by consideration dealing with this "range problem".

Antiproton-Nucleon and Antiproton-Nucleus Interactions
Edited by F. Bradamante *et al.*
Plenum Press, New York, 1990

Annihilation range

Before any estimate of the interaction distance, it is necessary to remember that in quantum physics we are dealing with waves rather than with localized corpuscules. All sorts of interactions between quantum objects lead to distortions of the plane waves corresponding to the free motions. Therefore the so-called interaction range is really the size of the space region in which distorted wave substantially differs from the initial plane wave with momentum P. Any distorted wave is a Fourier superposition of plane waves with different momenta P. The momentum spread $\Delta P = \langle(p - P)^2\rangle^{\frac{1}{2}}$ in this superposition is connected with the distortion space range (in other words, with the interaction range) $\bar{r}$ by the uncertainty relation $\bar{r} \approx h/\Delta P$. On the other hand, the above mentioned distortion and the corresponding Fourier superposition generate the elastic scattering. Obviously the characteristic momentum transfer $\bar{q}$ of the elastic scattering process will be of the same order of magnitude as the mean momentum spread ΔP in the distorted wave. Therefore we have $\bar{r} = h/\bar{q}$ and this means that the right way to obtain the interaction range is the study of the dependence of the scattering amplitude on the momentum transfer square $q^2 = (P_i - P_f)^2$ where P_i and P_f are the initial and final particle's momenta in the c.m.s. Let us emphasize that for the statement formulated above, it is unimportant whether the interacting objects are composite systems or "point-like" particles without internal constituents. In any case only the alteration of the total momenta of the objects defines the interaction range. The explicit evaluation of the interaction range can be done by calculating the Born scattering amplitude. Its q-Fourier transformation will give the "interaction potential" as a function of the space variable r.

But sometimes even the calculation of the Born scattering amplitude in terms of the known analytical functions of q^2 may be very difficult or impossible. In these cases the interaction range can be obtained by finding only the singularities of the Born amplitude, considered as an analytical function of q^2. If the amplitude in question is presented by a Feynman graph, the general Landau rules for finding the singularities may be used.

Any elastic scattering process associated with annihilation include transitions to (and from) a virtual state which contains only bosons and no fermions. The simplest example of annihilation induced elastic scattering is a process going through a one-boson virtual state. The corresponding graph for the Born amplitude is shown in Fig. 1. Here P_1 and P_3 are the 4-momenta of the baryon in the initial and final states, whereas P_2 and P_4 are those quantities for the antibaryon. Also, the standard kinematical variables $s = (P_1 + P_2)^2 = (P_3 + P_4)^2$ and $t = (P_1 - P_3)^2 = (P_4 - P_2)^2$ are used. In the c.m.s. s is the square of the total energy and $-t = q^2$ is the square of the momentum transfer. The amplitude for this graph is a function M(s) depending only on the variable s.

We have (for nonrelativistic fermions and a pseudoscalar boson):

$$M(s) = -\frac{g^2}{s - \mu^2} \tag{1.1}$$

and where μ is the mass of the boson, g being the coupling constant.

The elastic scattering cross section, corresponding to the Born amplitude (1.1) will be

$$\sigma_e^{Born} = \left(\frac{g^2}{4\pi}\right)^2 \times \frac{4\pi\, m^2}{(s - \mu^2)^2} \approx \frac{g^2}{4\pi} \times \frac{\pi}{m^2} \tag{1.2}$$

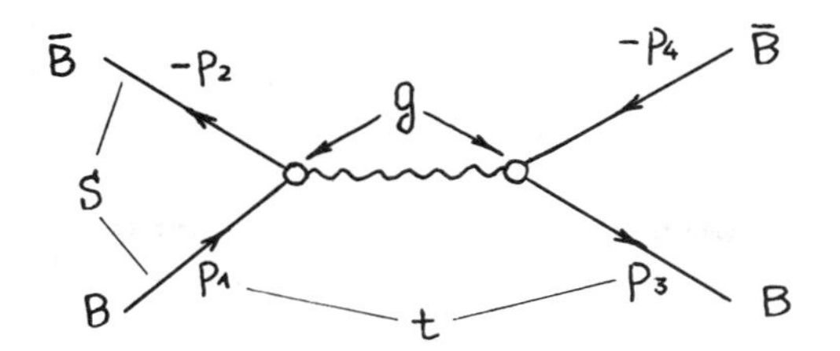

Fig. 1 The one-boson s-channel exchange.

The latter approximate equality is valid of $s \approx 4m^2 >> \mu^2$ (this condition is realized in the case when B = N and the boson is a pion).

Using the well known relation between the Born amplitude and the potential

$$V(r) = - \int e^{iqr} \times M \frac{d^3q}{(2\pi)^3} \tag{1.3}$$

we obtain:

$$V(r) = M(s)\, \delta(r) \approx g^2 \times \left(\frac{1}{2m}\right)^2 \delta(r) \tag{1.4}$$

This means that one-boson virtual annihilation is a zero-range process. The value of the observable quantities for instance - the elastic scattering cross section - will be defined in the order of magnitude by the characteristic length 1/2m which will enter into calculations being a coefficient in the δ-potential (1.4). Let us emphasize that the latter is equally valid for point-like and composite fermions B and $\bar{B}$, because for deriving the zero-range potential (1.4) from Eqs. (1.1) and (1.3), it is needless to make any assumption on the structure of these particles.

For composite fermions the "coupling constant" g in the potential (1.4) will in fact be a function of the variable $R\sqrt{s - \mu^2}$, where R is the radius of the composite system. The dependence on R arises because of diffraction of the boson wave emitted (or absorbed) at one point on the remaining part of the extended object. But we can see from what was said above that the size R of the annihilating composite systems have nothing in common with the annihilation range: the latter is strictly speaking zero, as it follows from (1.4) or, a least from a pragmatic point of view, (taking into account the previously mentioned values of the cross sections) equals in order of magnitude to 1/2m.

The s-channel one-boson exchange corresponds to one-boson annihilation which can take place only as a virtual process (for a short time $t \simeq \hbar/2mc$). Let us now discuss the range for a real annihilation with two or more bosons in final state. To clear up the question on the annihilation range in this case it is enough to analyse the two-boson annihilation process. In accordance with the general principles described before we have to investigate the q-dependence of the elastic scattering amplitude created by the two-boson s-channel exchange. The associated Feynman graph is shown in Fig. 2, where k, k' are the boson's 4-momenta and p, p' those of the fermions in the virtual states. The amplitude in Fig. 2 is an integral over the 4-momenta of the virtual particles (internal lines on the graph).

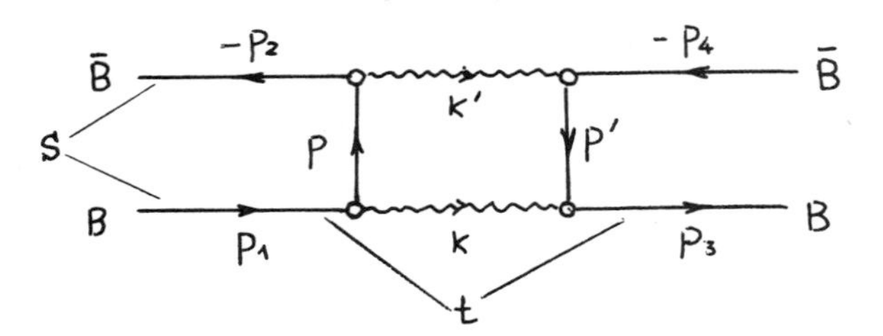

Fig. 2 The two-boson s-channel exchange.

The result of the integration will be a function depending on the variables s,t and also on the masses μ,m of the bosons and fermions. To make transparent the t-dependence of the amplitude it is necessary to connect the variable t to the 4-momenta of the virtual particles. Because of the 4-momenta conservation law at each vertex of the graph, we have

$$k = p_1 - p = p_3 - p' \,, \quad k' = p_2 + p = p' + p_4 \tag{1.5}$$

From these equations it follows

$$q = p_1 - p_3 = p_4 - p_2 = p - p' \tag{1.6}$$

Therefore, remembering that $t = q^2$, we obtain

$$t = - q^2 = (p - p')^2 \tag{1.7}$$

where q as before designates the c.m.s. 3-momentum transfer. From the Eq. (1.7) we learn that the t-dependence of the amplitude is directly connected with the two-fermion exchange in the t-channel. Moreover, the Eq. (1.7) coincides exactly with those for the "contracted" graph, presented below in Fig. 3. The contracted graph corresponds to creation of a virtual fermion-antifermion pair at one world point and to absorption of this pair at an another world point.

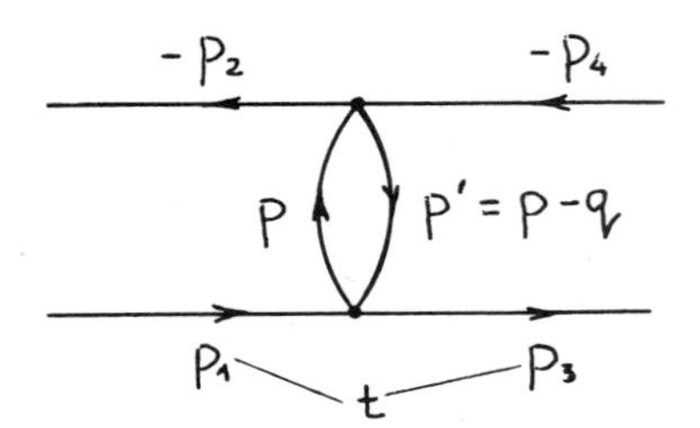

Fig. 3 The graph obtained by contraction of the initial boson lines.

The described physical situation makes it possible to obtain the interaction range for the contracted graph without performing explicit calculation of the loop in Fig. 3. It follows from the time-energy uncertainty relation that the created virtual fermion-antifermion pair would be existing only for a time interval not longer than $\hbar/2mc$. Therefore the distance between the creation and absorption points must be not larger than $c\hbar/2mc^2 = \hbar/2mc$. This gives for the interaction range in question $1/2m$ (coming back to the units with $\hbar = c = 1$). As it was mentioned above, the same will be true (because of the Eq. (1.7) for the annihilation range r defined by the two-boson s-channel exchange graph (Fig. 2). We have therefore

$$r_a = 1/2m \tag{1.8}$$

This means that the q^2-dependence of the annihilation induced Born scattering amplitude will be of the type

$$M(q^2) \sim \frac{1}{q^2 + 4m^2} \tag{1.9}$$

which in accordance with the general relation (1.3) leads for low energy fermions to an effective Yukawa potential

$$V(r) \sim e^{-2mr}/r \tag{1.10}$$

Let us mention that this "potential" is complex and energy dependent, because annihilation process into two bosons is allowed, and therefore the Born scattering amplitude corresponding to the two-boson s-exchange graph possesses an imaginary part:

$$\mathrm{Im}\, V(r) \sim \left| \quad \right|^2 \tag{1.11}$$

If the fermions $B,\bar{B}$ are composite particles, we shall have instead of the coupling constant g an amplitude $g(Q^2, \Pi_1^2, \Pi_2^2)$, where Q, Π_1 and Π_2 are the 4-momenta related with the lines adjoining to the vertices of the graph in Fig. 2. The momentum Q belongs to the boson ($Q = k, k'$). As it can be seen from Eq. (1.5), Q does not contain the momentum transfer g. The momentum Π_2 is that of a free fermion ($\Pi_2 = p_1, p_3, -p_2, -p_4$), therefore $\Pi_2^2 = m^2$ is a constant. The momentum Π_1 is associated with an internal fermion line ($\Pi_1 = p, p'$). The momenta p and p' are connected by the relation (1.6). If p is chosen as an independent vector variable for the integral related with the graph in Fig. 2, then $p' = p - q$ and the vertex amplitude s $g(Q^2, (p - q)^2, m^2)$ in general would be functions of the momentum transfer q. For this reason the whole amplitude in Fig. 2 will obtain an additional (compared with the Eq. (1.9) q-dependence. But it is important to mention that this mass off-shell effect is not related with the fermion's spatial size. The latter defines the dependence of the vertex amplitudes on the boson's momentum as the boson wave serves as a probe of the extension of its fermion source. On the other side, it is obvious that the fermion wave can not be in the same way a probe of its own internal structure. The physical nature of the vertex amplitude dependence on the fermion momenta is of the same kind as the energy off-shell effect for a particle moving in a potential. In the phenomenological approach, the simplest graph for the vertex amplitude giving its Π^2-dependence is the one boson exchange between the initial and final fermions (see Fig. 4). The amplitude $g(\Pi^2)$ as an analytical function of Π^2 has a singular point at $\Pi_0^2 = (m + \mu)^2$, corresponding to the normal threshold for a virtual inelastic process $B \to B + \pi$. The dimensionless amplitude $g(\Pi^2)$ is in fact a function of the variable $\Pi^2/(m + \mu)^2$. Therefore the q-dependence of the vertex amplitude should be really measured in the scale of the variable $q^2/(m + \mu)^2$, i.e. the characteristic parameter will be again of the order of the baryon Compton length $1/m$.

In the QCD approach it is hardly possible to present any reliable derivation for the analytical properties of the vertex amplitude $g(\Pi^2)$ as a function of Π^2 because of the difficulties arising from confinement. Concerning the quark bag models it must be mentioned that they are non-analytic in Π^2 (and not only on this variable) from the beginning.

Fig. 4(a) The one-boson exchange term in the vertex amplitude $g(Q^2,\Pi^2,m^2)$.
(b) Graph with one fermion line contracted giving the singular point $\Pi_0^2 = (m + \mu)^2$.

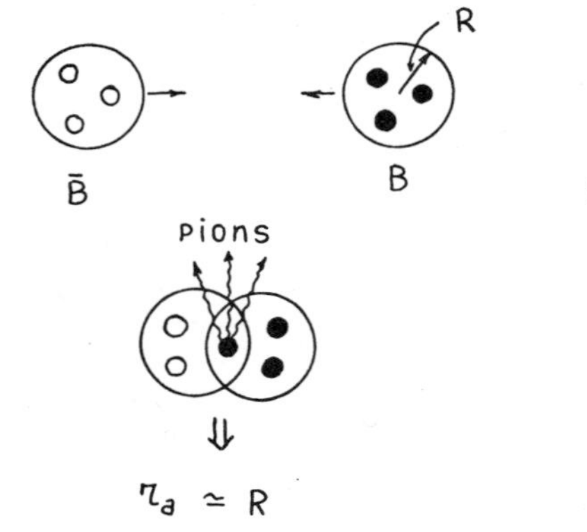

Fig. 5 Two view points on the annihilation range:

(a) A popular but not adequate corpusculare picture.

From the above discussion we conclude that the annihilation range is of the order of the baryon Compton length. It seems reasonable to emphasize once more two points: first, that concerning the range problem, we have to deal with elastic $B\bar{B}$ scattering induced by annihilation in the framework of the wave picture (diffraction) and second, that only full virtual annihilation without any fermion in the s-channel intermediate state) must be considered. A graphic illustration is given in Fig. 5.

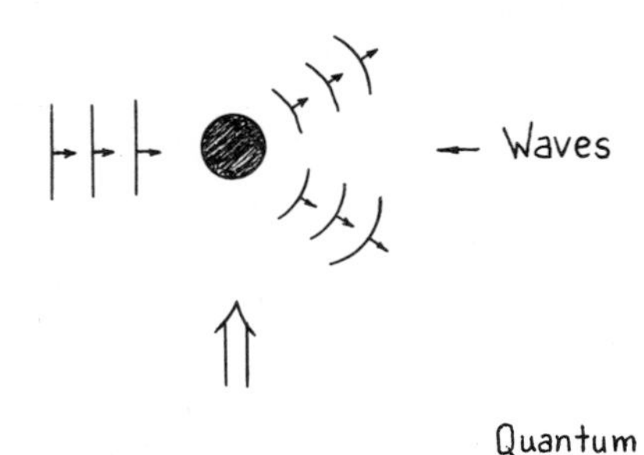

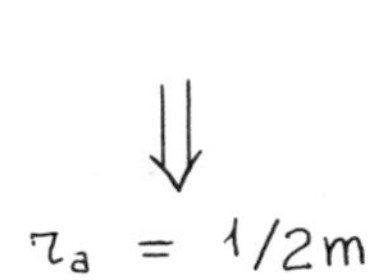

$\tau_a = 1/2m$

Fig. 5 (b) The quantum field wave approach (diffraction on an absorbing body).

During the discussions on the annihilation range problem this Summer at Erice, the annihilation of muonium (μ^+e^-) and antimuonium (μ^-e^+) into 4γ was treated. The constituents of these composite systems are point-like particles and the cross section of the above mentioned annihilation process may be calculated in the frame of QED perturbation theory. A simple estimation was given by S. Brodsky. He considered the "re-arrangement" graph (Fig. 6). Neglecting inter-atomic velocities it is easy to obtain for the total annihilation cross section σ_a:

$$v\,\sigma_a = v\,\sigma_a^{(e)} \quad v\,\sigma_a^{(\mu)}/\pi\, R_B^2 \tag{1.12}$$

Here v is the relative velocity of the colliding bodies ($1 \gg v \gg \alpha = 1/137$), R_B is the Bohr radius of the (μe)-atom:

$$R = \frac{1}{\alpha\,\bar{m}} \quad , \quad \bar{m} = \frac{m_e\, m_\mu}{m_e + m_\mu} \tag{1.13}$$

and

$$v\,\sigma_a^{(e)} = \frac{\pi\,\alpha^2}{m_e^2} \quad , \quad v\,\sigma^{(\mu)} = \frac{\pi\,\alpha^2}{m_\mu^2} \tag{1.14}$$

where m_e and m_μ are the electron and muon masses.

It follows from the Eqs. (1.12-1.14):

$$v\,\sigma = \pi\,\alpha^6/(m_e + m_\mu)^2 \tag{1.15}$$

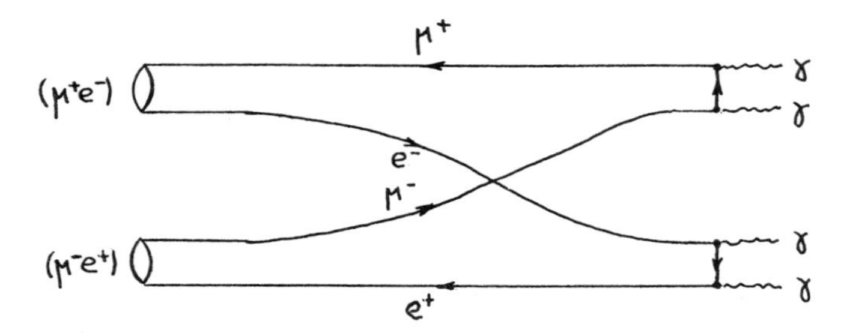

Fig. 6 Rearrangement graph for the annihilation process $(\mu^+e^-) + (\mu^-e^+) \rightarrow 4\gamma$.

This example shows that the annihilation range r_a for such extended systems as (μe)-atom is in order of magnitude equal to the Compton length $1/(m_e = m_\mu \approx 1/m_\mu$ of the atoms but not to their size $R \approx \frac{1}{\alpha\, m_e}$ which is 10 times greater.

$B\bar{B}$ nuclear potentials

These potentials $V(B\bar{B})$ can be obtained from the BB potentials V(BB) by G-conjugation in the frame of the OBEP model (Fig. 7). This approach (first proposed in 1968 by Bryan and Phillips[1]) is helpful because in the important case of NN interaction some information on $V(N\bar{N})$ may be extracted from the conventional nuclear physics (NN-scattering at low energies, few body nuclei). In any case, at low energies it is useful to have, at first glance, the possibility for interconnection of quite different physical phenomena in the BB and $B\bar{B}$ systems. There are some variants of the OBEP models (see for instance Refs. 1, 2, 3). They differ in the choice of some boson-baryon coupling constants, cut-off parameters of the singular (at small distances) terms and in the type of heavy mesons participating in the exchange. But all OBEP models obligatory include π-, η-, ρ- and ω-exchanges practically with the same set of coupling constants. Of particular importance are the ω-exchange forces. The quantum numbers J^{PC} of the ω-meson are the same as for photon

$$J^{PC} = 1^{--} \tag{1.16}$$

Therefore, just as for Coulomb interaction between $e^- - e^-$ and $e^- - e^+$ (Fig. 8a), the ω-exchange forces are repulsive in the BB systems and attractive for the $B\bar{B}$ pairs (Fig. 8b). Further, the ω-exchange is an isoscalar and also possesses no any other flavours. This means that ω-exchange is allowed for all baryons. In addition, the coupling constant

g_ω = [vertex diagram: B B line with wavy ω line]

is very large

$$\frac{g_\omega^2}{4\pi} = 10$$

Because of properties mentioned above, the ω-exchange gives strong repulsive forces between B-B and strong attraction between B-$\bar{B}$. Combining with other OBE it leads to a substantial distinction between V(BB) and $V(B\bar{B})$: the resulting $B\bar{B}$ nuclear potential well is much deeper than that for BB.

The existence of the repulsive core in NN nuclear interaction is a well known fact in conventional nuclear physics. It is the main reason for the relatively weak mean nuclear attraction even in the S-states of NN-system: only one loosely bound triplet state $^3S_1$ (the deuteron) and one singlet $^1S_0$ virtual state exist there (manifesting themselves in n-p scattering at very low energies). Bound or resonant states with non-zero orbital momenta (L = 0) are absent in the NN-systems. We do not consider here still doubtful experimental indications on the so-called di-baryon resonances. Their existence does not follow from the nuclear interaction (OBEP), but is predicted by some of the quark bag models.

In contrast, the potential $B\bar{B}$ well is deep enough for appearance of bound and resonant states with non-zero orbital momenta (this statement was published about 20 years ago (see Ref. 4 and also the survey [5]). The nuclear-type (quasi-nuclear) bound or resonant $B\bar{B}$ states are almost all (if not all) nodeless (because the potential well in question is not so deep and extended as it should be for existance excited states with nodes). Therefore the position of quasi-nuclear levels with $L \neq 0$ will be higher than that of the S-levels or, in other words, it is expected that near the threshold (2M) must be located the levels with non-zero orbital momenta and among them the P-levels (L = 1). As it will be shown below (see section 2.3) the numerical calculations with realistic OBEP forces and annihilation taken into account indeed confirm this qualitative predictions. This conclusion is an important point of the low-energy antiproton physics. It is the main reason of the recently discovered at LEAR P-wave enhancement phenomena in low energy $N\bar{N}$ scattering and in the process $p\bar{p} \to \Lambda\bar{\Lambda}$ near threshold. A graphic summary of the situation described above is presented in Fig. 8.

Effective $B\bar{B}$ interaction with annihilation included

The Optical Model. This is a well known semi-empirical approach for calculations of the elastic (σ_{el}), annihilation (σ_a) and total (σ_T) scattering cross sections σ_{el} taking into account the inelastic processes. The optical model potential $V_{opt}(r)$ is complex. Its imaginary part $W(r) = \text{Im } V_{opt}(r)$ is negative (corresponding to "absorption" of the initial particles due to inelastic processes). The "depth" and range parameters of W are fitting variables (the set of their values reproducing the experimental data is not unique). In the conventional optical models used in nuclear physics for the description of the particle-nucleus collisions the ranges of Re V(r) and Im V(r) are the same (equal to the size of the nucleus). In the case of the $B\bar{B}$ interaction the range of $W(B\bar{B})$ must be of the order of r_a whereas the range R of the main part of Re $V_{opt}(BB)$ is that of OBEP, r, with, as shown previously, $R \gg r_a$).

The annihilation cross section for the low energy $p\bar{p}$ collisions is close to the unitary limit. It seems therefore at first glance that we have to deal simply with a textbook picture for wave diffraction on a black body. But in the latter case one has $\sigma_a = \sigma_{el}$, whereas the $N\bar{N}$ experiments give $(\sigma_a/\sigma_{el})_{exp} = 1.5 - 1.8$. This is a phenomenological indication that the range of W must be smaller than that for Re V_{opt} = OBEP. In one of the first variants of the $V_{opt}(N\bar{N})$ [1] the range r_a of W was stated close to its physically reasonable value ($r_a = 1/2M \simeq 0.1$ fm) discussed above. But in a number of recent papers on this item for the range of W were chosen values significantly greater than the baryon Compton length (between 0.5 fm and 0.8 fm - see for instance Ref. 2). This can hardly be accepted from a physical point of view. It was also shown by numerical calculations that a better fit to the experimental data on the $B\bar{B}$ cross sections can be obtained in the frame of the optical model, introducing in addition to OBEP a short range (the same as for W) attractive potential U(r). The physical source for additional attraction between B and $\bar{B}$ is the reannihilation processes considered previously (see Figs. 1 and 2). But again, it follows from the physical reasons that the range of U must be equal to $r_a = 1/2M$, that is 5 - 8 times smaller than the fitting values used for U and W in the current optical model calculations of the $N\bar{N}$ cross sections. As it was described above, the defective $B\bar{B}$ potential in the optical model may be written in the form

$$V_{opt}(B\bar{B}) = V_{OBEP}(B\bar{B}) + U(r) + iW(r) \tag{1.17}$$

where both U and W are negative and possess the same effective range (smaller than the range of OBEP, but substantially larger than the baryon

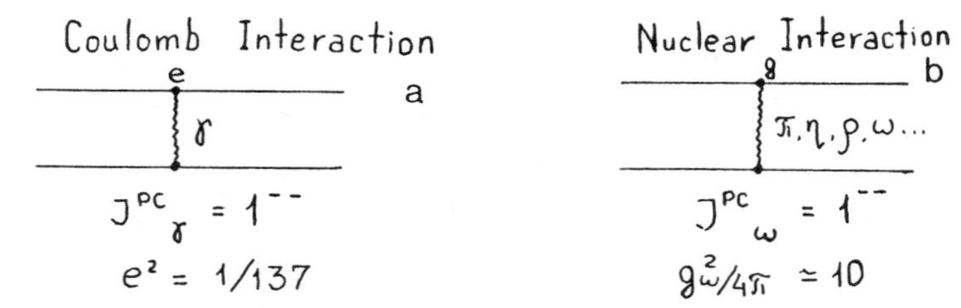

Fig. 7 Coulomb interaction (a) and OBEP models (b) are compared. In the latter case ω-exchange forces are of particular importance.

Compton length). The OBEP part in Eq. (1.17) represents the $B\bar{B}$ nuclear forces, whereas the complex potential U + iW is introduced for taking into account the annihilation phenomena (the range and the depth of U and W are energy- and L-independent fitting parameters). (Usually U(r) and W(r) are taken in the Woods-Saxon form. In order of magnitude U(r), W(r) are adjusted to be about 1 GeV. Let us mention that the possible choice of depth parameters - especially that for W - is not unique.)

The main defect of the optical model is its lack of unitarity. This makes difficulties connected with the non-orthogonality of the eigenfunctions. For instance, the apparatus of the perturbation theory is not applicable (at least in its conventional form). Further, the values of the optical model wave-functions at finite ($r < \infty$) distances are non-reliable. This can be seen even qualitatively: the optical model potential takes into account the absorption (annihilation) but does not reproduce the effect of reappearence of the initial particles. Therefore the values of the $B\bar{B}$ wave-function $|\Psi(r)|^2$ are usually underestimated in the optical model approach (the additional real potential U(r) which was mentioned above takes only partially into account the consequences of the reannihilation transitions). This defect and some other weaknesses make the applicability of the optical model in the low energy antiproton physics at least doubtful and therefore very limited.

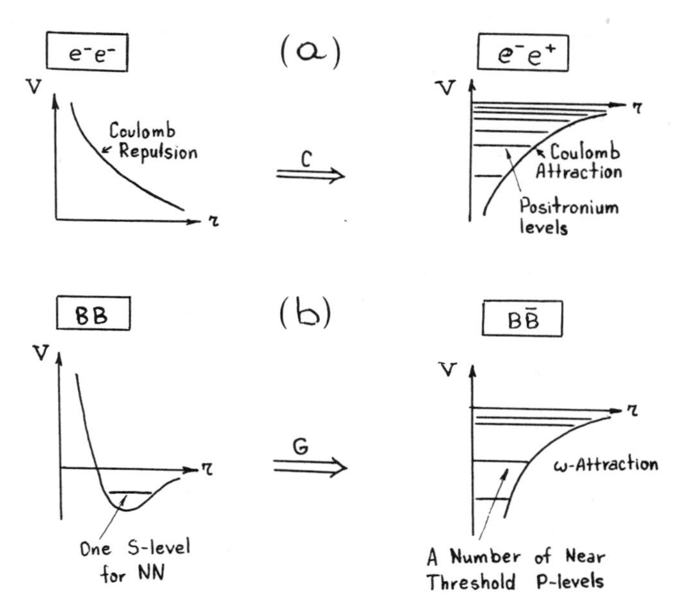

Fig. 8 Coulomb potential for C-conjugated e^-e^- and e^+e^- interactions (a) and for G-conjugated BB and $B\bar{B}$ interactions (b),

The Coupled Channel Model. This approach is selfconsistent from a theoretical point of view. The wave function contains n components, where n is the number of coupled channels. The Hamiltonian interaction V is a $n \times n$ matrix. Its non-diagonal matrix elements V_{ij} $(i,j = 1,2,\ldots,n)$ correspond to the coupling between channels i and j. In contrast to the optical model V is hermitian and therefore the CCM is a unitary theory with a complete set of orthogonal eigen-functions.

For the low energy antiproton physics the minimal number of channels is $n = 2$: (1) one $B\bar{B}$ channel and (2) one annihilation channel (describing the mesons). In the simplest CCM used in the theory of $N\bar{N}$ low energy scattering and annihilation[6-8] the $B\bar{B}$ channel corresponds to N and $\bar{N}$ with nuclear interaction between them. The annihilation channel represents two free (non-interacting with each other) mesons. The Hamiltonian interaction may be written in the form:

$$\hat{V} = \begin{pmatrix} V & V_{21} \\ V_{12} & 0 \end{pmatrix} \quad , \quad V_{12} = V_{21} \quad , \quad \hat{V}^{+} = \hat{V} \tag{1.18}$$

Here $V(r) = B\bar{B}$ - OBEP and $V_{12}(r)$ corresponds to the coupling between the $B\bar{B}$ and annihilation channels:

$$V_{12} = \lambda_L \frac{e^{-r_a/r}}{r} \quad , \quad r_a = 1/2M \tag{1.19}$$

The λ_L is an L-dependent fitting parameter (we assume for simplicity that the L-mixing due to the tensor $B\bar{B}$-forces is not very important at low energies because the D- and F-wave contributions are small). The $\Psi_{B\bar{B}}(r)$ (that is the $(B\bar{B})$-component of the CCM wave function) will be an eigen-function of an integro-differential "Shrödinger" equation with an effective interaction "potential" $V_{eff}(BB)$ which can be symbolically represented in the following form

$$V_{eff}\ (B\bar{B}) = V + \int V_{12}(r)\ G(r,r';E)\ V_{12}(r') \tag{1.20}$$

The second term in this equation is an integral operator (its action on $\Psi_{B\bar{B}}(r')$ includes integration over r') which contains the Green function of the Shrödinger equation for a system with kinetic energy $\varepsilon = 2(M - \mu) + E$ (M, μ are the baryon and meson masses, E is the kinetic c.m. energy of the $B\bar{B}$ pair). This term is complex and can be considered as an effective "optical potential"

$$\int V_{12}(r)\ G(r,r';E)\ V_{12}(r') = U(r) + iW(r) \tag{1.21}$$

$$U^+ = U \ , \quad W^+ = W \tag{1.22}$$

But in contrast with the optical models described above this "potential" is

(i) Non-local,
(ii) L- and E-dependent,

and, beside, U and W are rigidly interconnected. The calculations of Im V_{eff} performed by A. Voronin (Lebedev Physical Institute, Moscow, 1988; unpublished) show that the L- and E-dependence of the effective annihilation interaction are very significant (see Fig. 9). These dependences lead to quasi nuclear level's width smaller than given by optical model[9]. The real part of the effective reannihilation interaction for realistic (with right value for $r_a = 1/2M$) CCM is always negative. That means that the influence of the annihilation pushed down the quasi-nuclear bound or resonant $B\bar{B}$ levels also giving them the natural annihilation widths. However, it must be emphasized that these effects do not destroy the main features of the $B\bar{B}$ nuclear levels spectra for the reason that the annihilation range r_a is much smaller than the size of a quasi-nuclear state (which is equal in order of magnitude to the range R of the $B\bar{B}$ nuclear forces). As it will be shown in the next section, this is the most important fact creating the P-wave enhancement phenomena observed in low energy baryon-antibaryon physics.

A graphic summary of the results of the CCM approach is given in Fig. 10.

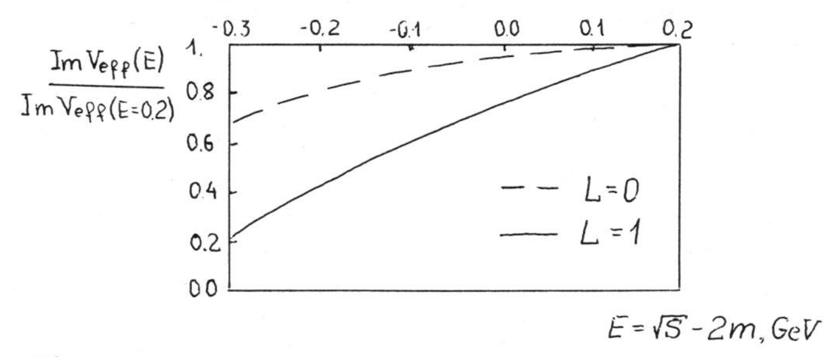

Fig. 9 The L- and E-dependence for Im V_{eff} (NN) in coupled channel model.

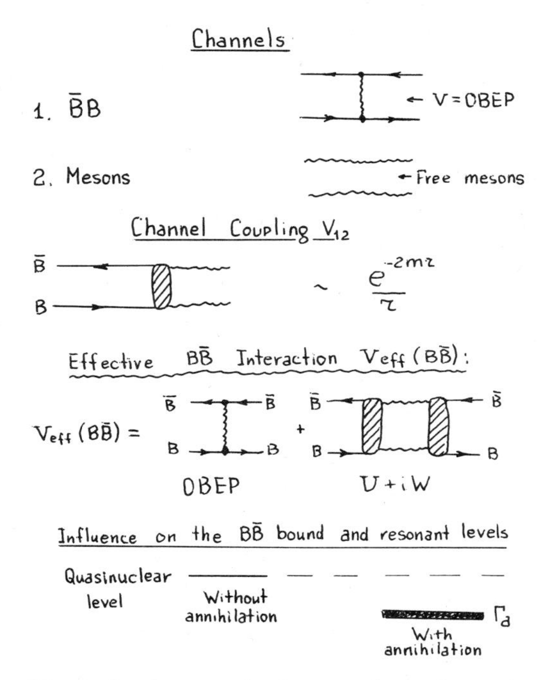

Fig. 10 The main items of the coupled channel approach in the low energy $B\bar{B}$ physics.

II INFLUENCE OF NUCLEAR INTERACTION ON NUCLEON-ANTINUCLEON SCATTERING AND ANNIHILATION AT LOW ENERGIES

The main experimental result obtained recently in scattering experiments at LEAR is the P-wave enhancement at low energies. Whereas the pp scattering at c.m. momenta $k \simeq 100$ MeV/c is isotropic, the $p\bar{p}$ elastic and charge-exchange ($p\bar{p} \to n\bar{n}$) scatterings are strongly forward directed, showing that the contributions of S- and P-waves are comparable in contrast to the usual range and L-barrier estimations for such low momenta. The annihilation from the P-states of the protonium is also enhanced. This follows from the $p\bar{p}$-atomic data: the annihilation width $\Gamma_a(2P)$ for the 2P protonium states is approximately 40 times larger than it would be expected from estimation mentioned above ($\Gamma_a(2P) = 40$ MeV instead of the "normal" value 1 MeV). Also the P-wave contribution to $p\bar{p}$ annihilation in flight at low energies is abnormally high. Another quantity showing an abnormal behaviour at low energies is the ratio $\rho = \mathrm{Re}\, f(0)/\mathrm{Im}\, f(0)$ ($f(0)$ is the forward $p\bar{p}$ scattering amplitude). This ratio grows from $\rho - -1$ at $k = 0$ to $\rho \simeq 0$ at $k = 70$ MeV/c. Such rapid growth can be explained only as a result of interference of S- and P-waves[10].

The physical reason for the P-wave enhancement at low energy $N\bar{N}$ collisions

The main reason is the existence of near-threshold bound or resonant states due to the strong attractive nuclear forces between N and $\bar{N}$. The energies E_B or E_R of these states are poles of the scattering amplitude $f(E,q^2)$ considered as an analytical function of the energy variable E. This means that

$$f(E = E_B) = \infty$$

The E_B (or E_R), being complex number, is located outside the physical region for the variable $E(0 \leqslant E)$. But if $|E_B|$ (or $|E_R|$) are small enough, the scattering amplitude $f(E)$ will be enhanced at low energies (this, for instance, is the reason why the n-p scattering cross-sections for low energy neutrons is enhanced both in triplet and singlet S-states: in the first case because of the loosely bound deutron state; in the latter, due to the existence of a very close to the threshold "antibound" or "virtual" pole).

In the case of $B\bar{B}$ systems all bound or resonant states are nodeless. Therefore the positions of the levels with L = 0 will be higher than those for L = 0. On the other hand at $kR \leqslant L$ the partial cross sections

$$\sigma_L \sim (kR)^{2L} \,/\, [(2L + 1)!!]^2$$

and therefore at low energies the contributions of waves with L > 1 even enhanced will be much smaller than the P-wave enhancement effect.

As the P-wave enhancement exists in the elastic scattering amplitude, it must also manifest itself in all inelastic collisions because of initial or final states interactions (the latter takes place in the near-threshold reaction $p\bar{p} \to \bar{\Lambda}\Lambda$): in fact, distorted by these interactions wave functions contain off-shell elastic scattering amplitudes (Fig. 11). A well known example demonstrating the influence of the near-threshold levels on an inelastic process is the annihilation cross section of e^+e^- pair into two photons. For "fast" positrons ($v >> 2\pi\,\alpha$; $\alpha = 1/137$) the annihilation cross section will be

$$v\, \sigma_a = \alpha^2/m_e^2$$

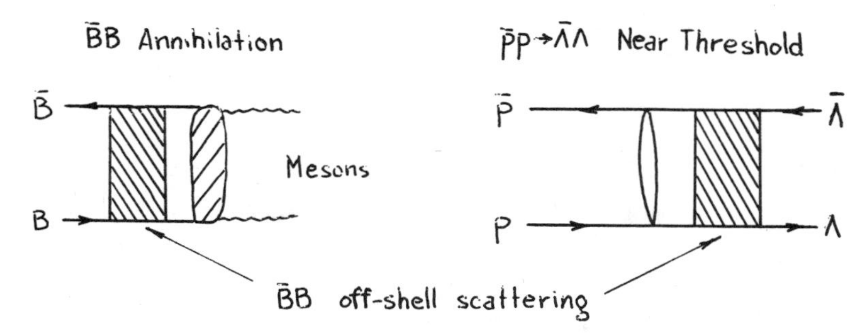

Fig. 11 Rescattering graphs corresponding to interactions in the initial and final states.

whereas for slow particles ($v << 2\pi\ \alpha$) the probability of the annihilation is enhanced:

$$v\ \sigma = \frac{\alpha^2}{m_e^2} \times \frac{1}{v} = \frac{\alpha^2}{m_e^2}\ \frac{\sqrt{m_e E}}{E}$$

This enhancement $1/E$ appears from the existence of a positronium level just at the threshold $E = 0$. The factor $\sqrt{E}$ corresponds to the usual threshold branching point $E = 0$ coinciding in the special case of Coulomb interaction with the pole $E = 0$ of the scattering amplitude.

It follows from the above consideration that the P-wave enhancement phenomena are connected with the theory of quasi-nuclear $B\bar{B}$ states from the beginning (or, in other words, on a qualitative level). Therefore, it is expected that even a rough model based on right physical principles may explain the phenomena observed at LEAR listed above. As it will be shown, this is indeed the situation which we have.

Adjustment of the CCM fitting parameters for $N\bar{N}$ interactions

A simplest CCM with minimal number of mesons in the annihilation channels was considered by the Lebedev Physical Institute group. To have non-relativistic kinematics all coupled channels, it was assumed that the masses μ of the two mesons are the same and equal to the mass of the ρ-meson.

There are two kinds of the fitting parameters which can not be obtained by G-conjugation from the conventional NN nuclear OBEP and therefore must be settled by fitting the experimental data. Firstly, there is the set of annihilation constants (see Eq. 1.19) for $L = 0,1,2$. As it follows from the fitting procedure

$$\lambda_S \simeq 4\ ,\quad \lambda_P \simeq 18\ ,\quad \lambda_D \simeq 44 \tag{2.1}$$

Second is the set (also for different L) of cut-off distances r_c for the singular OBEP terms. The nuclear NN data are not sensitive to the choice of r_c because of strong ω-repulsion at $r \geqslant r_c$. A zero cut-off for BB potential was used: $V(r < r_c) = 0$. The main reason for this statement is the repulsive character of the relativistic corrections. Depending on the quantum numbers $^{2S+1}L_J$ the r_c is varied within the limits 0.50 - 0.72 fm. (The cut-off distance depends on the quantum numbers because the singular terms come from the spin-orbital and tensor forces; for details see Ref. 8.) It is important that the fitting procedure do not give a unique set for the annihilation constants λ_L. Figure 12 shows a typical dependence of the partial (for given L) annihilation cross-section σ_a on λ. The

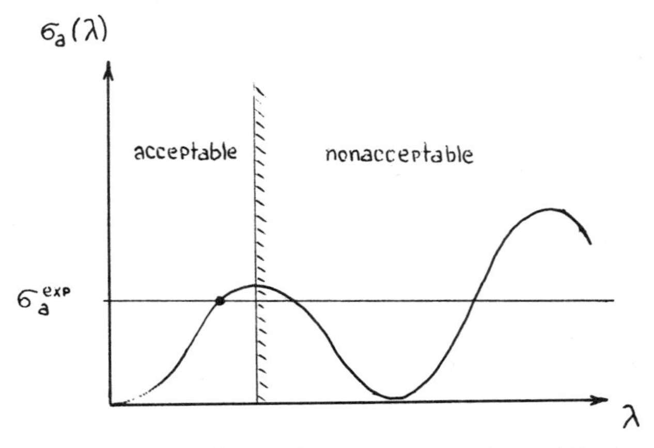

Fig. 12 Partial annihilation cross section $\sigma(\lambda)$ as a function of the channel coupling constant λ.

oscillation of $\sigma_a(\lambda)$ occur because bound states appear in the reannihilation attractive potential well U(r) at large λ's. Of course the use of such λ values is practically unacceptable in the potential approach (the $\Psi_{B\bar{B}}$ for these bound states will be localized at very small - relativistic - distances equal to 1/M in order of magnitude). Therefore only those sets of λ_L are possible for which the potential well U is not very deep. With this condition the set (2.1) is unique.

The fitting parameters λ_L and r_c were settled by reproducing the annihilation cross-sections, and also total (not differential) elastic and charge-exchange cross-sections. The OBEP coupling constants are taken from Ref. 3.

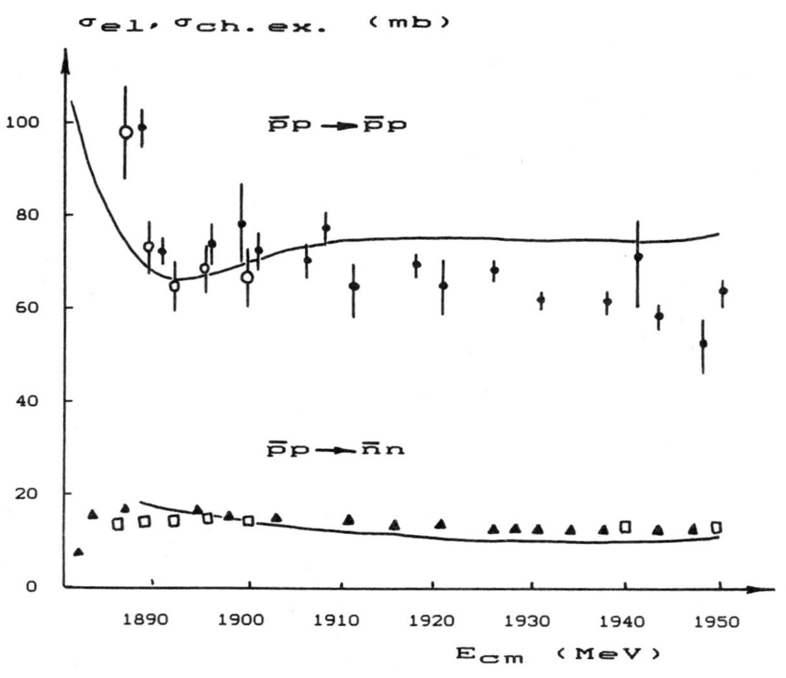

Fig. 13 Elastic and charge-exchange cross sections for $p\bar{p}$ interaction.

Theory and experiment

I would like to present below the comparison of the main theoretical results on $N\bar{N}$ scattering and annihilation obtained recently by the Moscow group at Lebedev Physical Institute[8] with the experimental data. All calculations are based on the simple but physically transparent CCM described above. In Figs. 13 and 14 the elastic scattering, charge-exchange and annihilation $p\bar{p}$ cross sections as functions of energy or momentum are shown. Some of the experimental points shown in these figures are used to settle the fitting parameters.

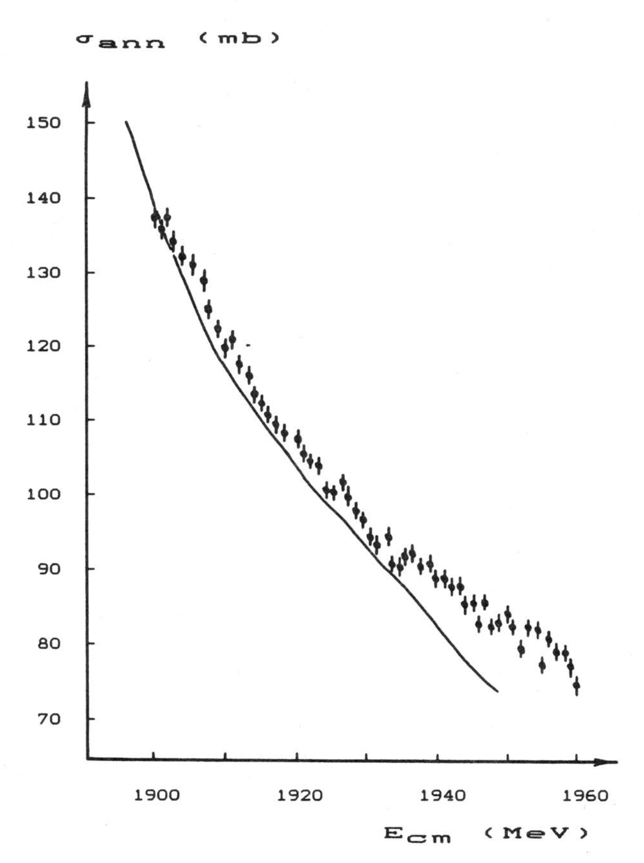

Fig. 14 Annihilation cross section of $p\bar{p}$ interaction.

Figure 15 illustrates one of the most important physical phenomena: a strong angular anisotropy of $p\bar{p}$ elastic scattering at low momentum (let us recall that pp scattering at this and even higher momenta is isotropic). From the theoretical point of view it is significantly not a fair agreement between theoretical curve and experimental points (which may be occasional for the used simple model) but itself the fact that the strong anisotropy in $p\bar{p}$ scattering at low energies follows from the calculations performed with fitting parameters settled only by the inclusive experimental data. The observed anisotropy results from the inteference

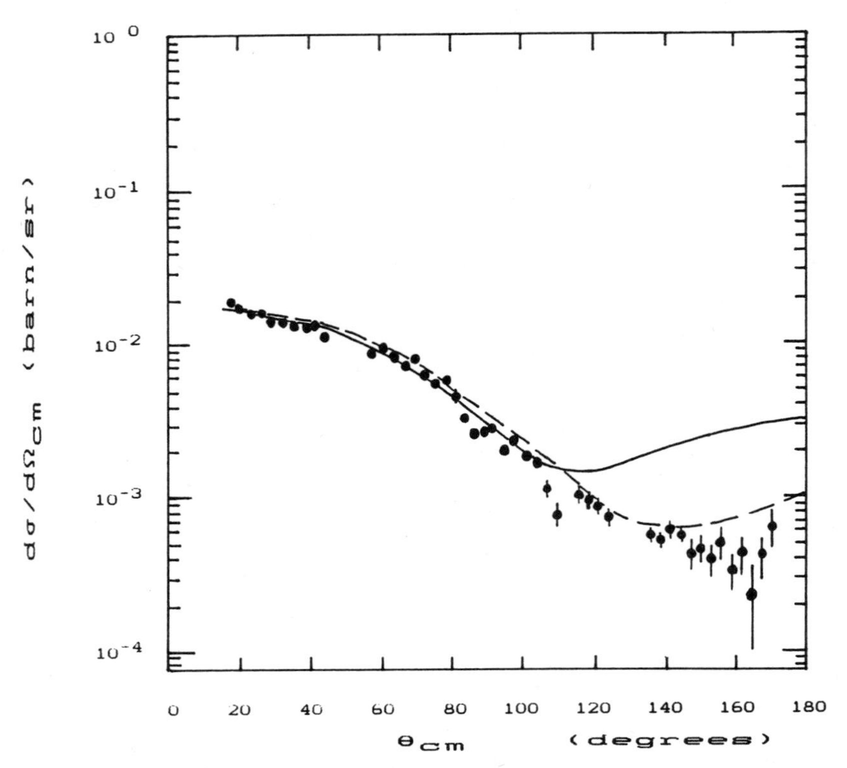

Fig. 15 Elastic differential cross sections for the $p\bar{p}$ scattering at the momentum of incident antiproton P = 287 MeV/c (c.m. momentum K = 144 MeV/c). The experimental data are taken from Ref. 11. The curves differ by the choice of the fitting paramter r_c. Solid curve is the calculation with $r_c(1^3S_1) = r_c(3^3S_1) = 0.50$ fm, $r(1^3P_2 = 0.72$ fm. Dashed curve is the same for $r_c(1^3S_1) = 0.47$ fm, $r_c(3^3S_1) = 0.60$ fm, $r_c(1^3P_2) = 0.65$ fm.

S- and P-waves. This means that an abnormal high P-wave contribution takes place in low energy pp scattering. Figure 16 makes clear the source of the P-wave enhancement in the low energy pp scattering. This is the nuclear interaction, but not the annihilation. As it can be seen on the figure, the P-wave contribution with annihilation switched out is even larger than that in the presence of annihilation. The latter effect is just what was theoretically expected: the annihilation interaction pushed down and broadened the near-threshold P-levels reducing their influence on the scattering amplitude. Let me emphasize that this picture and conclusion are quite opposite to the statements made in Ref. 1 on the basis of optical model calculations with annihilation range $r_a >> 1/2M$.

A fairly good agreement of the theoretical[8)] and experimental $p\bar{p}$-atomic data is demonstrated in Table 1. The same situation takes place for the momentum dependence of the quantity ρ = Re f(0)/Im f(0) in the low energy region (Fig. 17).

In Table 2 the calculated spectrum of the near-threshold $N\bar{N}$ quasi-nuclear levels is presented[6-8)]. As it can be seen from this table, there exist at least 5 near-threshold P-levels. This result confirms the main features of the theoretical expectations considered in the previous Section 1.

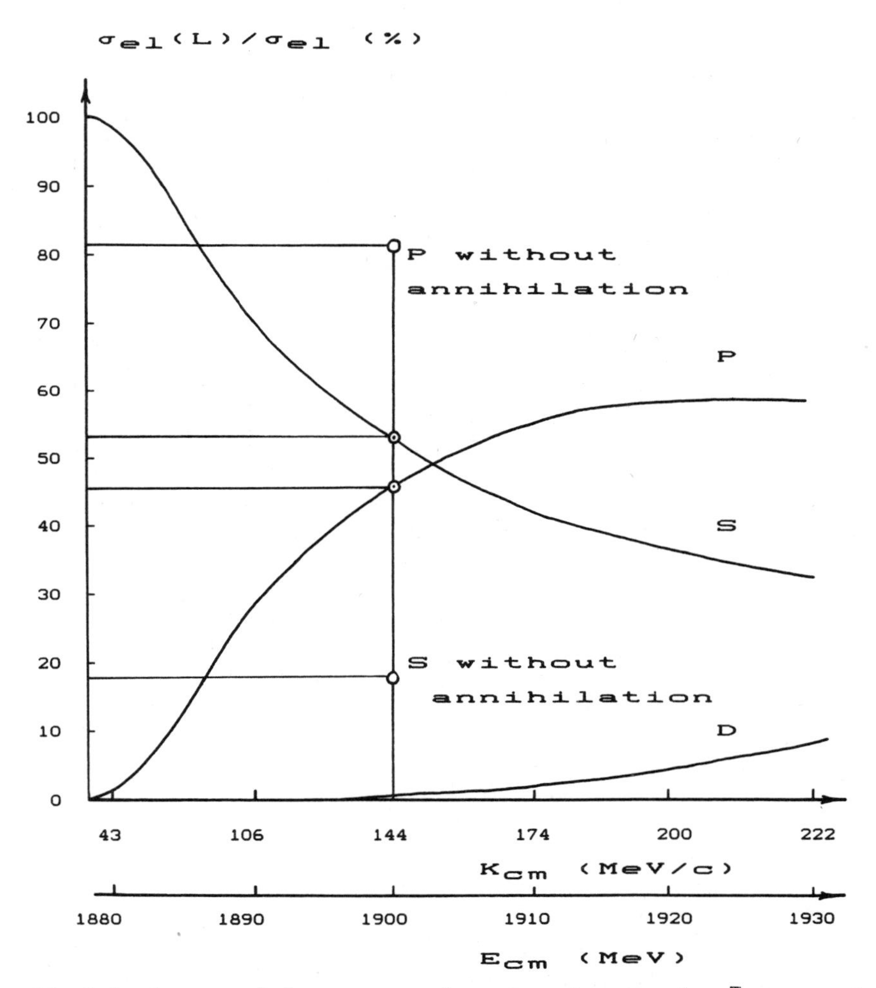

Fig. 16 Relative partial cross sections for the elastic $p\bar{p}$ scattering.

Table 1 Nuclear shifts and widths of 1S- and 2P-protonium states [12)]

	Experiment	Theory
$Re\ (\Delta E_{1S})$, KeV	0.70 ± 0.15 0.73 ± 0.15 0.70 ± 0.08	0.74
Γ_{1S}, KeV	1.60 ± 0.40 0.85 ± 0.39 1.13 ± 0.19	1.39
Γ_{2P}, meV	100 40 37 ± 11	39
$\rho\ (E = 0)$	- 1.24 ± 0.25 - 0.88 ± 0.31	- 1.08

Table 2 $N\bar{N}$ resonances in mass region 1700-1980 MeV

$^{2S+1}L_J$	$I^G\ (J^P)$	Mass, MeV	Total width, MeV	$\frac{\Gamma_{\bar{N}N}}{\Gamma}$
$^1P_1$	$1^+\ (1^+)$	1840	70	--
$^3P_0$	$1^-\ (0^+)$	1855	75	--
$^3P_1$	$1^-\ (1^+)$	1870	75	--
	$0^+\ (1^+)$	1800	35	--
$^3P_2$	$1^-\ (2^+)$	1870	85	--
	$0^+\ (2^+)$	1860	70	--
$^1D_2$	$1^-\ (2^-)$	1975	35	0.3
$^3D_1$	$1^+\ (1^-)$	1920	15	0.2
$^3D_2$	$1^+\ (2^-)$	1930	20	0.3
	$0^-\ (2^-)$	1955	35	0.6
$^3D_3$	$0^-\ (3^-)$	1975	35	0.3

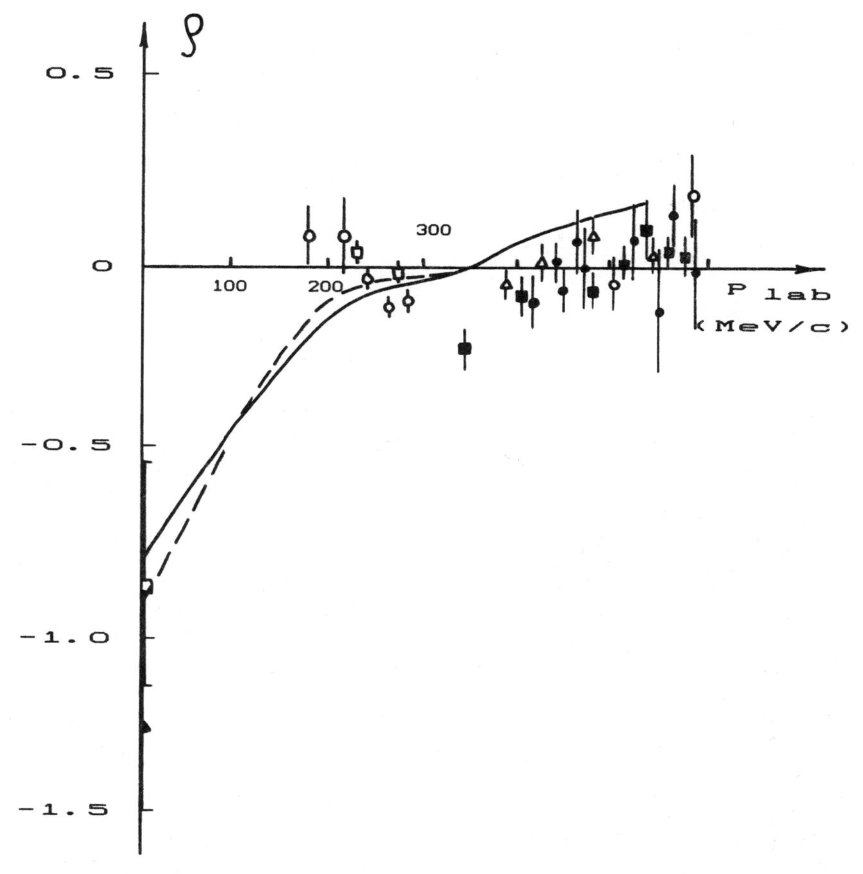

Fig. 17 Real-to-imaginary ratio for the forward elastic $p\bar{p}$ scattering amplitude as a function of incident antiproton momentum. The experimental data are taken from Refs. 11, 12 and 13. Solid curve is the calculation from Ref. 8, dashed - taking into account $p\bar{p} \to n\bar{n}$ channel using procedure proposed in Ref. 10.

III FINAL STATE INTERACTION AND THE NEAR-THRESHOLD HYPERON-ANTIHYPERON PRODUCTION IN PROTON-ANTIPROTON COLLISIONS

Modification of the simple coupled channel model

We shall consider here the near-threshold reaction $p\bar{p} \to \bar{\Lambda}\Lambda$ in the frame of CCM approach. For the case in question the two-channels model used above must of course be modified. The following complications are obvious:

(i) the existence of two $B\bar{B}$ channels ($N\bar{N}$ and $\Lambda\bar{\Lambda}$) instead of one in Section 2;

(ii) the appearing of a new coupling in the Hamiltonian - between the channels $N\bar{N}$ and $\Lambda\bar{\Lambda}$.

But besides these points there is also a non-trivial question whether the dominating annihilation modes into mesons for the $N\bar{N}$ and $\Lambda\bar{\Lambda}$ channels are the same or not? It is well known that the branching ratio for the $N\bar{N}$ annihilation mode containing strange mesons is smaller than 5%. As to the $\Lambda\bar{\Lambda}$ annihilation modes practically nothing is known. If we assume for the $\Lambda\bar{\Lambda}$ pair the same dominating annihilation modes as for $N\bar{N}$, then an additional coupling between the $N\bar{N}$ and $\Lambda\bar{\Lambda}$ channels would exist due to the set of virtual transitions:

$$N\bar{N} \rightleftarrows \text{mesons} \rightleftarrows \Lambda\bar{\Lambda}.$$

As these transitions are induced by strong interactions, it is natural to assume further that the coupling constants λ_L (see Section 2) between the baryon and meson channels are of the same order of magnitude for both $N\bar{N}$ and $\Lambda\bar{\Lambda}$ pairs. In this case it follows analytically from the CCM approach that the annihilation cross section for each $B\bar{B}$ pair will be reduced by a factor 1/n, where n is the total number of the $B\bar{B}$ channels[8]. This feature makes it impossible to obtain the experimental $p\bar{p}$ annihilation cross section even taking the greatest acceptable values for the constants λ_L. Therefore we conclude that the dominating annihilation modes for $N\bar{N}$ and $\Lambda\bar{\Lambda}$ channels must be quite different. Probably the dominant annihilation modes of $\Lambda\bar{\Lambda}$ pair are those containing strange mesons. It follows from the above consideration that for solving the problem in question we have to deal with four channels (two $B\bar{B}$ channels and two different meson channels) with an additional condition that $N\bar{N}$ and $\Lambda\bar{\Lambda}$ channels are not coupled through the virtual annihilation processes. We assume that the coupling between these channels occurs because of the K- and K*-exchange interaction (see Fig. 18). We take the coupling constants $g_{N\Lambda K}$ and $g_{N\Lambda K*}$ from Ref. 3 and introduce a cut-off fitting parameter r_{cpl} (using again the zero cut-off for $r < r_{cpl}$). The OBEP parameters for interaction inside each $B\bar{B}$ channel are also from Ref. 3. Each of the two annihilation channels contains two free mesons. The $N\bar{N}$ annihilation channel is the same as that used in Section 2. The masses of the mesons in the annihilation channel are equal and close to the mass of the K*. The numerical values of the constants λ_L are the same for both $B\bar{B}$ channels. Let us mention that the annihilation range for the $\Lambda\bar{\Lambda}$ channel is equal $r' = 1/2M_\Lambda$, where M_Λ is the mass of Λ. A graphic summary of this modified CCM is presented in Fig. 18.

The main results

From the previous calculations of the quasi-nuclear levels spectrum (without annihilation taken into account) it was expected that a number of near threshold P-states exists in the coupled $N\bar{N} - \Lambda\bar{\Lambda}$ system[14]. Nevertheless, the results obtained by the experimental collaboration PS185[15] for the $\Lambda\bar{\Lambda}$ production in the $p\bar{p}$ collisions just close to the threshold are very impressive and even surprising. On Fig. 19 the energy dependence of the total cross section of the reaction is shown. A very rapid grow of the cross section $\sigma_t(\varepsilon)$ can be seen with the rising of the c.m. kinetic energy ε of the Λ and $\bar{\Lambda}$. As $\sigma_t(\varepsilon) \sim \varepsilon^{3/2}$ it is clear that a large contribution from the P-waves takes place practically just at the threshold. Some of the experimental data presented in Fig. 19 were used for adjusting the values of fitting cut-off parameters r_{cpl} for different partial waves of the $\Lambda\bar{\Lambda}$ pair (the r_{cpl} is varied within the limits 0.67 - 0.96 fm[8]). The solid curve on Fig. 19 is calculated using the modified CCM described above. With the settled r_{cpl} the angular distribution of created $\Lambda\bar{\Lambda}$ was calculated for the energy ε = 3.6 MeV. The results are shown in Fig. 20 for two values of the $\Lambda\bar{\Lambda}$-OBEP cut-off radius r'_c = 0.64 fm and 0.58 fm (r'_c was assumed to be the same for all partial waves and was not used as a

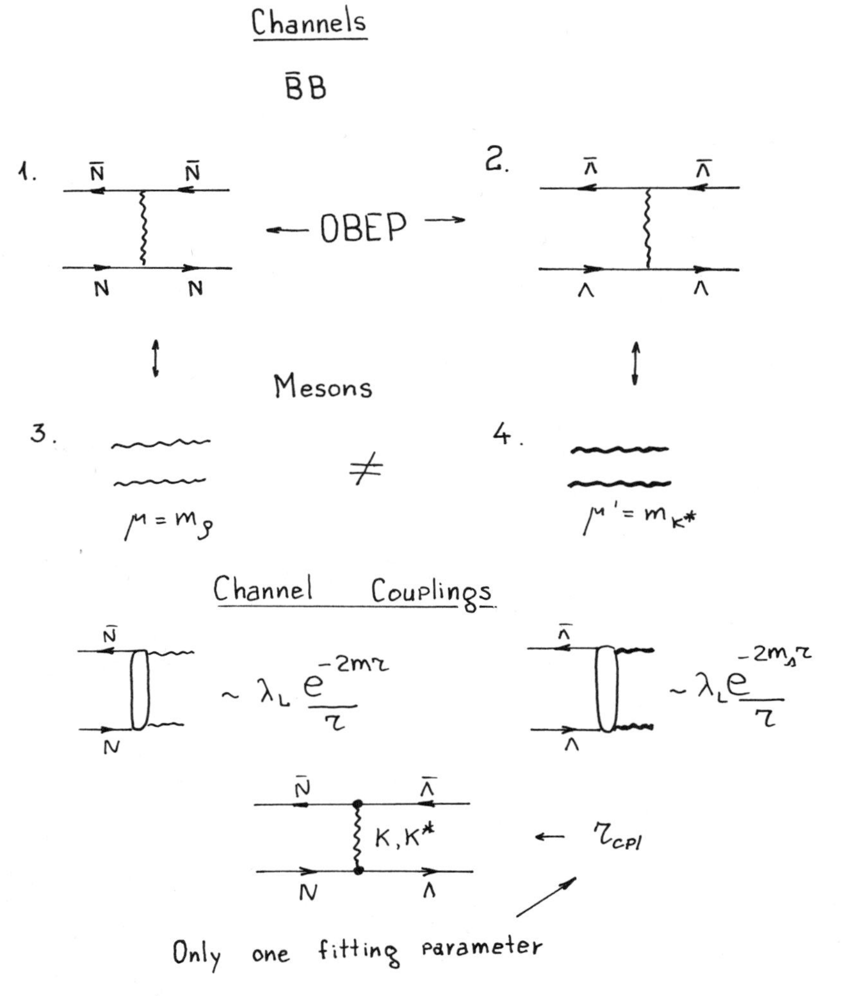

Fig. 18 The CCM approach to the theory of near-threshold reaction $p\bar{p} \to \Lambda\bar{\Lambda}$.

fitting parameter; the first of its two values was taken from Ref. 14, the other one was employed for the purpose of checking the stability of the theoretical results). With the same values of all parameters the polarization of $\bar{\Lambda}$ and Λ were calculated as a function of the c.m. emission angle $\nu^{\times}_{\bar{\Lambda}}$. The results are presented in Fig. 21. Both for the angular distribution and polarization the agreement between the theoretical and experimental data are fairly good.

To make clear the reason why the P-wave enhancement occurs in the considered reaction, the partial wave analysis of the theoretical curves were performed. It was ascertained that about 50% of the cross section at the energy ε = 3.6 MeV come from the triplet P-waves. For r' = 0.64 fm the main contribution (about 25%) gives the $^3P_1$-wave and for r' = 0.58 fm about 40% contributes to the $^3P_2$-wave. The investigations of the Argand plots for these waves clearly show the existence of the near-threshold

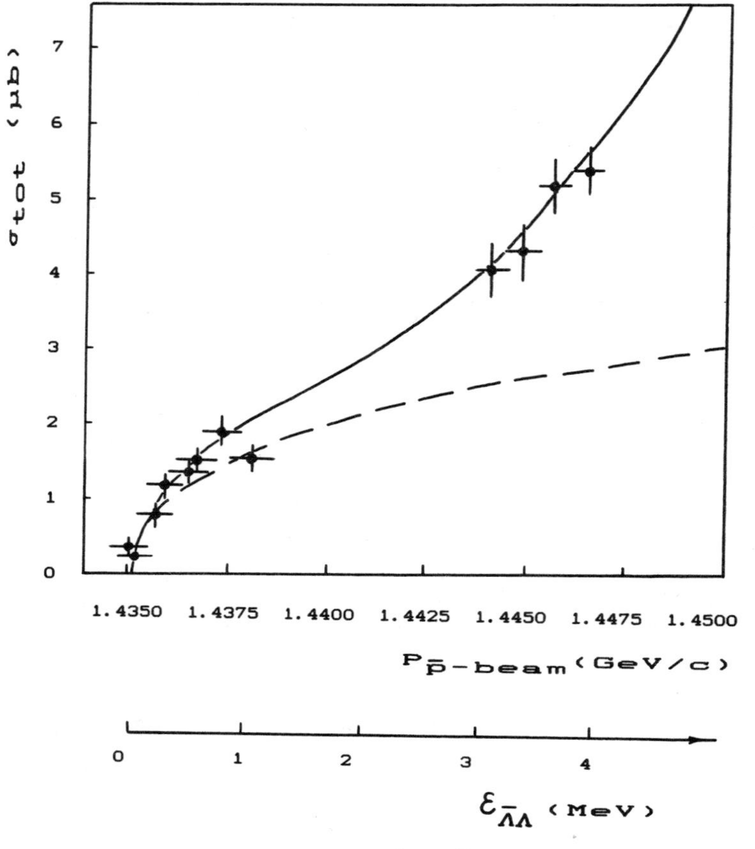

Fig. 19 Total cross section for $p\bar{p} \to \Lambda\bar{\Lambda}$ reaction as a function of incident antiproton momentum (ε is the kinetic c.m.s. energy of $\bar{\Lambda}$ and Λ). The experimental data are taken from Ref. 15. Solid curve is the calculation of this cross section; dashed - S-wave contribution.

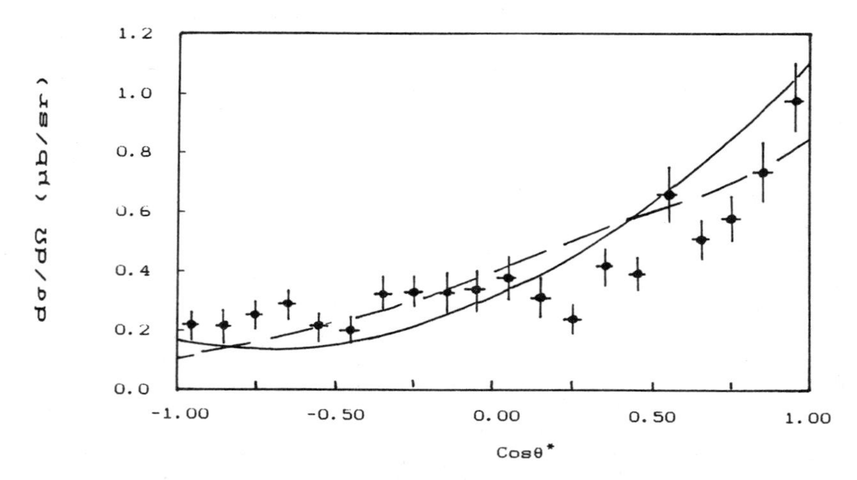

Fig. 20 Differential cross section for $p\bar{p} \to \Lambda\bar{\Lambda}$ reaction at the energy ε = 3.6 MeV. Solid curve corresponds to r_c' = 0.64 fm, dashed - r_c' = 0.58 fm (see text).

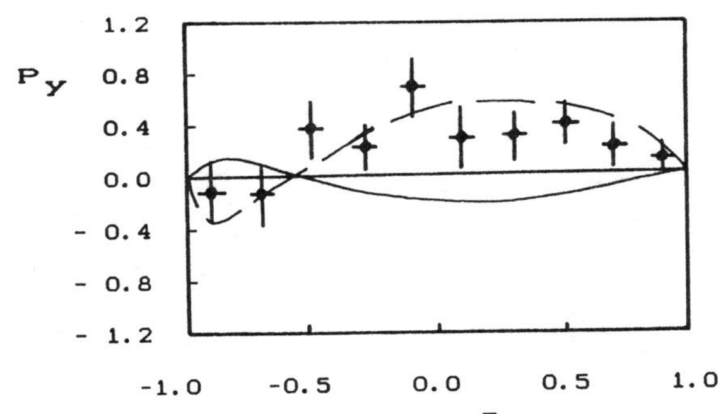

Fig. 21 Polarization of $\Lambda(\bar{\Lambda})$ at ε = 3.6 MeV. The type of the curve is the same as in Fig. 20.

resonant states in the $\Lambda\bar{\Lambda}$ system approximately at the energy ε_0 of several MeV with the total width Γ_t of about 10 MeV (for the $^3P_1$ resonance ε_0 = 6 MeV, Γ_t = 8 MeV (the $\Gamma_t = \Gamma_{\Lambda\bar{\Lambda}} + \Gamma_a$ include the "elastic" width $\Gamma_{\Lambda\bar{\Lambda}}$ = 5.5 MeV and the annihilation width Γ_a = 2.5 MeV). A typical Argand plot (for the $^3P_1$ amplitude) showing the "quality" of the resonance is presented in Fig. 22.

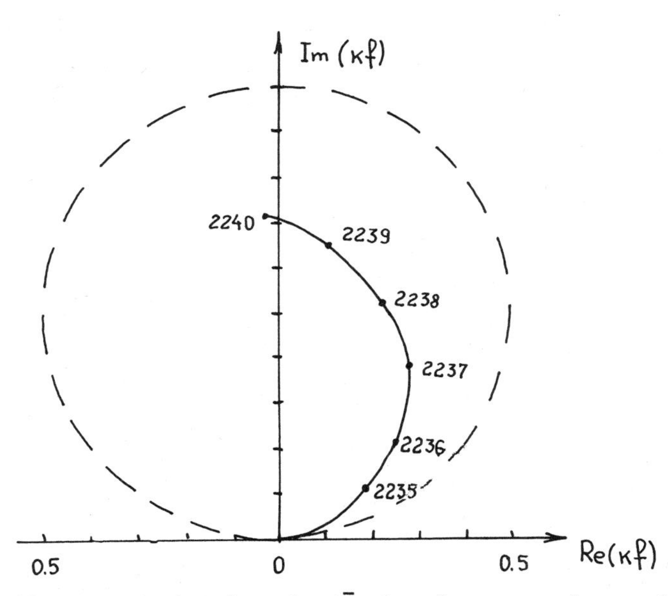

Fig. 22 Argand plot for the $\Lambda\bar{\Lambda}$ elastic scattering amplitude. Points on the curve corresponds to the total c.m. energy in MeV. Dashed curve is the unitary circle.

I should like to emphasize once more that not the fair agreement of the developed rough theory with the experiment is important but the fact that a physically transparent approach really gives a right qualitative picture explaining the low energy P-wave enhancement effect and its observable manifestations.

A number of calculations of the $p\bar{p} \rightarrow \Lambda\bar{\Lambda}$ reaction's features are published in the last years[16-20]. Only in Ref. 20 (besides Ref. 8 discussed above) the quantities of interest at very low energy $\varepsilon = 3.6$ MeV were considered. In this paper the calculations are performed in DWBA approach in the frame of optical model. The observed $\sigma_t(\varepsilon)$, angular distribution and polarization are reproduced by calculations, but the physical reasons of these theoretical results are hidden and the starting formulas for the reaction amplitude are in my opinion at least doubtful, because of items mentioned above in Section 1. Also, the author's statement about the "S-wave suppression" due to the annihilation is not confirmed by our calculations in which the optical DWBA method was used with parameters close to those in Ref. 20 (the S- and P-partial wave scattering amplitudes have their normal values; the "suppression" considered by the author corresponds to comparison of Born approximation with and without distortion; it has nothing in common with the problem in question).

Conclusion

(1) P-wave enhancement observed in $p\bar{p}$-scattering and annihilation and in $p\bar{p} \rightarrow \Lambda\bar{\Lambda}$ near-threshold production are of the same physical nature.

Bound and resonant states in the $B\bar{B}$ systems manifest themselves in phenomena discovered recently at LEAR.

(2) Processes $p\bar{p} \rightarrow X\bar{X}$ near threshold are unique tools to obtain believable information on matter-antimatter interaction. Therefore of great interest are:

(a) Further LEAR experiments as close to the thresholds as possible.

(b) Operation of Super-LEAR to study hyperon-antihyperon and charm-anticharm interactions at low energies.

ACKNOWLEDGEMENT

I should like to thank O.D. Dalkarov and K.V. Protasov for useful discussions. I am also very indebted to K.V. Protasov for his help in preparing the written version of these lectures.

REFERENCES

1. R.A. Bryan, R.J.N. Phillips; Nucl.Phys. 85 (1968) 201.
2. C.B. Dover an J.M. Richard; Ann.Phys. NY 121 (1979) 70.
3. M.M. Nagels, T.A. Rijken, J.J. de Swart; Phys.Rev. D20 (1979) 1633.
4. O.D. Dalkarov, V.B. Mandelzweig, I.S. Shapiro; Pis'ma v ZhETF 10 (1969) 402 and Nucl.phys. B21 (1970) 88.
5. I.S. Shapiro; Phys.Rep 35C (1978) 129.
6. I.S. Shapiro; Nucl.Phys. A478 (1988) 665c.
7. I.S. Shapiro; Proc. IV LEAR Workshop, Villars-sur-Ollon, Switzerland, 6-13 Sept. 1987, p. 377.
8. O.D. Dalkarov, K.V. Protasov, I.S. Shapiro; Preprint 37, Moscow, FIAN (1986).
9. M. Maruyama and T. Ueda, Preprint Osaka Univ., OUAM 80-11-5 (1980).

10. O.D. Dalkarov and K.V. Protasov; JETP Lett. 44 (1986) 638.
11. W. Bruckner et al.; Phys.Lett. 166B (1986) 113 and
Phys.Lett. 158B (1985) 180.
12. M. Ziegler et al.; Preprint CERN-EP/88-05,
T.P. Gorringe et al.; Phys.Lett 162B (1985) 71,
C.-J. Batty; contribution to this School.
13. L. Linssen et al.; Nucl.Phys. A469 (1987) 726.
14. I.S. Shapiro and R.T. Tyapaev; Sov.Phys.JETP 59 (1984) 21.
15. R. von Frankenberg; Proc. IV LEAR Workshop, Villars-sur-Ollon,
Switzerland, 6-13 Sept. 1987, p. 347 (Ed. C.Amsler et al.,
Harwood Academic Publishers, New York)
16. F. Tabakin and R.A. Eisenstein; Phys.Rev. C31 (1985) 1857.
17. J.A. Niskanen; Helsinki preprint HU-TFT-85-28 (1985).
18. S. Furui and A. Faessler; Nucl.Phys. A468 (1987) 669.
19. R.G.E. Timmermans, T.A. Rijken, J.J. de Swart;
Nucl.Phys. A479 (1988) 383c.
20. M. Kohno and W. Weise; Phys.Lett. 179B (1986) 15,
Nucl.Phys. A479 (1988) 433c,
Phys.Lett. 206B (1988) 584.

A REVIEW OF $N\bar{N}$ ANNIHILATION MECHANISMS

A.M.GREEN

Research Institute for Theoretical Physics
University of Helsinki
Siltavuorenpenger 20 C
Helsinki 17, Finland

1 INTRODUCTION

The $N\bar{N}$ interaction can be considered as an extension of the corresponding NN interaction, and so contains all of the uncertainties present in the latter. In addition to these problems the $N\bar{N}$ interaction has the complication of annihilation - the main topic of these lectures.

The effect of annihilation can be treated with various degrees of sophistication. These range from - a) simply adding a complex phenomenological interaction to an $N\bar{N}$ potential derived from some standard NN potential, upto– b) microscopic approaches involving quarks and gluons. A detailed review of this subject can be found in refs. [1] and [2], and so this written version of the lectures will concentrate on developments since the completion of the second review in December 1985.

As with many branches of physics, interest is stimulated by an ability to carry out appropriate experiments. In this case, it is fair to say that the conception and construction of the Low Energy Antiproton Ring (LEAR) at CERN in the early 1980's was the main catalyst. However, it should be added that the antinucleon work at Brookhaven (BNL) and Tokyo (KEK) has also played and continues to play an important role. The outcome of these experimental programmes has been a wide range of data covering many different aspects of antinucleon interactions with the spectrum going from the study of the basic $N\bar{N}$ interaction upto the study of collective $\bar{N}$-nucleus degrees of freedom. In these lectures the data of most direct interest is that of $\bar{N}N$ scattering and protonium, and the various $\bar{N}N$ annihilation channels into specific mesons. Immediately it must be said that much of this data is far from complete and also that there are inconsistencies between the results of different groups. The incompleteness is seen by the fact that the bulk of the new data does not involve spin observables and also that adequate antineutron beams are only just becoming available [3]. In comparison, NN scattering has for years been carried out with polarised beams of both protons and neutrons on polarised targets [4]. The inconsistencies are most apparent in the measurement of $N\bar{N}$ annihilation channels into specific mesons and can be seen in table 1 [5], e.g. $\bar{p}p \rightarrow \rho^0\omega$ and $\pi^{\pm}A_2^{\mp}$.

Antiproton-Nucleon and Antiproton-Nucleus Interactions
Edited by F. Bradamante *et al.*
Plenum Press, New York, 1990

Table 1 Branching ratios for $\bar{P}P$ annihilation at rest

Final State	Branching Ratio %	Reference
$\pi^\circ\gamma$	$(1.50 \pm 0.70) \times 10^{-2}$	a
	$(1.74 \pm 0.22) \times 10^{-3}$	a
$\pi^\circ\pi^\circ$	$(2.06 \pm 0.14) \times 10^{-2}$	a
	$(4.8 \pm 1.0) \times 10^{-2}$	a
	$(1.4 \pm 0.3) \times 10^{-2}$	a
	$(2.6 \pm 0.4) \times 10^{-2}$	b
$\pi^\circ\eta$	0.82 ± 0.10	a
	0.046 ± 0.013	b
$\pi^\circ\eta'$	0.050 ± 0.019	b
$\pi^\circ\omega$	< 1	a
	0.52 ± 0.05	b
$\pi^\circ f$	0.24 ± 0.07	a
$\pi^+\pi^-$	0.35 ± 0.02	a
	0.31 ± 0.03	c, PS183
	0.58 ± 0.07 (P-wave)	a
$\pi^\pm\rho^\mp + \pi^\circ\rho^\circ$	5.16 ± 0.81	a
	4.53 ± 0.45	c, PS183
$\pi^\circ\rho^\circ$	1.54 ± 0.21	b
$\pi^\pm\delta^\mp$	0.69 ± 0.12	c, PS183
$\pi^\pm B^\mp$	0.7 ± 0.1	a
	1.96 ± 0.27	c, PS183
$\pi^\pm A_2^\mp$	3.78 ± 0.87	a
	5.71 ± 0.85	a
	2.83 ± 0.32	c, PS183
$\rho^\circ\omega$	2.1 ± 0.2	a
	0.7 ± 0.3	a
	3.9 ± 0.6	c, PS183
$\eta\omega$	1.3 ± 0.2	a, PS182
	0.44 ± 0.14	b
$\eta\rho^\circ$	0.65 ± 0.16	a, PS182
	0.96 ± 0.16	b

$\rho^\circ\rho^\circ$	0.12 ± 0.12	a
$\eta'\rho^\circ$	0.11 ± 0.06	a
$\phi\rho^\circ$	$2.8 \pm 0.9 \times 10^{-2}$	a, PS171
$\omega\omega$	1.4 ± 0.6	a

a) Quoted by C. Amsler at Les Houches 1987. Some of these numbers were only preliminary. See also ref. [6]

b) Ref. [7]

c) Ref. [8]

However, it should be added that in many cases the extraction of such branching ratios is a difficult - if not impossible - task, since in the laboratory only the resultant photons and pions are observed. The above decays are then reconstructions, which can be unique if the mesons under consideration have narrow widths e.g. the ω. Unfortunately, in many cases the basic s- and p-wave mesons have widths $\approx 100\ MeV$, which can lead to ambiguities, since these widths are comparable to the kinetic energies involved.

The reason for the above criticism of present day $\bar{N}N$ data is not simply to urge experimental studies of the problems involved, but also to give an excuse for the present day theoretical confusion in attempting to explain these data. The latter is illustrated by fig. 1, where a series of quark mechanisms are drawn for annihilation into two or three mesons. In fig. 1a) the $q_3\bar{q}_6$ annihilation vertex is that of the $^3P_0$- or Pair Creation-model, in which the $q\bar{q}$ annihilate into the vacuum. On the other hand, in fig. 1b) the vertex is that of the $^3S_1$- or One-Gluon-Exchange-model, in which the $q\bar{q}$ annihilate into a gluon (or effective gluon). In figs. 1c, d, f) the $q\bar{q}$ vertices can also be of the $^3S_1$ form, and most combinations of these various possibilities can be found in the literature. In section 2 some of these combinations will be discussed in detail.

2 QUARK MECHANISMS FOR $N\bar{N}$ ANNIHILATION

2a. The $q\bar{q}$ vertex.

Before launching into a discussion of all the possibilities presented by fig. 1, when the $q\bar{q}$ vertex is either $^3P_0$ or $^3S_1$, it is sensible to see how reasonable each of these vertices really is. This can be approached in three ways.

Firstly, the two alternatives figs. 1a) and b), where the $^3P_0$- and $^3S_1$-vertices are taken to lowest order, can be compared directly with experimental data. This certainly favours [9,10,11,12] the $^3S_1$-model since it permits the decays, $N\bar{N}$ (S-wave) $\rightarrow M_1^S M_2^S$ i.e. $N\bar{N}$ in a relative S-wave can annihilate into two s-wave mesons e.g. $N\bar{N}(^{13}S_1) \rightarrow \pi\rho$.

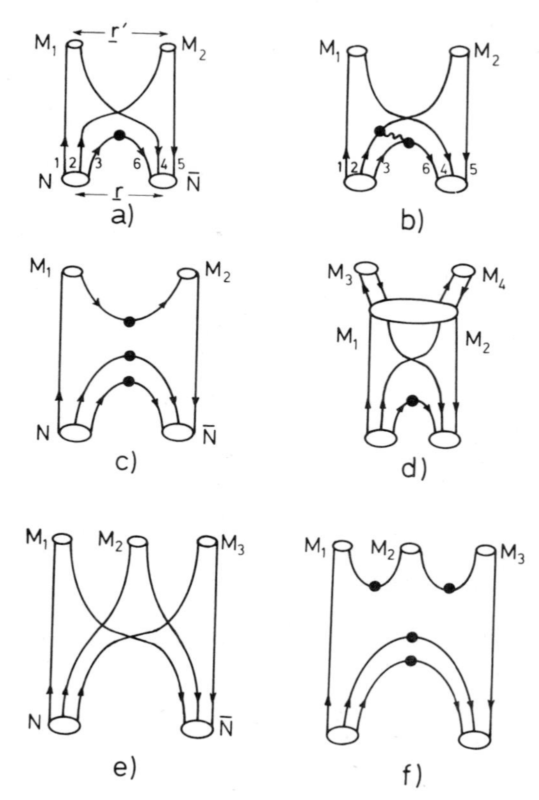

Fig. 1. Quark mechanisms for $N\bar{N}$ annihilation into two mesons a,b,c,d) and three mesons e,f). In the literature a,b,c and e) are often referred to as R2,S2,A2 and R3 respectively.

Furthermore, it inhibits the decay $N\bar{N}(^{31}S_0) \to \pi\rho$, so that the ratio

$$R_S = \frac{BR[N\bar{N}(^{31}S_0) \to \pi\rho]}{BR[N\bar{N}(^{13}S_1) \to \pi\rho]} \ll 1 \tag{2.1}$$

as is observed experimentally [13] and known as the $\pi\rho$-puzzle. On the other hand, the $^3P_0$-model does not permit decays such as $N\bar{N}(S) \to M_1^S M_2^S$ - when taken to lowest order. This follows directly from the form of the $^3P_0$-vertex which can be expressed as

$$\theta(^{13}P_0) = K_1[\nabla_c \chi(q_3\bar{q}_6, s=1)]^{J(q_3\bar{q}_6)=0}\delta(\mathbf{c}) \tag{2.2}$$

where $\mathbf{c} = \mathbf{r}_3 - \mathbf{r}_6$ in the notation of fig. 1a), and $\chi(s)$ is the spin triplet wavefunction. Because of the derivative ∇_c this operator requires q_3 and $\bar{q}_6$ to be in a relative P-wave. However, this situation is not possible for $N\bar{N}(S)$ unless there are *two* $q\bar{q}$ pairs in a relative P-wave. This then leads to the decay $N\bar{N}(S) \to M_1^S M_2^P$ i.e. a final s- and *p-wave* pair of mesons but not two s-wave mesons. This failure is one of the main arguments presented by the $^3S_1$-advocates. But it should be added that the $^3S_1$-model also has its problems, when treated as in fig. 1b). For example [12], it fails to allow the decay $N\bar{N}(^{13}S_1) \to \pi B(1233\text{ MeV})$, which has the non-negligible branching ratio $(0.7 \pm 0.1)\%$ or $(1.96 \pm 0.27)\%$. Furthermore, it must be remembered that the $^3S_1$-vertex, which can be expressed in the notation of fig. 1b) as

$$\theta(^3S_1) = K_2\delta(\mathbf{r}_{36})[\sigma(2) \times \sigma(36)].\nabla_{26}\delta(\mathbf{r}_{26}) \tag{2.3}$$

is a *two-body* operator - in contrast to $\theta(^{13}P_0)$ of eq. (2.2) which is a *one*-body operator. This difference permits $\theta(^3S_1)$ to have more annihilation possibilities than the restricted form of $\theta(^{13}P_0)$. However, it should be pointed out that the process under study involves *six* quarks (and antiquarks). Therefore, there could well be other one- or two-body quark operators that give similar results to the above $^3S_1$- and $^{13}P_0$-vertices, when immersed into a multiquark environment and the basic quark coordinates $\mathbf{r}_i$ integrated away to leave the more observable coordinates $\mathbf{r}$ and $\mathbf{r}'$ in fig. 1a). This leads naturally to consider the second way of comparing the two $q\bar{q}$ vertices.

Secondly, the $^{13}P_0$ and $^3S_1$-vertices can be used in simpler situations - in particular for the study of meson decays as in fig. 2. Here the $^{13}P_0$-model has a

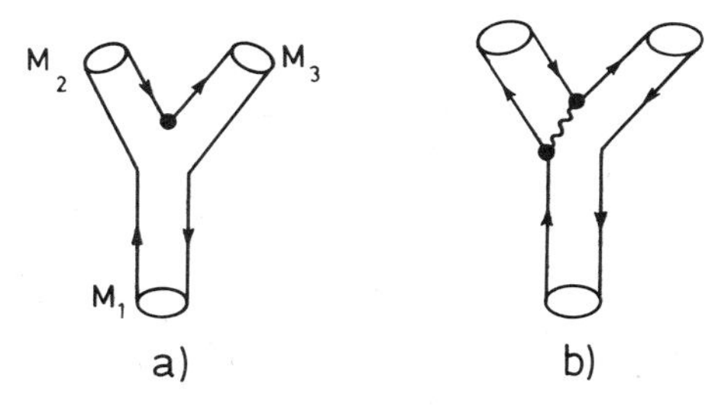

Fig. 2. Meson decay $M_1 \to M_2M_3$ using a) the $^{13}P_0$-vertex and b) the $^3S_1$-vertex.

resounding success [14] and to quote these authors "We confidently expect that the hundreds of additional amplitudes which we have predicted here will confirm the applicability of the model". On the other hand, the $^3S_1$-model fails [15] and to quote the authors of ref. [15] themselves - "We acknowledge that the $^3P_0$-model is better able to reproduce experimental data". This confession by the $^3S_1$-advocates should be sufficient to drop the $^3S_1$-vertex as a serious possibility in fig. 1. Furthermore, additional support for the $^{13}P_0$-vertex comes from the third approach.

Thirdly, as will be discussed in section 3, when attempts are made to construct a more basic theory involving gluon degrees of freedom, the $^3P_0$-vertex emerges as a natural operator.

Having come out strongly in favour for the $^3P_0$-vertex, it is only fair to say that in some situations vestiges of One-Gluon-Exchange (OGE) remain to give a $^3S_1$-vertex. Such a case could be that of strange meson production [16]. This involves internal kinematics where the perturbative QCD limit in the form of OGE is possibly applicable. In the non-perturbative limit of interest for $N\bar{N}$ annihilation into non-strange mesons many gluons are involved. The above preference for the $^{13}P_0$-vertex can then be expressed by saying that the multipole expansion of the $q\bar{q}$ annihilation vertex

$$\theta(q\bar{q}) = a\theta(^{13}P_0) + b\theta(^3S_1) \tag{2.4}$$

suggests that $a \gg b$. However, in ref. [17] it is pointed that, if $b \approx a$, then the $\pi\rho$-puzzle in S-waves -see eq. (2.1) - can be understood by the b-term and the corresponding P-wave $T = 1$ inhibition, $N\bar{N}(^{33}P_{1,2}) \not\rightarrow \pi\rho$, can possibly be understood as a cancellation between the a-and b-terms.But in ref. [18] such an approach is shown to be not successful.

2b. The annihilation $N\bar{N}(S) \rightarrow M_1^S M_2^S$ using the $^3P_0$ model.

If the $^3S_1$-vertex is rejected as the main mechanism for $q\bar{q}$ annihilation, then most $^3P_0$-advocates resort to the process involving *three* $^3P_0$-vertices - as depicted in fig. 1c). This readily gives $N\bar{N}(S) \rightarrow M_1^S M_2^S$ but the ratio in eq. (2.1) becomes in Born approximation $R_S \approx 1$ compared with the experimental value $R_S \ll 1$. Of course, this can be resolved if the initial state interaction (ISI) for $N\bar{N}(^{13}S_1)$ is much larger than that for $N\bar{N}(^{31}S_0)$. In fact, there are indications that this is not an unreasonable possibility. As shown in ref. [19] the ISI for $N\bar{N}(^{13}S_1)$ can enhance specific annihilation channels by easily an order of magnitude compared with the ISI for $N\bar{N}(^{31}S_0,\ ^{11}S_0$ or $^{33}S_1)$. The reason for this effect in $N\bar{N}(^{13}S_1)$ is probably due to the special character of the OBE potential in this particular channel. For example, in ref. [20] it is shown that for $N\bar{N}(^{13}S_1)$ there is a coherence of the π, ρ and ω exchange terms in the tensor component of the potential, which makes the interaction effectively more attractive in this channel.

Another alternative for enhancing $BR_1 = BR[N\bar{N}(^{13}S_1) \rightarrow \pi\rho]$ with respect to $BR_2 = BR[N\bar{N}(^{31}S_0) \rightarrow \pi\rho]$ is to include other competing decay modes. In refs. [21,22] it is $BR_3 = BR[N\bar{N}(^{31}S_0) \rightarrow \pi f(1273)]$ that is suggested as a likely competitor to BR_2 and this is supported by both experimental and theoretical arguments. Unfortunately both of these arguments have been disputed. Firstly, the experimental data quoted is that of ref. [23] which finds $BR_3 = 2.6\%$ for $\bar{p}n \rightarrow \pi^- f$ much larger than the $BR_3 = 0.24 \pm 0.07\%$ for $\bar{p}p \rightarrow \pi f$ [5]. However, the former number is based on $\bar{p}d$ data and so is the result of a more complicated analysis than the $\bar{p}p$ value. Furthermore, the two theoretical estimates [18,21,22] of BR_3 differ by a factor of 75 with refs. [21,22] favouring the larger number - hence getting the desired competition with BR_2 to make $R_S \ll 1$.

A third alternative for reducing R_S from ≈ 1 has been suggested in ref. [18]. There it is demonstrated that the strong tensor potential in the $^{13}S_1$-channel can generate a $^{13}D_1$ component that is comparable to the $^{13}S_1$ component *inside 1fm*– eventhough the overall D-state probability is only $\approx 10^{-5}$. Since, in Born approximation

$$BR(^{13}D_1 \to \pi\rho)/BR(^{13}S_1 \to \pi\rho) = 5 \sim 25 \tag{2.5}$$

depending on details, the whole transition is therefore governed by the D-wave component. There is clearly much more to be done both experimentally and theoretically before a direct application of fig. 1c) is understood.

In addition to the above problems with fig. 1c) there are other objections.

Firstly, the need for *three* $^3P_0$-vertices is somewhat disturbing, since each vertex is expected to be in some sense small. Qualitative evidence for this is the fact that the widths of mesons are smaller than their masses and these widths are well described by the single $^3P_0$ vertex as shown in fig. 2a). More quantitatively, by studying hadronization [24] in hadron-hadron and hadron-nucleus collisions, and assuming a simple annihilation amplitude of the form

$$\theta(Ann.) = \gamma\delta(c) \tag{2.6}$$

- see eqs. (2.2) and (2.3) for comparison - a value of $\gamma^2 \approx 0.25$ can be extracted. For the ratio of fig. 1c) and fig. 1a) this now gives a very crude estimate of

$$R_{S/P} = \frac{BR[N\bar{N}(S) \to M_1^S M_2^S]}{BR[N\bar{N}(S) \to M_1^S M_2^P]} = \frac{[\gamma^2]^3}{\gamma^2} = 0.06 \tag{2.7}$$

- a feature not seen experimentally for the relative branching into s-wave mesons and a p-wave meson. However, as will be discussed in section 2e), a more careful comparison suggests $R_{S/P} \approx 0.5$ –still favouring fig. 1a) over fig. 1c).

Secondly, in fig. 1c) the two successive annihilation vertices involve an intermediate state of unknown structure and energy. This means that the combined vertex for $q^2\bar{q}^2$ annihilation has a strength of $\lambda\gamma^2$, where γ is from eq. (2.6) and so is known, for example, from hadronization or meson widths. However, λ is essentially a free parameter with only rather wide limits imposed by the structure of the intermediate $q^2\bar{q}^2$ state between the two annihilation vertices.

Thirdly, as will be discussed in section 3, if the only interactions are the three $^3P_0$-vertices depicted in fig. 1c) then models involving explicit gluon degrees of freedom suggest that the final mesons will preferentially be hybrid i.e. contain gluon excitation.

2c. Two meson production involving meson-meson rescattering.

In an attempt to overcome the use of three $^3P_0$-vertices for the production of two s-wave mesons i.e. $N\bar{N}(S) \to M_1^S M_2^S$, the authors of refs. [25,26] have proposed a model involving meson-meson rescattering. This is depicted in fig. 1d) and can be expressed as the chain of events

$$N\bar{N}(S) \to M_1^S M_2^P(\ell = 0) \to M_3^S M_4^S(\ell = 1)$$

Here the first stage is simply the single $^3P_0$ mechanism of fig. 1a) generating an s-wave meson and a p-wave meson in a relative S-wave. These two mesons then rescatter to give two s-wave mesons in a relative P-wave. To estimate the reliability of this model the ratio

$$C(TS) = \frac{BR[N\bar{N}(ST, L=0) \to M_1^S M_2^S]}{M^2[N\bar{N}(ST, L=0) \to M_1^S M_2^S]} \tag{2.8}$$

is calculated. Here BR are the observed branching ratios and M the corresponding theoretical matrix elements including phase space. The constancy of $C(ST)$ - essentially the initial state interaction - is then a measure of the model's reliability. As seen in detail in ref. [25], this has both its successes and failures. For example, the spin singlet correlations - $C(T, S=0)$ - and some of the triplet correlations are roughly consistent for a given T, S. Furthermore, C's for different T, S are in line with those based on phenomenological potentials - in particular the need for $C(0,1)$ to be much greater than the other C's -see ref. [19] for more details. However, the model badly underestimates those M's involving pions. This same problem - the underestimation of non-resonant $\pi^+\pi^-\pi^0$ annihilation - is also seen in the pure rearrangement process of fig. 1e). - see for example ref. [27]. To a great extent this defect is remedied in ref. [26] by allowing the $q\bar{q}$ wavefunction of the pion to have a smaller RMS radius than that of other mesons. For example, by using a pion wavefunction which has $\approx \frac{1}{2}$ the RMS radius of other mesons the $C(TS)$'s shown in table 2 result. Some models for the pion [28] indeed suggest that the $q\bar{q}$ component is consistent with a *point-like* structure. The experimental RMS radius of 0.66 ± 0.01fm [29] is then described in terms of a vector meson within the meson cloud surrounding the $q\bar{q}$ component.A recent discussion of this can be found in ref. [30]. In table 2 the spin singlet $C(T,0)$'s are consistent as before. However, now the $C(01, \pi\rho)$ and $C(11, \pi\pi)$ are much more reasonable - being earlier in ref. [25] orders of magnitude larger. The main disturbing features are the large values of M for the η' cases. This seems unavoidable since they are so closely related to the corresponding η values.

Table 2. The $C(TS)$ from eq.(2.8) using the experimental BR's from table 1 - see ref. [26] for details.

$M_1^S M_2^S$	$\omega\omega$		$\rho^\circ\rho^\circ$
$C(00, M_1^S M_2^S)$	9.5 ± 4.0		8.3 ± 8.3
	$\rho\omega$		$\pi\rho$
$C(10, M_1^S M_2^S)$	2.9 ± 2.2		$[0.58 \pm 0.44\%]$
	$\pi\rho$	$\eta\omega$	$\eta'\omega$
$C(01, M_1^S M_2^S)$	90 ± 14	45.5 ± 7.0	$[4.0 \pm 0.6\%]$
	$\pi\omega$	$\rho\eta$	$\rho\eta'$
$C(11, M_1^S M_2^S)$	< 4	5.0 ± 1.2	0.3 ± 0.2
	$\rho\rho$	$\pi\pi$	
	$< 200 \pm 67$	12 ± 1	

The numbers in [ ] are BR predictions assuming the C's for a given T,S are indeed constant.

The use of a pion wavefunction that is smaller than that of other mesons also permits the chain of events

$$N\bar{N}(S) \to \pi M_2^P(\ell = 0) \to \pi M_4^P(\ell = 2)$$

i.e. annihilations such as $N\bar{N}(^{31}S_0) \to \pi f$ and $N\bar{N}(^{11}S_0) \to \pi A_2$ - both of which are forbidden by fig. 1a) since the final mesons have to be in a relative *D-wave*. The triple $^3P_0$-vertex model of fig. 1c) is also able to give these transitions. However, as mentioned earlier , the authors of refs. [18,21,22] differ by a factor of 75 in their estimate of $BR(^{31}S_0 \rightarrow \pi f)$ and so the actual prediction is at present uncertain. The meson-meson scattering model of fig. 1d) is not successful - with these transitions giving values of M that are far too small when judged by the constancy of $C(TS)$ from eq. (2.8) -see ref. [31].

One possible reason for these failures could be due to the simple model used for the meson-meson scattering mechanism. The latter is depicted in fig. 3a) and in the

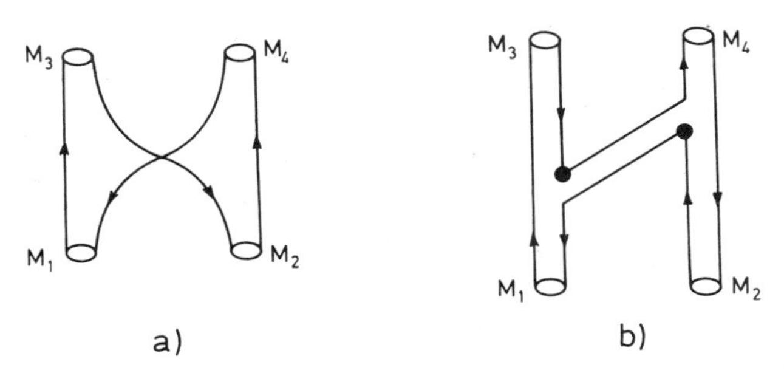

Fig. 3. Meson-Meson scattering generated by
a) a plaquette (P)– see section 3
b) two $^3P_0$ vertices (Q^2) to give an effective meson exchange

present work the assumption is made that fig. 3a) dominates over fig. 3b). In section 3 a qualitative comparison is made between these two possibilities. However, for practical reasons all quark wavefunctions for the mesons are taken to have *gaussian* form. Such wavefunctions can grossly underestimate matrix elements in which M_1 and M_2 are close to each other - as in the S-wave case - with M_3 and M_4 *not* close to each other - as in the D-wave case. This problem is further aggravated when using pions with small RMS radii. Very crude estimates made through replacing M_3, M_4 by exponential wavefunctions show that indeed the rescattering matrix element could well be increased by an order of magnitude.

2d. $N\bar{N}$ annihilation into three mesons.

The traditional mechanism for $N\bar{N}$ annihilation into 3 mesons is that depicted in fig. 1e) - the quark rearrangement model. This was originally proposed in 1966 [32] and developed over the years by the Osaka and Helsinki groups - see refs. [1,2] for a review. Recently [27] this process has been treated using the Resonating Group Method which calculates an explicit interaction for the $N\bar{N}(S) \rightarrow M_1^S M_2^S M_3^S$ transition. Usually this interaction is either taken to be proportional to the wavefunction overlap $(N\bar{N}|M_1M_2M_3)$ or estimated as $(N\bar{N}|U(confining)|M_1M_2M_3)$. In most cases these two are proportional to each other - see ref. [33]. Now in the RGM method the corresponding interaction is $(N\bar{N}|T(Kinetic\ energy) + U(confining)|M_1M_2M_3)$. As far as can be judged by the rather meagre and uncertain experimental data for these 3 meson channels the model is reasonable. However, an alternative model is that of fig. 1f) in which the number of $q\bar{q}$ vertices is essentially maximized - being 4 compared with 0 in fig. 1e). As shown in ref. [34] this model is equally reasonable. The question then arises concerning the relative magnitude of mechanisms that lead to the same final states. This is not easy to answer since most mechanisms involve at least one free parameter. For example, in fig. 1e) the strength of the overall interaction is difficult to estimate. Similarly, in fig. 1f) the strength of the $^3P_0$-vertex [K_1 in eq. (2.2)] is not necessarily the same as that extracted from meson decays. In addition to these basic uncertainties there is freedom with the parameters of the quark wavefunctions. For example, the N and $\bar{N}$ quark wavefunctions are usually taken to have a charge RMS radius of 0.6 fm eventhough the experimental value is $\approx$0.8 fm. The difference is then usually attributed to a meson cloud outside the region where the quarks are confined. However, this cloud is assumed to never play a direct role in the annihilation - eventhough this could well be a ready source of, in particular, pions.

2e. Large N_c arguments.

One way of attempting to select the most important mechanisms in $N\bar{N}$ annihilation is to use large N_c (number of colours) arguments. These favour planar versus non-planar diagrams e.g. fig. 1f) over 1e). Roughly speaking the more quark lines that must cross each other the more non-planar the diagram. In a recent article [35] these arguments have been applied to the mechanisms shown in fig. 1 with the result that all the cross sections involved can be expressed as

$$\sigma(N\bar{N} \rightarrow \text{Mesons}) \overset{N_c \to \infty}{\longrightarrow} e^{-AN_c}(\gamma^2)^{m+l} N_c^{2m} \tag{2.9}$$

Here the exponential factor arises due to the decreasing overlap of the $N\bar{N}$ and meson wavefunctions as N_c increases - a feature conjectured in ref. [36]. The γ^2 is the probability for creating or annihilating a $q\bar{q}$ pair - see eq. (2.6). The l and m are the number of $q\bar{q}$ pair creations and annihilations $-(m,l) = (0,0)$ and $(2,2)$ for figs. 1e) and 1f) respectively. Clearly as $N_c \rightarrow \infty$ fig. 1f) dominates over 1e) - the usual planar versus non-planar argument. However, in real life $N_c = 3$ and $\gamma^2 = 0.25$ [24] which then indicates that σ(fig. 1e)$\approx$ 3σ(fig. 1f) - a conclusion *opposite* to that of the $N_c \rightarrow \infty$ limit. Ref.[35] also shows that σ(fig. 1a)$\approx$ 2σ(fig. 1e) $\approx$ 2σ(fig. 1c). It must be admitted that these comparisons are marginal and that the inclusion of spin and isospin (so far neglected) could make changes. But taken at their face value they suggest that indeed fig. 1a) is the dominant mechanism for $N\bar{N}$ annihilation.

3 THE EFFECT OF GLUON DEGREES OF FREEDOM

The previous section shows that in $N\bar{N}$ annihilation the consideration of only quark and antiquark degrees of freedom leads to many alternative mechanisms, each of which is reasonable. Clearly some microscopic model is needed to "break this degeneracy". This instantly gives rise to models that introduce in some way the gluon degrees of freedom. Of course, the most direct and exact method would be to solve the QCD lagrangian that couples quarks and gluons. Unfortunately, such calculations involving $q^3\bar{q}^3$ configurations on a lattice are at present not practical - the limit being $q\bar{q}$ and q^3 systems which already require the largest available computers. However, it is possible that such exact calculations are unnecessary for $N\bar{N}$ annihilation since this only requires some average properties of the $q^3\bar{q}^3$ system that are on the scale of $\frac{1}{2} - 1$ fm - the size of nucleons and mesons. It is probably unnecessary to know how the quarks precisely behave when they are at distances much smaller than $\frac{1}{2}$ fm. This situation is analogous to the one encountered in shell model calculations of atomic nuclei. There the main interest involves configurations of nucleons in the *lowest* empty valence orbits - the effect of higher orbits being simulated by an effective NN interaction between nucleons in those lowest shells. The question is now reduced to knowing how accurately the gluon degrees of freedom need to be simulated. In the literature several suggestions with varying degrees of complexity have been put forward. The minimal model is the one based on "link operators" [37] which represent the gluon field in terms of links between the quarks in a given hadron. In this way the colour Van der Waals force problem is avoided for the interaction between two hadrons. However, the only variables associated with the links are those describing their end points - and so they are unable to contribute to the dynamics of the system. In spite of their rather inert nature, the model has been extended to the Baryon-Meson interaction [38] by associating a potential with the length of the link - see fig. 4). A similar idea is behind the flip-flop model of ref. [39], where the links rearrange to ensure their potential energy is minimized - see fig. 5. Going beyond these minimal models, by bestowing additional degrees of freedom on the links, leads rapidly to numerical complications.

Perhaps the best example of this next step is the Flux-tube model of ref. [40] in which the gluon flux, represented earlier by the links L_i in figs. 4 and 5, is discretized. If such a flux discretization is into N coupled pieces then the wavefunction of this flux-tube can be written as

$$\Psi_{\mathrm{FluxTube}} = \prod_{i=1}^{N} N_i \exp[-\frac{1}{2}\alpha_i^2 a_i^2] \tag{3.1}$$

where a_i and α_i are the transverse normal coordinates and the corresponding normal frequencies. So far the use of eq. (3.1) has been restricted [14,41] to the case of meson decay $M_1 \rightarrow M_2 M_3$ in which one flux tube breaks up into two flux tubes - the amplitude for the decay being the overlap

$$\gamma_{23}^{1} = < \Psi_{\mathrm{FT3}} \Psi_{\mathrm{FT2}} \mid \Psi_{\mathrm{FT1}} > \tag{3.2}$$

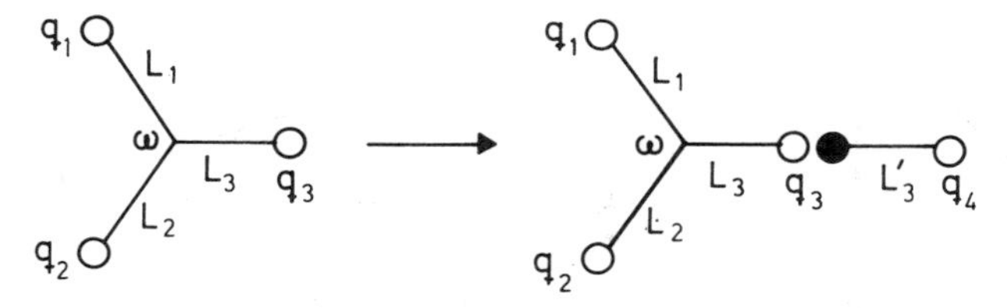

Fig. 4. a) A baryon described in terms of three quarks at coordinates $\mathbf{q}_i$ and three links $L_i = L_i(\mathbf{q}_i, \omega)$
b) The number 3 link has broken along a straight line to give the original L_3 and a link $L_3'(\mathbf{q}_3, \mathbf{q}_4)$ for the meson.

This is still a manageable calculation. But in more complicated situations such as $N\bar{N}$ annihilation, the complexity of treating an N body overlap in three dimensions is formidable. In view of this it would be desirable if some intermediate stage between the link model and flux tube model could be developed. Presumably such a compromise would involve average properties of the flux-tube and so require less than the N degrees of freedom needed above.

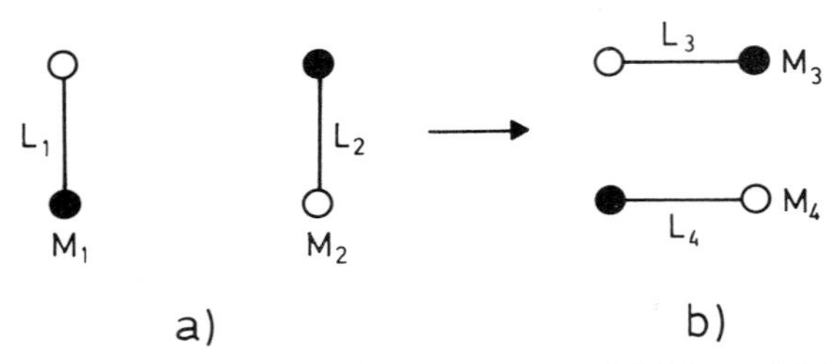

Fig. 5 The flip-flop model for meson-meson scattering $M_1M_2 \rightarrow M_3M_4$. The transition occurs when $L_1 + L_2$ becomes greater than $L_3 + L_4$ - where L_i represents the length of the link.

One possibility along these lines is to parameterize the overlap γ_{23}^1 in eq.(3.2) and use this as a building block in the evaluation of more complex situations. With the flux tube wavefunction of eq.(3.1) it has been shown [41] that γ_{23}^1 can in fact be written in the form

$$\gamma_{23}^1 = A_{23}^1(N, N_c)\sqrt{f(N, N_c)b/\pi}\ exp[-f(N, N_c)by_c^2/2] \tag{3.3}$$

Here $b \approx 1Gev/fm$ is the energy density of the flux tube, y_c is the distance shown in fig. 6a) and $N_c, N - N_c$ are the numbers of pieces in the final two flux tubes. The factor $A_{23}^1(N, N_c)$ is only weakly dependent on N and N_c. More precisely, A is independent of N_c to better than 1% for $0 < N < 20$ and decreases slowly with N in this range as $A \approx 1 - 0.01N$. The coefficient f has the form as $N \rightarrow \infty$

$$f = \frac{1}{\pi} ln[(1-\rho)^{-\frac{1}{\rho}}\rho^{-\frac{1}{1-\rho}}] \quad , \qquad \rho = N_c/N \tag{3.4}$$

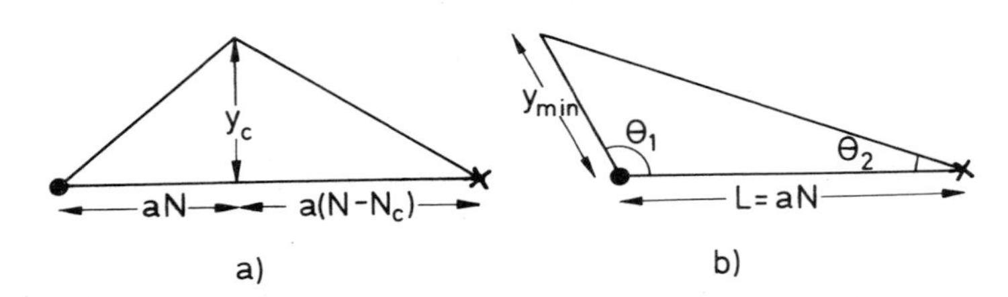

Fig. 6. a) The geometry associated with eq.(3.3).The distance y_c is the perpendicular from the tube breaking point to the line joining the quark (q) and antiquark ($\bar{q}$).
b) The geometry associated with eq.(3.5). The distance y_{min} is the minimum distance from the tube breaking point to the line joining the quark (q) and antiquark ($\bar{q}$). For θ_1 and $\theta_2 < \frac{\pi}{2}$, $y_{min} = y_c$

A more realistic flux tube wavefunction –closer to that expected for QCD on a lattice– indicates that a better form for γ^1_{23} is

$$\bar{\gamma}^1_{23} = \bar{A}^1_{23}(N, N_c)\sqrt{\bar{f}(N, N_c)b/\pi} exp[-\bar{f}(N, N_c)by^2_{min}/2] \tag{3.5}$$

where y_{min} is depicted in fig.6b) and permits the flux tube to make backward excursions at its ends.

3a. Interactions within the flux-tube model.

The end of the last section suggested that the flux-tube model in its present form is already too complicated for use in $N\bar{N}$ annihilation. However, in spite of this, it is important to understand the relationship between this model and QCD, so that simulations can be made as realistic as possible. The starting point is the strong coupling form of QCD on a lattice i.e.

$$H^{\text{lattice}}_{QCD} = H_{\text{Quark}} + H_{\text{Glue}} \tag{3.6}$$

where

$$H_{\text{Quark}} = \sum_q m_q \sum_n q^\dagger_n q_n + \frac{1}{a} \sum_{q, l_{ij}} q^\dagger_j U_{l_{ij}} \alpha_{ji} q_i \tag{3.7}$$

and

$$H_{\text{Glue}} = \frac{g^2}{2a} \sum_l C^2_l + \frac{1}{ag^2} \sum_{\text{plaquettes}} Tr[1 - U_{l_4} U_{l_3} U_{l_2} U_{l_1}] \tag{3.8}$$

In eq. (3.7) m_q are the masses of the quarks located on the lattice sites n, a is the lattice spacing, α_{ji} is the Dirac matrix in the direction of the link l_{ij} between neighbouring sites i and j, and $U_{l_{ij}}$ is the link variable related to the gluon field $A_\mu(x)$ by $U_{l_{ij}} = \exp \int_{x_i}^{x_j} dx_\mu A_\mu(x)$. In eq. (3.8) g is the QCD coupling constant. The first term is the energy in the gluon field on link l - C^2_l being a constant depending on the flux quantum numbers - and the second term corresponds to flux loops known as plaquettes made up from four U links. This notation is depicted in fig. 7.

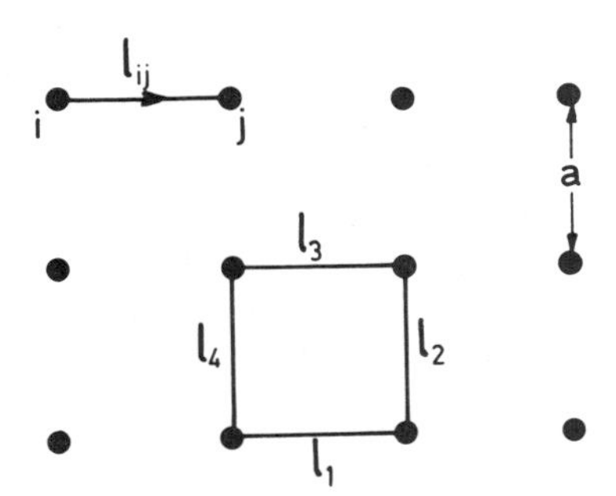

Fig. 7. A link l_{ij} and a plaquette $U_{l_4} U_{l_3} U_{l_2} U_{l_1}$ in a two dimensional lattice with spacing a.

The second term in eq. (3.7) is seen to give rise to an operator that is similar to the $^3P_0$-vertex of eq. (2.2). This follows, if the link direction is taken as $\hat{\varepsilon}$, so that

$$\frac{1}{a}q^\dagger(\mathbf{x}_j)\hat{\varepsilon}\cdot\alpha\, q(\mathbf{x}_i=\mathbf{x}_j+a\hat{\varepsilon})=\frac{1}{a}q^\dagger(\mathbf{x}_j)\hat{\varepsilon}\cdot\alpha\,[q(\mathbf{x}_j)+a\hat{\varepsilon}\cdot\nabla\, q(\mathbf{x}_j)+...] \qquad (3.9)$$

On averaging over all directions for $\hat{\varepsilon}$, since its orientation within a multiquark configuration is random, the first term in eq. (3.9) vanishes and the second one reduces to $\frac{1}{3}\,\alpha\cdot\nabla$ which is reminiscent of the $^3P_0$ -vertex. It is seen that the resultant operator is of $0(g^\circ)=0(1)$ compared with the $0(1/g^2)$ for the plaquette term in eq. (3.8) - in both of these terms the explicit $1/a$ dependence cancels.

At first sight it would, therefore, appear that plaquette effects, being of $0(1/g^2)$, must be small compared with mechanisms involving the $^3P_0$-vertex. However, it should be remembered that for the validity of the flux tube model $g\approx 1$, a value such that the corresponding lattice spacing a is smaller than hadronic sizes, but yet large enough so that topological mixing may possibly be treated as a perturbation [40,42]. This last point is already seen if the mean square radius of the flux tube described by eq. (3.1) is calculated. The result is

$$< y^2_{\rm centre} >^{1/2}\approx [\frac{4}{\pi b}\log N]^{1/2}$$

where y_{centre} is the midpoint of the flux tube and N the number of bits into which the flux tube is discretized. As discussed after eq. (3.3) a value of $N\approx 10$ seems to be the most appropriate.

In refs. [43,44] attempts have been made to estimate plaquette effects in meson-meson scattering as depicted in fig. 8. In both cases the outcome was that the mechanism appeared to be far too weak. The author of ref. [43] carried out the calculation using a simplified lattice procedure and came to the conclusion that the interaction was of very short range comparable to the appropriate lattice spacing of $\approx 0.2fm$. Only when the two quarks and two antiquarks were within this distance of each other did they interact significantly. In ref. [44] advantage was taken of the approximation for the flux tube breaking amplitude in eq. (3.3). The transition for the two dimensional rectangular geometry depicted in fig. 8 could then be expressed schematically in terms of

$$\int dxdy\gamma_1(x)\gamma_2(L_1-x)\gamma_3(y)\gamma_4(L_2-y) \qquad (3.10)$$

But as in ref. [43] this resulted in a transition matrix element that was much smaller than those expected from conventional meson exchange. However, it should be pointed out that – strictly speaking – the γ_i's of eq. (3.3) are being used in eq. (3.10) under conditions that violate the underlying small oscillation approximation on which they are based i.e.$x, L_1-x\ll L_2$ and $y, L_2-y\ll L_1$. An attempt [45] has been made to go beyond this small oscillation approximation by expressing the interaction in terms of surface area differences on the world sheet. This results for simple geometries in γ's that drop off much more slowly in y_c and y_{min} than the gaussian forms of eqs. (3.3) and (3.5). The indications are that this effect could well increase the value of the expression in eq. (3.10) by an order of magnitude. Even so, the resulting matrix elements still seem to be too small compared with conventional meson exchange.

The conclusion of the last paragraph prompted the author of ref. [43] to conclude that the scattering was dominated by t - and/or s-channel meson exchange which is generated by the $^3P_0$-model as in fig. 9. However, it has been pointed [46] out that there could be large flux loop correlations within the QCD vacuum - as shown in fig. 10. - and that these could mediate the scattering. This would suggest that it is more favourable to exploit the structure of the vacuum than to distort the flux tubes within the mesons. Such a conclusion would give some justification for the success of the flip-flop model of fig. 5. Also there would then be less need for the corresponding γ's in fig.10 to violate the small oscillation criterion–at the expense of being able to parameterize in a reliable way the large spacial deviations now forced on the vacuum correlations. In section 2c) and fig. 1d) it is the plaquette mechanism that is considered to be dominant.

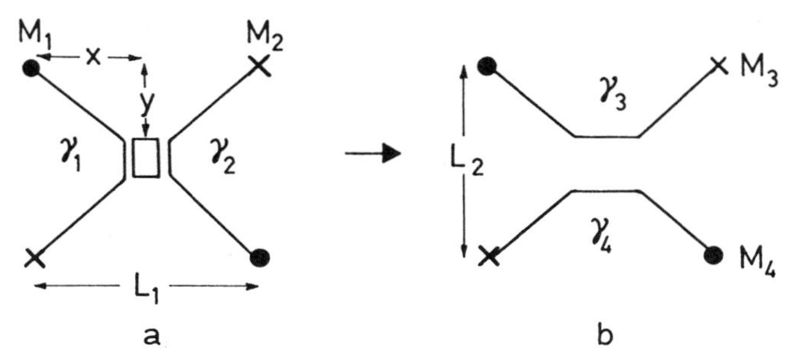

Fig. 8. Meson-meson scattering $M_1M_2 \rightarrow M_3M_4$ using a basic plaquette with links of length a. The geometry depicted is for 2-dimensions with the plaquette at distances x,y from the quark in meson M_1.The γ_i's are the appropriate flux breaking amplitudes.

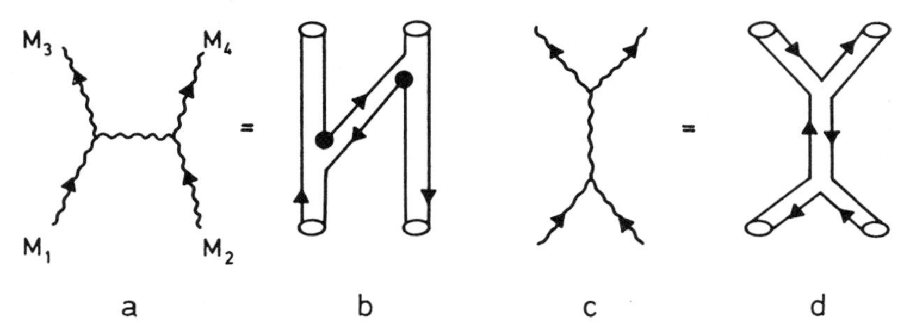

Fig. 9. Meson-meson scattering $M_1M_2 \rightarrow M_3M_4$ treated as t-channel exchange of a) a meson b) a $q\bar{q}$ state and s-channel exchange of c) a meson d) a $q\bar{q}$ state.

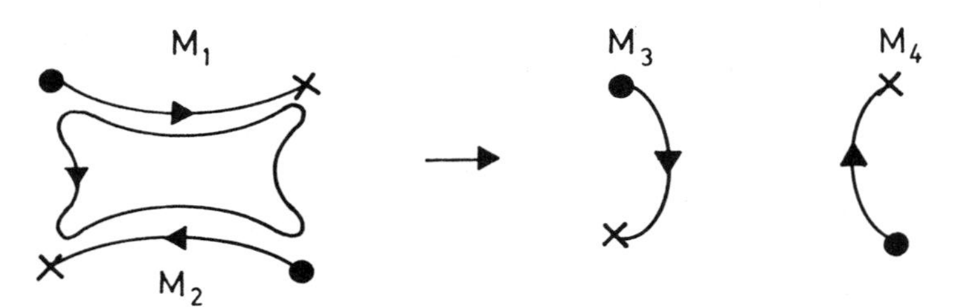

Fig. 10. Meson-meson scattering through a macroscopic flux-tube correlation in the QCD vacuum - ref. [46]

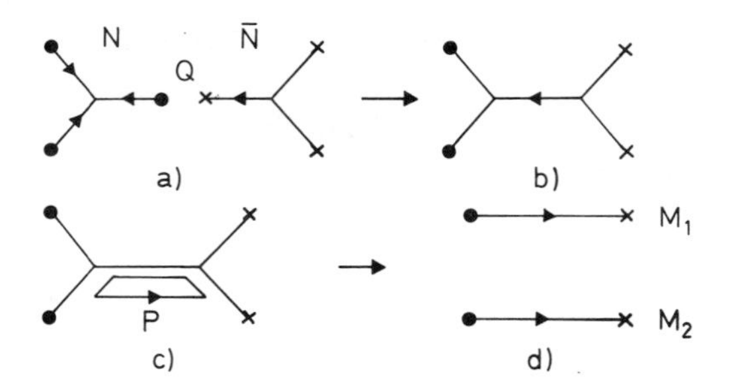

Fig. 11. The scenario for fig. 1a).
a) The $^3P_0$ operator(Q) acts to give b) a baryonium state.
c) A plaquette (P) acts to give the two mesons M_1 and M_2

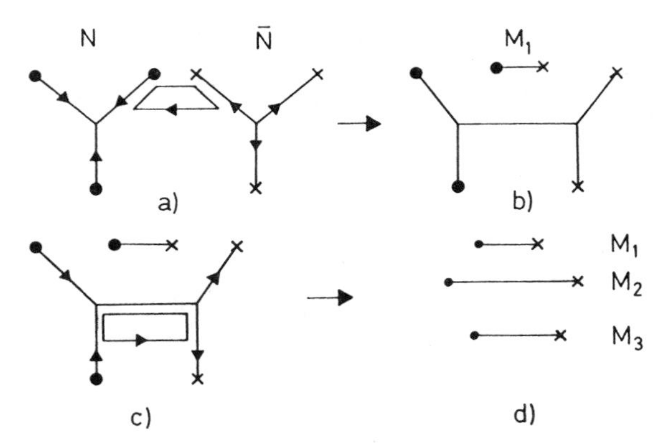

Fig. 12. The scenario proposed in ref. [33] as a model for the rearrangement mechanism of fig. 1e).
a) A plaquette generates b) a meson M_1 and a baryonium state.
c) A second plaquette generates d) mesons M_2 and M_3

In spite of these uncertainties in the much simpler case of meson-meson scattering, attempts are being made to apply the same techniques to $N\bar{N}$-annihilation. However, it is only fair to say that this is still at a very qualitative level. Possible ways of progressing have been outlined in refs. [25,27] . Both of these papers propose that the quarks are treated in a static approximation using standard $q\bar{q}$ or q^3 wavefunctions. For fixed quark positions the gluon fields in the mesons or nucleons (e.g. in the form of flux tubes) are allowed to interact (e.g. by means of plaquettes). Finally the positions of the quarks are integrated over. Naturally, this procedure is at its worst for light quarks - the ones of most interest in $N\bar{N}$ annihilation. In ref. [25] the scenario for fig. 1a) is depicted as the $O(QP)$ process in fig. 11, and that for fig. 1e) is the $O(P^2)$ process in fig. 12. The authors of ref. [27] propose for fig. 1e) the somewhat different scenario of fig. 13, which is only of $O(P)$. However, this possibility requires $\bar{q}_6$ to be near junction A and q_3 to be near junction B– giving an additional reduction factor not needed in the scenario of fig. 12. With the qualitative arguments given here, it is at present not clear which will dominate $N\bar{N} \to M_1M_2M_3$. Furthermore, even when the quarks are treated in the static approximation, incorporating the gluon degrees of freedom as illustrated in the above figures will be very difficult unless some simplified form of "gluon wavefunction" can be constructed. Such a wavefunction would presumably need to be less complicated than the discretized flux-tube of refs. [14,40,41]. Possibly the use of the flux tube breaking amplitudes parameterized in eqs. (3.3) and(3.5) can result in manageable numerics. However, as found in ref. [44] these are not easy to deal with for general geometries even in the simpler case of meson-meson scattering. In addition, the problem raised in ref. [46] and depicted in fig.10 concerning macroscopic flux loops in the vacuum needs to be resolved.

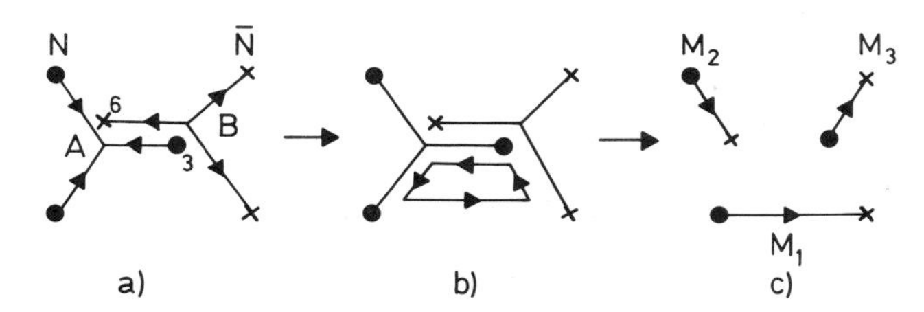

Fig. 13. The scenario proposed in ref. [27].
a) q_3 is near junction B at the same time as $\bar{q}_6$ is near junction A
b) The single plaquette produces three mesons as in c).

4 THE SKYRMION-ANTISKYRMION SYSTEM

The treatment of baryons as Skyrmions has received much attention over the last few years - see ref. [2] and also ref. [47] for a recent review. In their application [48,49] to $N\bar{N}$ scattering the most interesting result is the appearance of a short ranged ***repulsion*** which cancels to some extent the traditional attraction due to ω-meson exchange. The degree of this cancellation is at present uncertain since it requires the application of Skyrmion ideas in the region of small r where the model is at its weakest because –a) the Skyrme lagrangian is a large r (small q) expansion and b) the product ansatz could be unreliable for small r – see refs. [49,50,51].

The extension of the model to $N\bar{N}$ annihilation has been attempted in ref. [52]. This work uses the original Skyrme lagrangian but with the baryon stabilization achieved by introducing ω-mesons. This means that the baryons are chiefly made of π's and ω's i.e. $< \pi > \neq 0$, $< \omega > \neq 0$ but $< \rho > = 0$. The authors only attempt to predict selection rules in which certain annihilation channels are suppressed, since dominant modes would require more dynamical details. The outcome is that the channels $N\bar{N}(S) \to \rho\rho, \pi\pi$ and $\omega\pi\pi$ are all suppressed. The first could well be affected by the use of a more elaborate lagrangian [53] in which the ρ-meson also enters into the stabilization to give $< \rho > \neq 0$. As seen in earlier pages, experimentally the above $\bar{p}p$ branching are $< 9 \pm 3$, 0.4, $\approx 3\%$ and $\bar{p}n \to \rho^- \rho^\circ \sim 5.8\%$. However, these numbers have their uncertainties, for example, the amount of P-wave annihilation is not clear.

The author wishes to acknowledge useful correspondence and discussions with Drs.C.Amsler, N.Isgur, G.Liu and J.Paton.

REFERENCES

[1] A.M.Green and J.A.Niskanen, Int. Rev. Nucl. Phys. Vol. 1–Quarks and Nuclei (World Scientific,Singapore,1984) p.570

[2] A.M.Green and J.A.Niskanen, Prog. in Particle and Nuclear Physics Vol. 18, ed. A.Faessler , (Pergamon 1987) p.93

[3] T.Armstrong et al., Phys. Rev. D36 (1987) 659

[4] C.Lechanoine-Leluc, F.Lehar, P.Winternitz and J.Bystricky, Journal de Physique 48(1987) 985

[5] C.Amsler, private communication and compilation presented at the Les Houches Workshop, March 1987

[6] L.Tauscher, Proc. of IV LEAR Workshop, Villars-sur-Ollon, Switzerland, ed. C. Amsler et al. (Harwood Academic Publishers, 1988) p. 397

[7] G.Smith, Proc. of Workshop on the Elementary Structure of Matter, Les Houches, March 1987 , eds. J.-M.Richard, E.Aslanides and N.Boccara (Springer-Verlag Proc. in Phys. 26) p.197

[8] M.Chiba et al., Proc. of IV LEAR Workshop, Villars-sur-Ollon, Switzerland, ed. C. Amsler et al. (Harwood Academic Publishers, 1988) p. 401

[9] M.Kohno and W.Weise, Nucl. Phys. A454 (1986) 429

[10] M.Maruyama and T.Ueda, Phys. Lett. 149B (1984) 436 and Prog. Theor. Phys. 74 (1985) 526
M.Maruyama, Osaka University Ph.D. Thesis (1984) unpublished

[11] E.M.Henley, T.Oka and J.Vergados, Phys. Lett. 166B (1986) 274

[12] M.Maruyama and T.Ueda, Prog. Theor.Phys. 73(1985) 1211

[13] G.B.Chadwick et al., Phys. Rev. Lett. 10 (1963)62
M.Foster et al., Nucl. Phys. B6 (1968) 107

[14] R.Kokoski and N.Isgur, Phys. Rev. D35 (1987) 907

[15] E.M.Henley, T.Oka and J.D.Vergados, Nucl. Phys. A476(1988)589

[16] H.R. Rubinstein and H. Snellman, Phys. Lett. 165B (1985) 187

[17] E.M. Henley, L. Wilets and M.A. Alberg, Seattle preprint 40048-40-N7
– A new model for $N\bar{N}$ annihilation.

[18] M. Maruyama, T. Gutsche, G. Strobel, A. Faessler and E.M. Henley, Tübingen preprint 1988
- Present status of the description of the $\rho\pi$ puzzle with explicit consideration of the initial state interaction.

[19] J.A.Niskanen, V.Kuikka and A.M.Green, Nucl.Phys. A443 (1985)691

[20] C.Dover and J.-M.Richard, Phys.Rev. C21 (1980) 1466

[21] C.Dover, P.Fishbane and S.Furui, Phys. Rev. Lett. 57 (1986) 1538

[22] C.Dover, Proc. 2nd. Conf. on the Intersection between Particle and Nuclear Physics, Lake Louise, Canada,1986 ed. D.F.Geesaman (AIP conference proceedings No.150) p.272

[23] D.Bridges et al., Phys. Rev. Lett. 56 (1986) 215

[24] W.Q. Chao and H.J. Pirner, Z. Phys. C14 (1982) 165

[25] A.M. Green and G.Q. Liu, Helsinki preprint HU-TFT-88-4
- to appear in Nucl.Phys.A
- $N\bar{N}$ annihilation into two mesons - effect of final state interactions.

[26] A.M. Green and G.Q. Liu, Helsinki preprint HU-TFT-88-13
- to appear in Z.Phys.A–Atomic Nuclei
- $N\bar{N}$ annihilation into two mesons - the effect of the pion radial wavefunction.

[27] G. Ihle, H.J. Pirner and J.M. Richard, Heidelberg preprint HD-TUP-87-13
–S-wave $N\bar{N}$ interaction in the constituent quark model.

[28] G.E.Brown, Nucl. Phys. A358 (1981) 39c

[29] S.R. Amendolia et al., Phys. Lett. 178B (1986) 435, Nucl. Phys. B277 (1986) 168

[30] V.Bernard, B.Hiller and W.Weise, Regensburg preprint TPR-88-2
–Pion electromagnetic polarizability and chiral models.

[31] A.M.Green and G.Q.Liu, in preparation

[32] H. Rubinstein and H. Stern, Phys. Lett. 21 (1966) 447

[33] A.M.Green, V.Kuikka and J.A.Niskanen, Nucl.Phys. A446 (1985)543

[34] M. Maruyama, S. Furui, A. Faessler and R. Vinh Mau, Nucl. Phys. A473 (1987) 649

[35] H.J. Pirner, Heidelberg preprint HD-TUP-88-2
- $N\bar{N}$-annihilation in the large N(colour)-limit

[36] E. Witten, Nucl. Phys. B160 (1979) 57

[37] O.W.Greenberg and J.Hietarinta, Phys. Rev. D22 (1980) 993 and The Proc. of the 17th. Winter School of Theoretical Physics, Karpacz, Poland 1980 (Harwood Acad. Pub. Co.,Chur,Switzerland) p.411

[38] O.W.Greenberg, R.J.Perry and J.Hietarinta, Phys. Lett. 131B (1983) 209

[39] F.Lenz et al., Ann. Phys.(N.Y.) 170 (1986) 65
K.Masutani, Nucl. Phys. A468(1987) 593

[40] N.Isgur and J.Paton, Phys. Rev. D31 (1985) 2910

[41] N.Dowrick, J.Paton and S.Perantonis, J. Phys.G 13 (1987) 423

[42] J.Merlin and J.Paton, Phys. Rev. D36 (1987) 902

[43] G.A.Miller, University of Washington preprint 40048-38-N7, December 1987
–Flux tube rearrangement and meson-meson scattering

[44] A.M. Green and J. Paton , in preparation.

[45] S.Perantonis, Oxford university thesis–unpublished

[46] K. Zablocki and J. Paton - private communication.

[47] E.M.Nyman and D.O.Riska, Int. Jour. of Mod. Phys. A3 (1988) 1535

[48] H.Yabu and K.Ando, Prog.Th.Phys. 74 (1985) 750

[49] J.J.M.Verbaarschot, Phys.Lett. B195 (1987) 237

[50] T.Otofui et al., Phys.Lett. B205 (1988) 145

[51] A.Schramm,Y.Dothan and L.C.Biedenharn, Phys. Lett. B205(1988) 151

[52] I. Zahed and G.E. Brown, Stony Brook preprint to appear in Phys. Rev.
–A nonperturbative description of $N\bar{N}$ annihilation

[53] C.Adamic, G.E.Brown and I.Zahed, in preparation

EXCLUSIVE PRODUCTION OF HEAVY FLAVOURS IN PROTON-ANTIPROTON COLLISIONS

P.Kroll and W.Schweiger
(presented by W.Schweiger)

Department of Physics
University of Wuppertal
Wuppertal, FRG

INTRODUCTION

Theoretically it is well founded that the large momentum-transfer region of exclusive processes like $p\bar{p} \to 2$ hadrons should be dominated by perturbative QCD[1]. Or, to be more precise, the corresponding scattering amplitude is known to be a convolution of a process specific, perturbatively calculable hard scattering amplitude - which is the sum of all tree diagrams - with process independent quark distribution amplitudes - which account for the nonperturbative formation of quarks into hadrons. For the reactions we are interested in, namely $p\bar{p} \to B_f\bar{B}_f$, where B_f is a baryon containing at least one heavy quark $f(= s, c, b)$ a full calculation along this so called "hard scattering scheme" would require an enormous effort due to the tremendous number of diagrams ($O(10^5)$) contributing. This huge number of diagrams can be reduced considerably by assuming baryons to be bound states of a pointlike quark and a spatially extended diquark. As an additional advantage one thereby includes non-perturbative (higher twist) effects which, as e.g. spin measurements have revealed[2], are still present in the available large momentum-transfer exclusive scattering data. The existence of diquarks, although not deduced from QCD, is also strongly suggested by many effects they can explain such as baryon production in hard collisions, scaling violations in the structure functions of deep inelastic lepton-hadron scattering, or static properties of baryons. Recently it has been shown[3] that a consistent description of the electromagnetic proton form factor and $\gamma\gamma \to p\bar{p}$ cross sections can be achieved within the hard scattering scheme including diquarks.

But even with diquarks the treatment of $p\bar{p} \to B_f\bar{B}_f$ along the hard scattering scheme would be an extensive undertaking. In our constituent scattering model for $p\bar{p} \to B_f\bar{B}_f$ we therefore assume only one constituent (quark or diquark) out of each baryon to participate in the elementary scattering processes and let the other one act as spectator. Since we are interested, first of all, in integrated cross sections (i.e. small angle scattering), this assumption essentially means that we neglect contributions with quarks or diquarks in far off shell intermediate states, or with large transverse momenta inside the baryons. What remains are contributions with a fairly hard gluon mediating the flavour changing processes ($q\bar{q} \to g \to f\bar{f}$ or $D\bar{D} \to g \to D_f\bar{D}_f$) and two relatively soft gluons which are absorbed in the parametrization of the baryon wave functions. For the production of heavy flavours the applicability of such a model to small angle scattering seems to be justified for two reasons. Firstly, with increasing mass m_f of the heavy constituents (quark or diquark) the lowest order perturbative

Antiproton-Nucleon and Antiproton-Nucleus Interactions
Edited by F. Bradamante *et al.*
Plenum Press, New York, 1990

treatment of the flavour changing elementary processes is more and more put on a sound basis, since the gluonic energy scale is set by $\hat{s} \geq 4m_f^2$. Secondly, the competing non-perturbative mechanism, namely t-channel baryon exchange, should become less important with increasing mass of the exchanged baryons.

In what follows we will present our constituent scattering model in more detail and compare it first with data for $p\bar{p} \to Y\bar{Y}(f = s)$. In doing so the influence of the relativistic quark-diquark bound state wave functions can be parametrized in a simple way. Adopting the same parametrization for charmed or even bottomed baryons, however, might be inadequate, since the wave functions of these baryons are expected to be much narrower than those of hyperons. Therefore we make also use of flavour dependent wave functions to be found in the literature[4] to give estimates of cross sections for various charmed and bottomed baryons.

THE CONSTITUENT SCATTERING MODEL

The dynamical mechanism of our model is visualized in Fig.1. The full hadronic helicity amplitude $T_{\{\lambda\}}$ thus is given by a product of wave functions with elementary helicity amplitudes $\hat{T}^i_{\{\lambda'\}}$ integrated over the intrinsic transverse momenta $k_\perp, l_\perp$ of the active constituents and the fractions x, y of the hadronic momenta carried by them. Finally it is summed over the various elementary processes i and appropriate sets of helicities $-\{\lambda'\}$ $(\Delta^2 = -t)$

$$T_{\{\lambda\}}(s,t) \sim \sum_{i\{\lambda'\}} C^i_{\{\lambda,\lambda'\}}(B_f) \int d^2k d^2l dx dy \hat{\Phi}^*_{B_f i}(x,(1-x)\vec{\Delta})\hat{\Phi}^*_{\bar{B}_f i}(y,(1-y)\vec{\Delta})$$
$$\times \hat{T}^i_{\{\lambda'\}}(\hat{s},\hat{t})\hat{\Phi}_{pi}(x,k_\perp)\hat{\Phi}_{\bar{p}i}(y,l_\perp)\Theta(\hat{s}-4m_f^2) \tag{1}$$

Thereby the elementary scattering amplitudes $\hat{T}^i$ are approximated by the Born amplitudes for $q\bar{q} \to f\bar{f}$, $S\bar{S} \to S\bar{S}$, $S\bar{S} \to V\bar{V}$, $V\bar{V} \to S\bar{S}$, $V\bar{V} \to V\bar{V}$, where S and V denote scalar and vector diquarks (of various flavours), respectively. For quarks they are given explicitely in Ref.5. For the treatment of diquarks (i.e. masses of diquarks, coupligs to gluons, diquark form factors, etc.) we refer to Ref.6, where the basic ideas of the diquark model for exclusive reactions are discussed in more detail.

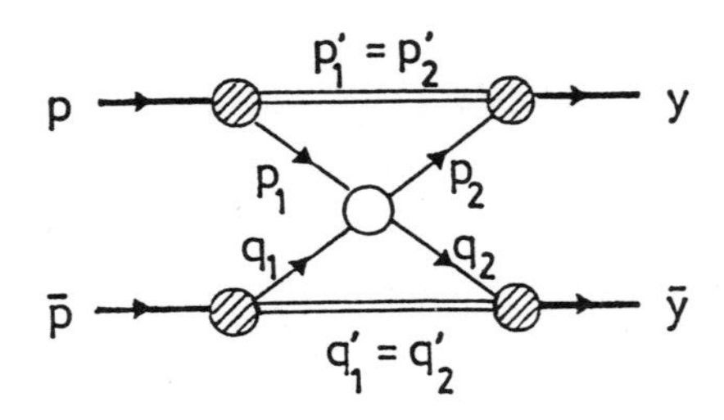

Fig. 1. The basic diagram of the constituent scattering model

For the reactions to be studied the coefficients $C(B_f)$ occurring in Eq.(1) can be found in Ref.5. They are obtained from the spin- flavour part of the baryon wave functions $\psi_{B_f\lambda}(x, k_\perp)$ for which we choose

$$\psi_{B_f\lambda}(x,k_\perp) = \sqrt{\frac{2}{3}}\sin\gamma \sum_{ij\mu\mu'\alpha\bar{\alpha}} C_{ij}^{\mu\mu'}(B_f,\lambda)\delta_{\alpha\bar{\alpha}}V_{i\mu\alpha}q_{j\mu'\bar{\alpha}}\hat{\Phi}_{B_f}^V(x,k_\perp)$$
$$+\sqrt{\frac{2}{3}}\cos\gamma \sum_{ij\alpha\bar{\alpha}} C_{ij}^{0\lambda}(B_f,\lambda)\delta_{\alpha\bar{\alpha}}S_{i0\alpha}q_{j\lambda\bar{\alpha}}\hat{\Phi}_{B_f}^S(x,k_\perp) \tag{2}$$

in analogy to SU(6). The first and the second quark have been coupled together to scalar and vector diquarks (i,j flavour, $\lambda, \mu, \mu'(=\lambda-\mu)$ helicities, $\alpha, \bar{\alpha}$ colour indices). The Clebsch-Gordan coefficients $C_{ij}^{\mu\mu'}(B,\lambda)$ of the baryons we are interested in may also be found in Ref.5. With the angle γ we allow for some flavour symmetry breaking. Because relativistic quark-diquark wave functions of baryons are not well known at present, we have assumed throughout $\hat{\Phi}^V = \hat{\Phi}^S$. There is still a comment in order about the arguments of the final state wave functions in Eq.(1). They are a consequence of the spectator conditions, namely $\vec{p_1}' = \vec{p_2}'$ and $\vec{q_1}' = \vec{q_2}'$ (see Fig.1) and hold approximately for small intrinsic transverse momenta. The step function in Eq.(1) guarantees that the incoming $q\bar{q}$ (diquark-antidiquark) pair has enough energy to produce the corresponding heavy final state constituents.

Eq.(1) can be cast into an even simpler form noticing that for small momentum transfer the dominant elementary amplitudes are only weakly dependent on x and y. The simplified expression then reads

$$T_{\{\lambda\}}(s,t) \sim \sum_{i\{\lambda'\}} C^i_{\{\lambda,\lambda'\}}(B_f)\hat{T}^i_{\{\lambda'\}}(<\hat{s}>,<\hat{t}>)F_i^2(t) \tag{3}$$

where p is the cms initial state momentum and F_i denotes the overlap integral

$$F_i(t) = \int d^2k dx \Theta(xp - m_f)\hat{\Phi}^*_{B_f i}(x,(1-x)\vec{\Delta})\hat{\Phi}_{pi}(x,k_\perp) \tag{4}$$

The constituent Mandelstam variables $<\hat{s}>, <\hat{t}>$ are calculated for the mean values $<x>=<y>$, the precise values of $<x>$ and $<y>$, however, being not so crucial.

In exploiting Eq.(3) the strategy we persued in Ref.5 was to parametrize the overlap integral (4) instead of choosing specific wave functions. For the quark ($i=1$) and diquark processes ($i \neq 1$) it was written as

$$A: \qquad F_i(t) = \begin{cases} \sqrt{c_0}F(t) & i=1 \\ \sqrt{c_0 c}\sqrt{F(t)} & i\neq 1 \end{cases} \tag{5}$$

The function F(t) was chosen to be a sum of two exponentials. The constant c determines the strength of the diquark contribution relative to the quark one.

Since in Ref.[1] we were interested only in the hyperon channels this strategy was most convenient and in fact sufficient. However, when extending this model to charmed or even bottomed baryons one may expect a flavour dependence of the overlap integral in addition to the kinematical effects which disappear with increasing s. As an alternative we therefore take a flavour dependent wave function from the literature and adapt it appropriately to the situation of quark and diquark.

In Ref.4 a meson wave function has been proposed which was obtained by transforming the harmonic oscillator wave function to the light cone. Adapting it to the case of a baryon (p, Y, B_c, B_b) made of quark and diquark, one finds for an active quark[7]

$$B: \quad \hat{\Phi}_{B_f1}(x,k_\perp) = N_f(1-x)^2\exp[-b^2(\frac{m_f^2}{x}+\frac{M_f^2}{1-x})]\exp[-b^2\frac{k_\perp^2}{x(1-x)}] \tag{6}$$

where m_f is the mass of the active quark, M_f that of the remaining spectator diquark, and N_f the normalization constant. The masses have to be understood as constituent masses. The wave functions of several baryons are shown in Fig.2.

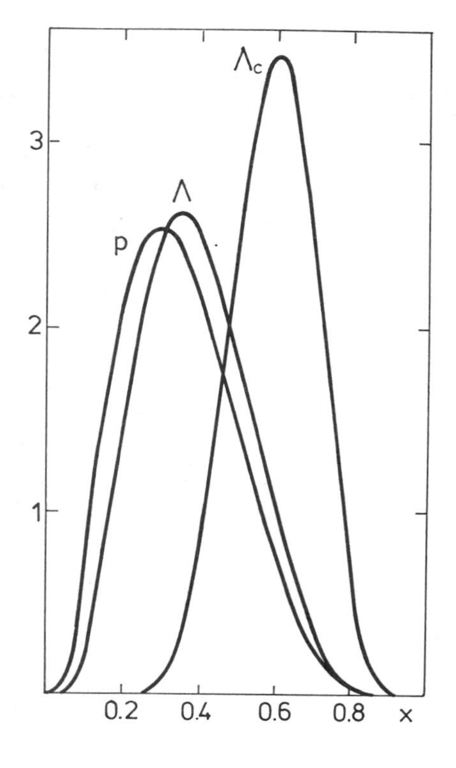

Fig. 2. The wave function B integrated over $k_\perp$ for p, Λ, and Λ_c. x is the fraction of the hadronic momentum carried by the quark.

Taking the diquark parameters from Ref.8, where the diquark model was proposed for an explanation of the puzzling spin effects in pp scattering[2], we are left with only few parameters, namely the overall normalization c_0, the flavour symmetry breaking angle γ, and the parameters occuring in the parametrization of the overlap integral (A) and wave function (B), respectively. To determine these open parameters we fit our model to the differential cross section for $\Lambda\bar{\Lambda}$ at $p_L = 6GeV/c$ [9] and integrated cross sections for $\Lambda\bar{\Lambda}$ and $\Sigma^0\bar{\Lambda} + cc$ in the p_L range from 2.7 to $7GeV/c$. The resulting value of 50° for the flavour symmetry breaking angle γ thereby occurs to be in nice agreement with that one which has been deduced from an analysis of meson-nucleon reactions[10]. For the oscillator parameter b^2 of model B we find the reasonable value of $1.39GeV^{-2}$ which, for protons, leads to a value of $0.267GeV/c$ for the mean intrinsic transverse momentum $< k_\perp^2 >^{1/2}$. At the end of this section we want to emphasize that, in contrast to case A, in case B the t-dependence of the quark and diquark overlap functions (4) as well as their relative strength are fully determined by the wave function.

RESULTS AND CONCLUSIONS

In Figs.3 and 4 we display the fit for the $\Lambda\bar{\Lambda}$ and $\Sigma^0\bar{\Lambda}$ integrated cross sections and corresponding predictions for the other hyperon channels. Both the versions of the model - the overlap integral parametrization (A), as well as the wave function approach (B) - reproduce the available data quite well down to $s \simeq 7GeV^2$, the second one doing even better at low energies. Whereas the diquark contributions play only a minor role in single annihilations reactions (Fig.3), they are essential for double annihilation reactions (Fig.4). As a consequence

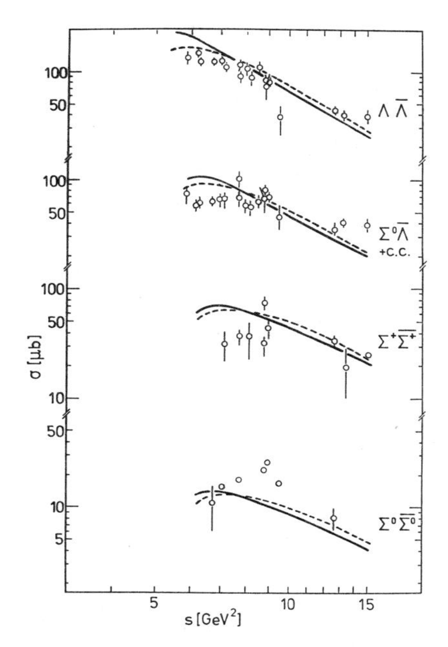

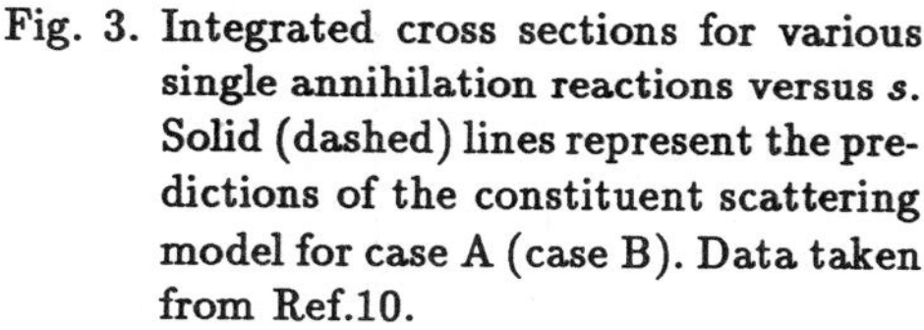

Fig. 3. Integrated cross sections for various single annihilation reactions versus s. Solid (dashed) lines represent the predictions of the constituent scattering model for case A (case B). Data taken from Ref.10.

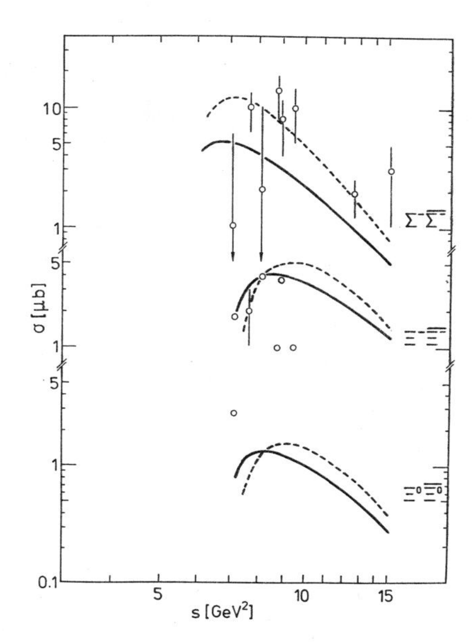

Fig. 4. Same as Fig.3 for double annihilation reactions. Note, that for some of the data no errors are quoted by the experimentalists.

our model asymptotically exhibits an s^{-2} (modulo α_s) behavior for integrated cross sections of single annihilation reactions and an s^{-6} behavior for that of double annihilation reactions. This difference is visible in Figs.3 and 4 and seems to be compatible with the trend of the data.

Keeping all the parameters fixed, we now extend this model straightforwardly to $p\bar{p} \to B_c\overline{B_c}$. In Fig.5 we present the predicted integrated cross sections for $p\bar{p} \to \Lambda_c\bar{\Lambda}_c$ and $p\bar{p} \to \Sigma_c^{++}\overline{\Sigma_c^{++}}$ and for comparison $p\bar{p} \to \Lambda\bar{\Lambda}$. The cross sections obtained with the overlap integral parametrization are rather large, suppressed as compared to $\Lambda\bar{\Lambda}$ only by kinematical effects which go away rapidly with increasing s. The results obtained with the wave function approach, on the other hand, are strongly suppressed even for large s, because the mismatch between the wave function for $p(\bar{p})$ and $\Lambda_c(\bar{\Lambda}_c)$ (see Fig.2) leads to a smaller overlap integral for $p\bar{p} \to \Lambda_c\bar{\Lambda}_c$ than for $p\bar{p} \to \Lambda\bar{\Lambda}$. We think of the overlap integral parametrization as an overestimate of the $B_c\overline{B}_c$ cross section since it is implausible that the charmed channels are merely suppressed by kinematics. Rather we expect the predictions with wave function (B) to be more realistic suggesting a suppresion of $\sigma(\Lambda_c\bar{\Lambda}_c)$ by a factor of ≈ 100 as compared to $\sigma(\Lambda\bar{\Lambda})$. Our results are obtained with a flavour dependence as given by Eq.(2). One, however, may think of other possibilities. Assuming the Λ_c wave function to be unsymmetrized in the c quark one gets a cross section for $\Lambda_c\bar{\Lambda}_c$ about 3 times larger than that obtained with Eq.(2). Other single annihilation channels like $\Sigma_c^{++}\overline{\Sigma_c^{++}}$ or $\Lambda_c\overline{\Sigma_c^+}$ + *c.c.* are suppressed as compared to $\Lambda_c\overline{\Lambda_c}$ by about the same factors as for the corresponding hyperon channels. Double annihilation reactions - which procced only via diquarks - are strongly damped by the diquark form factors. For instance $\sigma(\Sigma_c^0\overline{\Sigma_c^0}) : \sigma(\Lambda_c\bar{\Lambda}_c) \simeq 0.01$. From the same argument it is obvious that single annihilation reactions are totally dominated by the $q\bar{q} \to f\bar{f}$ contribution (($s \geq 20GeV^2$)). We have also estimated the cross section for $p\bar{p} \to \Lambda_b\bar{\Lambda}_b$ and found

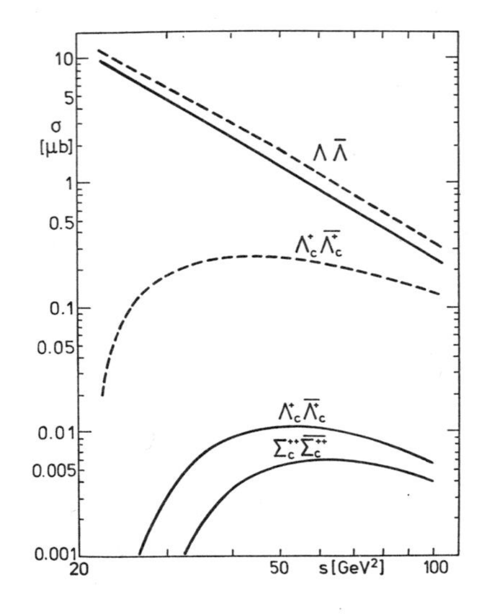

Fig. 5. Integrated cross sections for the reactions $p\bar{p} \rightarrow \Lambda\bar{\Lambda}$, $\Lambda_c\bar{\Lambda}_c$, and $\Sigma_c^{++}\overline{\Sigma_c^{++}}$ vs. s. Solid (dashed) lines are predictions for the wave function approach (B) (overlap integral parametrization (A)).

$2.2 \times 10^{-12}\mu b$ at $s \approx 500 GeV^2$. This reaction is likely too small to be measured. For predictions of observables different from integrated cross sections and production of heavy flavour meson pairs we refer to Ref.11.

In conclusion it seems not to be unrealistic to measure exclusive pairs of e.g. $\Lambda_c\bar{\Lambda}_c$ at a future facility like SuperLEAR. Such experiments and perhaps better and more complete data on hyperon production well away from the corresponding thresholds could serve as a valuable information for clarifying the interplay of perturbative QCD with nonperturbative effects.

REFERENCES

1. G.P.Lepage and S.J.Brodsky, Phys.Rev.**D22**,2157(1980).
2. P.R.Cameron et al., Phys.Rev.**D32**,3070(1985).
3. M.Anselmino, F.Caruso, P.Kroll, and W.Schweiger, preprint CERN-TH4941/87(1987).
4. S.J.Brodsky, T.Huang, and G.P.Lepage, SLAC-PUB2540(1980).
5. P.Kroll and W.Schweiger, Nucl.Phys.**A474**,608(1987).
6. P.Kroll, preprint CERN-TH4938/88, talk presented at the Adriatico Research Conference on "Spin and Polarization Dynamics in Nuclear and Particle Physics", Trieste (1988).
7. T.Huang, private communication.
8. M.Anselmino, P.Kroll, and B.Pire, Z.Phys.**C36**,89(1987).
9. H.Becker et al., Nucl.Phys.**B141**,48(1978).
10. R.T.van der Walle, Erice Lectures 1979, ed. A.Zichichi, Plenum Press, New York, pp.477 (1979).
11. P.Kroll, B.Quadder, and W.Schweiger, preprint WU B 88-17(1988).
12. V.Flaminio et al., CERN-HERA 84-01(1984).

$p\bar{p}$ ANNIHILATION INTO VECTOR MESONS AND INTO $\pi\pi\pi$, $\bar{K}K\pi$, $\pi\pi$ AND $K\bar{K}$ IN THE $^3P_0$ MODEL

Sadataka Furui

Institut für Theoretische Kernphysik
Universität Bonn
5300 Bonn, F.R.Germany

INTRODUCTION

Meson nucleon coupling form factors and coupling constants in the space like and in the time-like regions are calculated in the $^3P_0$ model[1]. In this model $q\bar{q}$ pairs are created/annihilated in the quantum numbers of the vacuum. I apply the same model to $p\bar{p}$ annihilation into $\pi\pi\pi$ and $K\bar{K}\pi$ in which two π are correlated to make a ρ and K and π ($\bar{K}$ and π) are correlated to make a K^* ($\bar{K}^*$) and study the selection rule on $p\bar{p}$ annihilation into $\pi\rho$ or $K\bar{K}^*$ ($\bar{K}K^*$). I study also $p\bar{p}$ annihilation into two or three mesons by solving the Schrödinger equation with the annihilation potential derived from the quark model combined with the G-parity transformed meson exchange potential of the Bonn group and the Paris group. I compare branching ratios of $\pi\pi$ and $K\bar{K}$ production from S- and P-wave antiprotonic atom calculated with the meson exchange potential and the quark model annihilation potential.

NUCLEON-VECTOR MESON COUPLING IN THE $^3P_0$ MODEL

In the $^3P_0$ model the meson baryon antibaryon coupling is calculated from the amplitude

$$T= \int d^3q_1 \dots d^3q_6 \ \psi_a(\vec{q}_1, \vec{q}_4)^\dagger \ O \ \psi_{N\bar{N}}(\vec{q}_1,\vec{q}_2,\vec{q}_3,\vec{q}_4,\vec{q}_5,\vec{q}_6) \tag{1}$$

where $\psi_a(\vec{q}_1,\vec{q}_4)$ and $\psi_{N\bar{N}}(\vec{q}_1,\vec{q}_2,\vec{q}_3,\vec{q}_4,\vec{q}_5,\vec{q}_6)$ are the wave function of a meson and that of the baryon-antibaryon system in the constituent quark model, the operator O is[2]

$$O=\lambda_A \frac{(-1)^{\mu+\nu}}{3} \ \sigma^{25}_{-\mu} \ y_\mu(\vec{q}_2-\vec{q}_5)\delta^{(3)}(\vec{q}_2+\vec{q}_5)\sigma^{36}_{-\nu} \ y_\nu(\vec{q}_3-\vec{q}_6)\delta^{(3)}(\vec{q}_3+\vec{q}_6) \tag{2}$$

where quark 2 and antiquark 5 and quark 3 and antiquark 6 are assumed to be pair created. The parameter λ_A is given by the strength of the quark pair creation γ and the spectrum of the four quark system. I take it here as a parameter and derive the coupling constant by comparing the result of the integral (1) with the effective lagrangian which is given by

$$L_{int}=i(g_{VNN}\bar{N}\gamma_{\mu}N-\bar{N}\frac{f_{VNN}}{2M_N}\sigma_{\mu\nu}q_{\nu}N)\ V_{\mu} \tag{3}$$

In the space-like region I obtain[2] the ratio of ρNN coupling and ωNN coupling as $g_{\omega NN}=3(m_{\omega}/m_{\rho})^{1/2}g_{\rho NN}$ and the ratio of the magnetic coupling and the electric coupling $f_{\rho NN}/g_{\rho NN}=3/10$ and $f_{\omega NN}/g_{\omega NN}=2/3$. The vector meson nucleon coupling in the time-like region or the nucleon anti-nucleon coupling to a vector meson consists of $p\bar{p}$ relative S- and D- wave terms. I obtain the S-wave $N\bar{N}$-vector meson coupling at the threshold is $G_{VNN}(-4M_N^2)=g_{VNN}+f_{VNN}$ which gives the ratio $G_{\omega NN}(-4M_N^2)/G_{\rho NN}(-4M_N^2)=15/13$. The form factor is proportional to

$$T\propto\left(\frac{4a^2(a^2+b^2/2)^2}{(2a^2+b^2/2)(3a^2+b^2)}k^2-\frac{6(a^2+b^2/2)}{3a^2+b^2}\right)$$
$$\times\exp\left(-\left(\frac{2}{3}a^2-\frac{a^2(a^2+b^2/2)}{(3a^2+b^2)(2a^2+b^2/2)}-\frac{a^4}{2a^2+b^2}\right)k^2\right) \tag{4}$$

which becomes zero around $Q^2=-4(k^2+M_N^2)=-5(GeV/c)^2$. In the case of D-wave $N\bar{N}$-vector meson coupling there is no zero in the form factor and I obtain $F_{\omega NN}(-4M_N^2)/F_{\rho NN}(-4M_N^2)=-3$. It is to be remarked that the $^3P_0$ model predicts the nucleon meson couplings not so different from the experimental values when the strength parameter γ is chosen to be 3.4 and that the model predicts the $N\bar{N}$-meson coupling which is different from the nucleon meson coupling.

$p\bar{p}$ ANNIHILATION INTO $\pi\pi\pi$ AND $K\bar{K}\pi$

In the production of $\pi\rho$ from the S-wave antiprotonic atom the contribution of $^3S_1$ dominates over $^1S_0$ and from the P-wave antiprotonic atom the contribution of $^1P_1$ dominates over $^3P_1$ or $^3P_2$[3]. The annihilation model A2 or the model which contains two $q\bar{q}$-pair annihilation and one $q\bar{q}$-pair annihilation which reproduces the general features of the branching ratios of $p\bar{p}$ annihilation into two mesons fails in explaining these data. To understand the phenomena we proposed the dynamical selection rule or the effective suppression of a certain channel due to the existence of competing decay channels[4]. In the above analyses the ρ meson is treated as a pure $q\bar{q}$ and the symmetry of the π in the decay product of ρ and the π which is produced before was not taken into account. The amplitude of S-wave $p\bar{p}$ annihilation into mesons in the $^3P_0$ model is obtained by contracting the two $^3P_0$ vertices to total angular momentum 0 and total spin 0. Thus the spin-flavour matrix element is proportional to that of $q\bar{q}$ in the intermediate state decaying into mesons, or $N\bar{N}$ decay into $\pi\rho$ from the $^1S_0$ channel is related to the decay of π. I observed that the amplitude of π decay into $\pi\rho$ vanishes due to the symmetry among the three π in the final state[2]. Thus the $^3P_0$ model that leads to three mesons or the A3 model predicts that the S-wave $p\bar{p}$ annihilation into $\pi\rho$ occurs only from the $^3S_1$ channel.

In order to clarify the symmetry among three mesons in the final state we study the $p\bar{p}$ annihilation into $K\bar{K}\pi$. The SU(6) spin-flavour matrix elements are characterized by the Isospin of the $p\bar{p}$; T, the isospin of the $K\bar{K}$; t, and the charge conjugation parity of the $p\bar{p}$; ε and we denote the amplitude as A^{ε}_{Tt} [5]. The charge conjugation parity means that we take the combination $p\bar{p}+\varepsilon\bar{p}p$ in the evaluation of the spin-flavour matrix element. In table 1 and 2 the spin-flavour matrix elements are shown. We observe that the in the $^1S_0$ channel $\varepsilon=1$ term contributes while in the $^3S_1$ channel $\varepsilon=-1$ term contributes. A calculation with the size parameter for a meson $a=3.1\ GeV^{-1}$ and for a baryon $b=4.1\ GeV^{-1}$ which corresponds to the mean square radius of a baryon is 0.61 fm and that of a meson is 0.50fm shows that the amplitude A^{-}_{11} is the largest in the S-wave $p\bar{p}$ annihilation into $K\bar{K}^*$ or $\bar{K}K^*$ [6].

Table 1. $^1S_0 N\bar{N} + \bar{N}N \to K\bar{K}\pi$ spin-flavour matrix elements. (LCR) means the assignment of the three mesons in the A3 diagram (see fig.1)

(LCR) a b c	(abc) (cba)	(cab) (bac)	A^{ε}_{Tt}	(abc) (cba)	(cab) (bac)	A^{ε}_{Tt}
$K^+K^-\pi^0$	9/32	9/32	A^{+}_{01}	-15/32	-15/32	A^{+}_{10}
$K^+\bar{K}^0\pi^-$	$9\sqrt{2}/32$	$9\sqrt{2}/32$	A^{+}_{01}	$-15\sqrt{2}/32$	$15\sqrt{2}/32$	A^{+}_{11}
$K^-K^0\pi^+$	$-9\sqrt{2}/32$	$-9\sqrt{2}/32$	A^{+}_{01}	$15\sqrt{2}/32$	$-15\sqrt{2}/32$	A^{+}_{11}
$K^0\bar{K}^0\pi^0$	-9/32	-9/32	A^{+}_{01}	-15/32	-15/32	A^{+}_{10}

Table 2. $^3S_1$ $N\bar{N}-\bar{N}N \to K\bar{K}\pi$ spin flavour matrix elements

(LCR) a b c	(abc) -(cba)	(cab) -(bac)	A^{ε}_{Tt}	(abc) -(cba)	(cab) -(bac)	A^{ε}_{Tt}
$K^+K^-\pi^0$	25/96	-25/96	A^{-}_{01}	$-25\sqrt{3}/192$	$25\sqrt{3}/192$	A^{-}_{10}
$K^+\bar{K}^0\pi^-$	$25\sqrt{2}/96$	$-25\sqrt{2}/96$	A^{-}_{01}	$25\sqrt{6}/192$	$25\sqrt{6}/192$	A^{-}_{11}
$K^-K^0\pi^+$	$25\sqrt{2}/96$	$-25\sqrt{2}/96$	A^{-}_{01}	$25\sqrt{6}/192$	$25\sqrt{6}/192$	A^{-}_{11}
$K^0\bar{K}^0\pi^0$	-25/96	25/96	A^{-}_{01}	$-25\sqrt{3}/192$	$25\sqrt{3}/192$	A^{-}_{10}

In the case of $p\bar{p}$ annihilation into $\pi\pi\pi$ with two pions correlated to form a ρ, we find the competing resonances in each partial waves as shown in the table 3. In the case of P-wave $p\bar{p}$ annihilation, the $^1P_1$ channel does not have a competing channel, while the $^3P_1$ and $^3P_2$ channels have competing πf decay mode. Thus the contribution of the $^1P_1$ channel will be dynamically enhanced. A study of the Dalitz plot distribution shows that in the $^3P_1$ channel the constant term and the term proportional to the square of the momentum between π and ρ cancell and thus the contribution of the $^1P_1$ becomes larger than that of the $^3P_1$.[6] For more quantitative understanding we need more theoretical as well as experimental study.

Table 3. Correlation of pions in the amplitudes of $N\bar{N} \to \pi^+ \pi^0 \pi^-$ for the partial waves $^{2T+1\ 2S+1}L_J^{\varepsilon}$

$^{31}S_0^+$	$((\pi\pi)_s\pi)_{L=0}$ $\varepsilon\pi$ $((\pi\pi)_d\pi)_{L=2}$ $f\pi$		
$^{13}S_1^-$	$((\pi\pi)_p\pi)_{L=1}$ $\rho\pi$		
$^{11}P_1^-$	$((\pi\pi)_p\pi)_{L=0,2}$ $\rho\pi$		
$^{33}P_1^-$	$((\pi\pi)_p\pi)_{L=0,2}$ $\rho\pi$	$^{33}P_1^+$	$((\pi\pi)_s\pi)_{L=1}$ $\varepsilon\pi$ $((\pi\pi)_d\pi)_{L=1}$ $f\pi$
$^{33}P_2^-$	$((\pi\pi)_p\pi)_{L=2}$ $\rho\pi$	$^{33}P_2^+$	$((\pi\pi)_d\pi)_{L=1}$ $f\pi$

$p\bar{p}$ ANNIHILATION INTO $\pi\pi$ AND $K\bar{K}$

Recently ASTERIX group at CERN published the branching ratios of S-wave and P-wave $p\bar{p}$ annihilation into $\pi^+\pi^-$ and $K\bar{K}$.[7] They observed that the branching ratios $BR(p\bar{p} \to K^+K^-)/BR(p\bar{p} \to \pi^+\pi^-)$ for the S-wave $p\bar{p}$ is about 1/3 and for the P-wave $p\bar{p}$ is about 1/16. We expect the $p\bar{p}$ annihilation into two mesons is dominated by the annihilation model A2 and the suppression of the $K\bar{K}$ pair production is due to the suppression of the $s\bar{s}$ pair creation as compared to the $u\bar{u}$ or $d\bar{d}$ pair creation in the $p\bar{p}$ system[8] One could

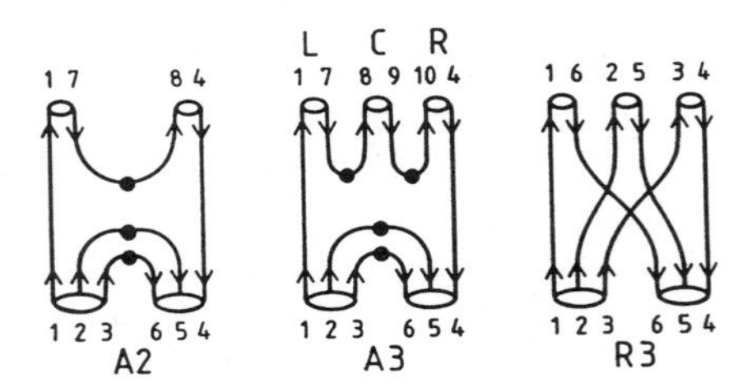

Fig. 1. The quark model for the $N\bar{N}$ annihilation into two mesons A2, into three mesons in the annihilation model A3 and in the rearrangement model R3

reproduce the branching ratios of K^+K^- and $\pi^+\pi^-$ by including the rearrangement model which contribute only in the $\pi^+\pi^-$. In this kind of model the $\pi^+\pi^-$ production from the $^3P_0$ channel becomes much larger than that from the $^3P_2$ channel while an extrapolation of the phenomenological helicity amplitude of $p\bar{p} \to \pi^+\pi^-$ suggests that the contribution of the $^3P_2$ channel is comparable to that of the $^3P_0$ channel.[9] In the annihilation model A2 I observe that the amplitude for the $^3P_0$ channel is suppressed as compared to the $^3P_2$ channel due to a term proportional to the square of the relative momentum between the π and ρ[8]. The term has the same origin as the recoil term in the meson baryon coupling[1].

In Table 4 I show the spin-flavour matrix element for S- and P-wave $p\bar{p}$ annihilation into $\pi^+\pi^-$ and $K\bar{K}$. From this table I observe that the $K^0\bar{K}^0$ production from the S-wave $p\bar{p}$ is suppressed due to the destructive interference of the isospin 0 and isospin 1 amplitudes while in the D-wave K^+K^- and $K^0\bar{K}^0$ are produced with comparable weights.

Table 4. The spin-flavour matrix element for $N\bar{N}\to\pi\pi$ and $K\bar{K}$ multiplied by 12

m_1m_2	$^{13}S_1$	$^{13}D_1$	$^{33}S_1$	$^{33}D_1$	$^{13}P_0$	$^{33}P_0$	$^{13}P_2$	$^{33}P_2$
$\pi^+\pi^-$	0	0	$13\sqrt{6}/18$	$2\sqrt{3}/9$	$15\sqrt{2}/2$	0	$18\sqrt{5}/10$	0
K^+K^-	$15\sqrt{3}/18$	$-\sqrt{6}/3$	$13\sqrt{3}/18$	$\sqrt{6}/9$	15/2	5/2	$9\sqrt{10}/10$	$7\sqrt{10}/10$
$K^0\bar{K}^0$	$15\sqrt{3}/18$	$-\sqrt{6}/3$	$-13\sqrt{3}/18$	$-\sqrt{6}/9$	15/2	-5/2	$9\sqrt{10}/10$	$-7\sqrt{10}/10$

The initial state interaction mixes the isospin and also it makes the decay amplitude complex. In order to study this effect we solve the coupled Schrödinger equation in momentum space. The $N\bar{N}$ interaction is described by the meson exchange potential and the annihilation potential which is derived from the quark model. In the meson exchange potential we use the Bonn potential without explicit Δ component, and one π, two π and ω exchange potential of the Paris group[10]. We multiply a cut off factor on the potential in the momentum space[8]. In the annihilation potential we take the combination of R3 and A2 although there are indications that A3 reproduces the branching ratios better than the R3 model[11]. The strength of the annihilation potential is fixed from the branching ratios of $p\bar{p}$ annihilation in flight calculated in combination of the Bonn potential.

From numerical calculations I found that the decay width and shifts are sensitive to the difference in the meson exchange potential. In the case of the Bonn potential I found large width and shift in the $^3P_0$ channel which is incompatible with the experiment. It could be due to the special cut-off scheme in the momentum space, but it is important that the shifts and widths depend on the intermediate and short range part of the meson exchange potential. In the case of Paris potential, I found the decay widths and shifts qualitatively in agreementrent with the results of the Dover-Richard model[12)] In table 5 I show the partial decay widths of $\pi^+\pi^-$ and K^+K^- in the combination of the potential of the Paris group and the quark model. I assume that the $s\bar{s}$ pair creation is suppressed as compared to $u\bar{u}$ or $d\bar{d}$ by a factor α=0.28. The relative magnitude of the width of $\pi^+\pi^-$ decay from the $^3P_0$ and from the $^3P_2$ is consistent with the phenomenological helicity amplitude approach[9)]. In the $^3SD_1$ channel, $K\bar{K}$ are produced mainly from the D-wave. The ratio of K^+K^- and $K^0\bar{K}^0$ production from $^3SD_1$ channel is 0.42 (exp 0.34±0.04) and that of the average of $^3P_0$ and $^3P_2$ is 0.051 (exp 0.060±0.018) in this model. The decay width from the $^3P_2$ channel is smaller than the Dover-Richard model, but it may be improved by replacing the annihilation potential from R3+A2 to A3+A2.

Table 5. The widths of $K\bar{K}$ and $\pi^+\pi^-$ production from the antiprotonic atom in the quark model with the meson exchange potential of the Paris group.

$^{2S+1}L_J$	Γ_{tot}	$\Gamma_{\pi^+\pi^-}$	$\Gamma_{K^+K^-}$	$\Gamma_{K^0\bar{K}^0}$
$^3SD_1$	1.4 keV	0.45 eV	$2.24\alpha^2$=0.19 eV	$1.27\alpha^2$=0.11 eV
$^3P_0$	25 meV	2.2 meV	$1.4\alpha^2$=0.12 meV	$0.51\alpha^2$=0.043 meV
$^3P_2$	4.6 meV	0.07meV	$0.06\alpha^2$=0.005meV	$0.0016\alpha^2$=0.00013meV

ACKNOWLEDGEMENTS

I thank G.Strobel and M.Maruyama for the help in the numerical calculation, Prof. A. Faessler and Prof. R.Vinh Mau for helpful discussions and Dr.B.Loiseau for providing valuable information on the Paris potential.

References

1) A. Le Yaouanc et al., Phys. Rev. D8 (1973) 2223
2) S.Furui, Nucl. Phys. A(1988) in press
3) S.Ahmed et al., Proc. seventh European Symp. on Antiproton Interaction, Durham, England, 1984, ed. by M.R. Pennington,(Hilger 1985) p.283
4) C.B.Dover, P.M.Fishbane and S.Furui, Phys. Rev. Lett. 57 (1986) 1358
5) B.Conforto et al., Nucl.Phys. B3 (1967) 469
6) S. Furui, Univ. Bonn preprint (1988)
7) M. Doser et al., preprint CERN/EP 88-42
8) M.Maruyama, S.Furui and A. Faessler, Nucl. Phys. A472 (1987) 643
9) G.C.Oades et al., Nucl.Phys. A464 (1987) 538
10) S.Furui, G.Strobel, M.Maruyama, A.Fessler and R.Vinh Mau, in preparation
11) M.Maruyama,S.Furui,A.Faessler and R.Vinh Mau, Nucl.Phys. A473 (1987) 649
12) J.M.Richard and M.E. Sainio, Phys. Lett. 110B (1982) 349

PROTONIUM ANNIHILATION INTO TWO MESONS

L.Mandrup and the Intermediate Energy Theory Group
Institute of Physics, University of Aarhus
DK-8000 Aarhus C, Denmark

INTRODUCTION

The annihilation of a proton-antiproton atomic bound state into a pseudoscalar meson and its antiparticle have recently been analyzed in terms of threshold values of the helicity amplitudes for this process[1]. Measurements[2] of such reactions particularly when made in coincidence with X-ray emission provide experimental constraints on nucleon-antinucleon annihilation models which are in general very simple and specific, in contrast to those obtained in scattering conditions, where the energy is higher and many partial waves contribute.

Two-particle final states are the simplest to treat but unfortunately such states account for a very small fraction (about 10^{-3}) of the total annihilation cross section. The largest number of decays end up with more than three pions as annihilation products. However, in the case of three pions there is experimental indication[2] that about half of the cross section arises from a correlated intermediate state of a π and a ρ meson. It is therefore quite likely that a substantial part of the annihilation process proceeds via two-particle intermediate states containing σ, ρ, ω or π mesons.

Simple experimental constraints on annihilation models are clearly very desirable. Since a large fraction of the cross section conceivably arises from the emission of two different mesons, it is necessary to consider annihilation to such final states. Extending the work of ref.1 to include unequal masses is straightforward and simple. The major difficulty is due to changes in the spin structure of the T-matrix elements. Both the ρ and the ω mesons are vector particles with three possible helicities compared to one for the pseudoscalar π mesons. A larger number of amplitudes are therefore needed to describe the annihilation process.

MESON EMISSION PROBABILITY

The proton-antiproton atomic bound state, $|\alpha\rangle = |n\,^{2S+1}L_J\rangle$, has principal quantum number n, total spin S, relative orbital angular momentum L and total angular momen-

Antiproton-Nucleon and Antiproton-Nucleus Interactions
Edited by F. Bradamante *et al.*
Plenum Press, New York, 1990

tum J. We work in the $p\bar{p}$ center of mass coordinate system and choose the z-axis in the direction of the emitted X-ray accompanying the final transition leading to the state $|\alpha\rangle$. One meson is emitted with momentum k in the direction defined by the spherical coordinates (θ,φ). The other meson is emitted with the same momentum in the opposite direction.

The probability W per unit time per unit solid angle for emission of the pair of mesons from the state $|\alpha\rangle$ is now ($\hbar$=c=1)

$$W(\alpha) = \frac{k}{32\pi^2 E_0} \sum_{\lambda_1\lambda_2} \sum_m g_m(\alpha)|M|^2 \tag{1}$$

$$M \equiv \langle k\theta\varphi;\lambda_1\lambda_2;T_1t_1T_2t_2|V|\alpha m\rangle \tag{2}$$

where m is the angular momentum projection along the z-axis, g_m is the occupation number for that substate, E_0 is half the protonium mass, λ_1 and λ_2 are helicities of the mesons, (T_1,t_1) and (T_2,t_2) are the total and third components of the isospins of the mesons and V is the interaction causing the reaction.

Expanding both initial and final states in states of given total isospin I and subsequently Fourier expanding the relative $p\bar{p}$ wave function we get by using the (relativistic) helicity representation

$$M = \sum_I \langle T_1t_1T_2t_2|I0\rangle\langle \tfrac{1}{2}\,\tfrac{1}{2}\,\tfrac{1}{2}\,-\tfrac{1}{2}|I0\rangle\cdot M_I \tag{3}$$

$$M_I = \sum_{\lambda_p\lambda_{\bar{p}}} \int\frac{d^3\vec{p}}{(2\pi)^3}\left(\frac{M_0^2}{E_pE_{\bar{p}}}\right)^{1/2}\tilde{R}_\alpha(p)\, M_I^{(h)}\,\langle\vec{p};\lambda_p\lambda_{\bar{p}}|LSJm\rangle \tag{4}$$

$$M_I^{(h)} \equiv \langle k\theta\varphi;\lambda_1\lambda_2;IT_1T_2|V|\vec{p};\lambda_p\lambda_{\bar{p}}\rangle \tag{5}$$

where $\vec{p}$ is the proton momentum, M_0 is the proton mass, E_p and $E_{\bar{p}}$ are the total energies and λ_p and $\lambda_{\bar{p}}$ are the p and $\bar{p}$ helicities. The function $\tilde{R}_\alpha(p)$ is the "radial" part of the Fourier transform of the protonium wave-function[1]. It is related to the radial part of the wave-function in coordinate space by

$$R_\alpha(r) = \frac{i^L}{2\pi^2}\int_0^\infty \tilde{R}_\alpha(p)j_L(pr)p^2\,dp \tag{6}$$

where j_L is the spherical Bessel function of order L and $R_\alpha(r)$ is the radial wave-function of the state $|\alpha\rangle$.

The results of Jacob and Wick[3] (θ_p and φ_p are the angular coordinates of the proton)

$$\langle\vec{p};\lambda_p\lambda_{\bar{p}}|LSJm) = \left(\frac{2L+1}{4\pi}\right)^{1/2} D^{J*}_{m\lambda}(\varphi_p,\theta_p,-\varphi_p)\langle L0S\lambda|J\lambda\rangle\langle\tfrac{1}{2}\lambda_p\tfrac{1}{2}-\lambda_{\bar{p}}|S\lambda\rangle \tag{7}$$

and from Frazer[4] $(\lambda=\lambda_p-\lambda_{\bar{p}})$

$$M_I^{(h)}=\sum_{J'm'}\frac{2J'+1}{4\pi}D^{J'*}_{m'\mu}(\varphi,\theta,-\varphi)D^{J'}_{m'\lambda}(\varphi_p,\theta_p,-\varphi_p)\langle\lambda_1\lambda_2|V_{IJ'}|\lambda_p\lambda_{\bar{p}}\rangle \tag{8}$$

are now used in eq.(4). The angular part of the integration is carried out giving

$$M_I=\frac{1}{(2\pi)^3}\left(\frac{2L+1}{4\pi}\right)^{1/2}D^{J*}_{m\mu}(\varphi,\theta,-\varphi)\sum_{\lambda_p\lambda_{\bar{p}}}\langle LOS\lambda|J\lambda\rangle\langle\tfrac{1}{2}\lambda_p\tfrac{1}{2}-\lambda_{\bar{p}}|S\lambda\rangle$$

$$\cdot\int_0^\infty\frac{M_0}{E_p}p^2\tilde{R}_\alpha(p)\langle\lambda_1\lambda_2|V_{IJ}|\lambda_p\lambda_{\bar{p}}\rangle dp \tag{9}$$

where $\mu=\lambda_1-\lambda_2$ and V_{IJ} is the part of V with isospin I and total angular momentum J.

Inserting eq.(9) into eqs.(1) and (3) we arrive at the general expression for the meson emission probability

$$W(\alpha)=\frac{k}{8\pi^4E_0}\frac{2L+1}{4\pi}\sum_{\lambda_1\lambda_2}\sum_m g_m(\alpha)|d^{J*}_{m\lambda_1-\lambda_2}(\theta)|^2$$

$$\cdot\left|\sum_I\langle T_1t_1T_2-t_1|IO\rangle\langle\tfrac{1}{2}\,\tfrac{1}{2}\,\tfrac{1}{2}\,-\tfrac{1}{2}|IO\rangle\cdot\Lambda^I_{LSJ}\right|^2 \tag{10}$$

$$\Lambda^I_{LSJ}\equiv\frac{1}{16\pi^2}\sum_{\lambda_p\lambda_{\bar{p}}}\langle LOS\lambda|J\lambda\rangle\langle\tfrac{1}{2}\lambda_p\tfrac{1}{2}-\lambda_{\bar{p}}|S\lambda\rangle\cdot h^{\lambda_1\lambda_2}_{\lambda_p\lambda_{\bar{p}}} \tag{11}$$

$$h^{\lambda_1\lambda_2}_{\lambda_p\lambda_{\bar{p}}}\equiv\int_0^\infty\frac{M_0}{E_p}p^2\tilde{R}_\alpha(p)\langle\lambda_1\lambda_2|V_{IJ}|\lambda_p\lambda_{\bar{p}}\rangle dp \tag{12}$$

The number of independent helicity-amplitudes can be reduced by using parity conservation which requires[4]

$$\langle\lambda_1\lambda_2|V_{IJ}|\lambda_p\lambda_{\bar{p}}\rangle=\eta_g\langle-\lambda_1-\lambda_2|V_{IJ}|-\lambda_p-\lambda_{\bar{p}}\rangle \tag{13}$$

$$\eta_g=\eta_1\eta_2\eta_p\eta_{\bar{p}}\cdot(-1)^{s_1+s_2+s_p+s_{\bar{p}}} \tag{14}$$

where the η's and the s's are the parities and spins of the mesons, the proton and the antiproton.

PSEUDOSCALAR PARTICLES IN THE 2P STATE

For spin zero negative parity particles like pions or kaons we have only two independent helicity amplitudes. The momentum distribution in the atomic bound state falls off exponentially over a 4 MeV/c scale. As this is small compared to the scale of the helicity amplitudes we may use the threshold values (p → 0). They are given by

$$\langle 00|V_{IJ}|\lambda_p\lambda_{\bar{p}}\rangle \equiv 16\pi^2(pk)^{J-1}\cdot g_{IJ}(\lambda_p,\lambda_{\bar{p}}) \tag{15}$$

$$\sqrt{J}\, g_{IJ}(+,+) + \sqrt{(J+1)}\, g_{IJ}(+,-) \to \sqrt{(2J+1)}\, a_1^{(IJ)} \tag{16a}$$

$$\sqrt{(J+1)}\, g_{IJ}(+,+) - \sqrt{J}\, g_{IJ}(+,-) \to p^2\sqrt{(2j+1)}\, a_2^{(IJ)} \tag{16b}$$

where a_1 and a_2 are constants. The threshold values of Λ^I_{LSJ} can then be calculated and the angular distribution of one of the mesons in the center of mass system obtained from eq.(10). When $g_m(\alpha)$ is independent of m, i.e. all substates are equally populated, the angular distribution is isotropic as expected. To break this spherical symmetry some anisotropy must be introduced in the population of the m-states. The bound states we are interested in are produced in an X-ray cascade process following the capture of the antiproton into some high lying initial state. We assume that electric dipole transitions dominate this cascade and we use the direction of the final X-ray to define the z-axis. Since most of the annihilation takes place[2] from the 2P-state, we choose this important example. The resulting g_m values turn out to be

$$g_0(^3P_2):g_{\pm1}(^3P_2):g_{\pm2}(^3P_2):g_0(^3P_1):g_{\pm1}(^3P_1):g_0(^3P_0):$$

$$g_0(^1P_1):g_{\pm1}(^1P_1) = 38:39:42:42:39:40:36:42 \tag{17}$$

and the normalized values are then obtained by division by 480 corresponding to unit probability for <u>all</u> the P-states.

The total angular distribution is now easily calculated for the different P-states. We find

$$W(^1P_1) = W(^3P_1) = 0 \tag{18a}$$

$$W(^3P_0) = \frac{3}{2^4\pi k M_0} K(\alpha)|a_2^{(0)}|^2 \tag{18b}$$

$$W(^3P_2) = \frac{123k^3}{2^7\pi M_0} K(\alpha)|a_1^{(2)}|^2[1-\frac{3}{41}\cos^2\theta] \tag{18c}$$

$$a_i^{(J)} = \sum_{I=0}^{1} \langle T_1 t_1 T_2 -t_1|I0\rangle\langle\tfrac{1}{2}\,\tfrac{1}{2}\,\tfrac{1}{2}\,-\tfrac{1}{2}|I0\rangle\cdot a_i^{(IJ)} \tag{19}$$

$$K(\alpha) \equiv \lim_{r\to 0} |R_\alpha(r)/r|^2 \tag{20}$$

where $K(\alpha)$ may be different for the various states. The total angular distribution, obtained by adding eqs.(18), is then of the form

$$W(2P) \equiv C_0 + C_1\cos^2\theta \tag{21}$$

with C_0 and C_1 expressed in terms of the threshold values of the partial wave helicity amplitudes.

The partial widths, obtained by integrating over all angles and renormalizing each of the states to unity, are then given by

$$\Gamma(^1P_1) = \Gamma(^3P_1) = 0 \tag{22a}$$

$$\Gamma(^3P_0) = \frac{9}{kM_0} K(^3P_0) \, |a_2^{(0)}|^2 \tag{22b}$$

$$\Gamma(^3P_2) = \frac{9k^3}{M_0} K(^3P_2) \, |a_1^{(2)}|^2 \tag{22c}$$

The total annihilation width Γ_{an} of a given state $|\alpha\rangle$ is given in terms of W(r), the imaginary part of the two body potential, by

$$\Gamma_{an}(\alpha) = -8\pi \int |R_\alpha(r)|^2 W(r) r^2 dr \tag{23}$$

COMPARISON WITH MODELS AND EXPERIMENTS

Experimental information on annihilation from P-states of protonium is essentially limited to the measurements of the Asterix group. As an example of the information which can be obtained, we consider the final states $\pi^+\pi^-$, K^+K^- and $K^0\bar{K}^0$ where the spin averaged branching ratios are measured to be $(4.81\pm0.49) \times 10^{-3}$, $(2.87\pm0.51) \times 10^{-4}$ and $(8.2\pm1.3) \times 10^{-5}$ respectively[2]. If we make use of eq.(22) we obtain the spin averaged branching ratios in the form

$$B(P) = \frac{9[K(^3P_0)|a_2^{(0)}|^2 + 5k^4 K(^3P_2)|a_1^{(2)}|^2]}{kM_0[\Gamma_{an}(^3P_0) + 3\Gamma_{an}(^3P_1) + 5\Gamma_{an}(^3P_2) + 3\Gamma_{an}(^1P_1)]} \tag{24}$$

Using the Dover-Richard parameter set 2 to calculate K and Γ_{an} the branching ratios quoted above give the spin averaged values

$$\begin{aligned} \sqrt{[|a_2^{(0)}|^2 + 1.41k^4 |a_1^{(2)}|^2]} &= 0.074 \text{ fm} \quad \pi^+\pi^- \\ &= 0.017 \text{ fm} \quad K^+K^- \\ &= 0.0089 \text{ fm} \quad K^0\bar{K}^0 \end{aligned} \tag{25}$$

The reliability of these values depend on how well the Dover-Richard potential can reproduce the total annihilation widths for the P-states. Here there are two measurements[5,6] of the spin averaged 2P width, 40 ± 11 meV and 45 ± 10 meV. The Dover-Richard potential with parameter set 2 which we use gives a spin averaged total annihilation width of 32 meV (27 meV without tensor mixing and coupling to $n\bar{n}$) so we conclude that we can obtain reasonable quantitative estimates of the threshold values of the partial wave helicity amplitudes for specific annihilation processes.

Martin and Morgan[7] have analyzed higher energy in flight data for annihilation to $\pi^+\pi^-$. Using their results for the individual helicity amplitudes and extrapolating down to threshold gives an average 2P threshold value which is about a factor of 1.5 too large. Bearing in mind the considerable uncertainties associated with the extrapolation from Martin and Morgan's lowest energy, which corresponds to an antiproton lab. momentum of 800 MeV/c, the agreement is very satisfactory.

The quark model calculations of Kohno and Weise[8] can also be extrapolated down to threshold to give values for the moduli of the helicity amplitudes, both for annihilation to $\pi^+\pi^-$ and to K^+K^-. The agreement is satisfactory for annihilation from the 2P state to K^+K^- whereas the predicted value for the average 2P threshold value is roughly a factor of 7 too large in the case of $\pi^+\pi^-$.

Another quark model calculation by Maruyama et al.[9] leads to threshold values which are quite similar to the values obtain by extrapolating Martin and Morgan's results and are in fair agreement with the preliminary experimental values.

CONCLUSION

The present analysis of protonium annihilation into two mesons relates measurements with the threshold values of partial wave helicity amplitudes. These in turn can be used to constrain and test theoretical models for the annihilation. Results from applications where one of the mesons has spin 1 are also very interesting and expected to be published soon.

REFERENCES

1) G.C. Oades, A. Miranda, L. Mandrup, A.S. Jensen and B.I. Deutch, Nucl.Phys.A464 (1987) 538.
2) S. Ahmad et al. p.287 in "Antiproton 1984", ed. M.L. Remington, Inst. of Physics Conference Series, number 73, Adam Hilger Ltd., 1986, Phys.Lett. 157B (1985) 333, Proceedings III LEAR Workshop, ed. Frontières, Tignes, France (1985) p353.
   M. Doser et al., CERN preprint CERN/EP 88-42.
3) M. Jacob and G.C. Wick, Ann.Phys. 7 (1959) 404.
4) W.R. Frazer, Elementary Particles, Prentice-Hall, Inc. 1966.
5) R. Bacher et al., Proceedings of Antiproton 86 (Thessaloniki, 1986), ed. S. Charalambous et al., World Scientific, Singapore, 1987, p.223.
6) C.A. Baker et al., Nucl.Phys. A483 (1988) 631.
7) B.R. Martin and D. Morgan, in Proc. of the IV European Symp. on Antinucleon-Nucleon Annihilation, Strasbourg, 1978, p.101.
8) M. Kohno and W. Weise, Nucl.Phys. A454 (1986) 429.
9) M. Maruyama, S. Furui and A. Faessler, Nucl.Phys. A472 (1987) 643.

POLARIZATION IN THE $p\bar{p}$-ELASTIC SCATTERING NEAR THRESHOLD

O.D. Dalkarov and K.V. Protasov *

Lebedev Physical Institute
Leninsky Prospect 53
Moscow 117924, USSR

ABSTRACT

Polarization of antiprotons in the $p\bar{p}$ elastic scattering at low incident momenta (< 300 MeV/c) near threshold is calculated in coupled channel model. Smallness of the polarization is shown to be due to existence of the P-wave quasi-nuclear states, which enhance all of the P-waves in the $p\bar{p}$ elastic scattering, but compensate each other in the polarization.

Very interesting results (e.g. large P-wave enhancement, unusual behaviour of the ρ-parameter - real-to-imaginary ratio for the forward elastic $p\bar{p}$-scattering amplitude[1]) were obtained at LEAR on the $p\bar{p}$ interaction. It seems desirable, however, to make more complete measurements of the spin observables of $p\bar{p}$-scattering and to perform experiments at very low $\bar{p}$ momenta. We shall present here predictions for the polarization corresponding to antiproton momenta $P_{lab} \leqslant 300$ MeV/c.

These calculations were done in the coupled channel model[2], within which the experimental data on $p\bar{p}$-interaction (total, annihilation, elastic, charge-exchange cross sections, etc.) were described in the energy range from the threshold to 1950 MeV in c.m.s. In Fig. 1 an example of the description of the differential $p\bar{p}$ elastic cross section at the antiproton incident momenta $P_{lab} = 387$ MeV/c is presented. Two curves (solid and dashed) correspond to the different sets of the fitting parameter - the range of the cut-off of the OBEP potential singular terms (see the caption of Fig. 1). The angular anisotropy of the differential cross section, being the indication of the large P-wave contribution, is seen clearly on the two curves. In Fig. 2(a) the polarization of antiprotons in the elastic $p\bar{p}$-scattering, which corresponds to the differential cross section on Fig. 1, is shown. First of all, it is necessary to pay atten-

*) Presented by K.V. Protasov

Antiproton-Nucleon and Antiproton-Nucleus Interactions
Edited by F. Bradamante *et al.*
Plenum Press, New York, 1990

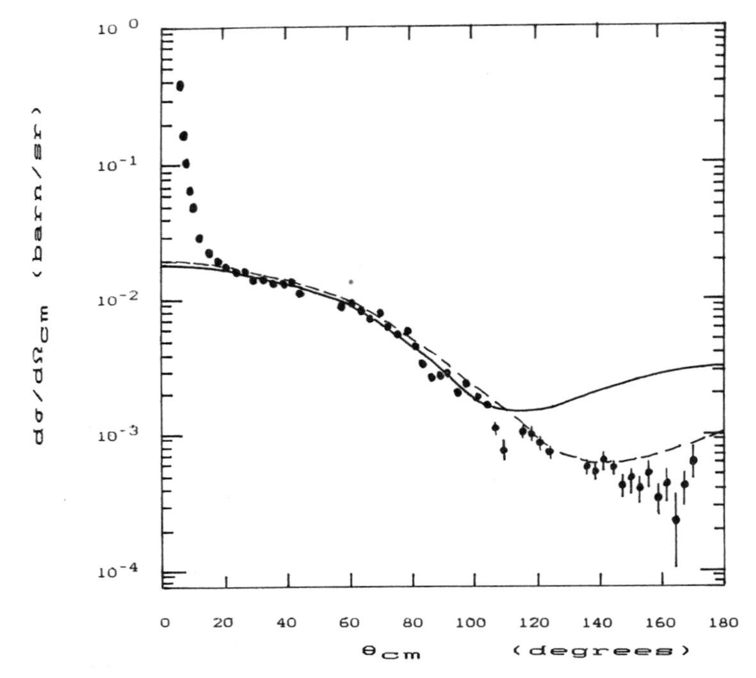

Fig. 1 Elastic differential cross-section for the $p\bar{p}$ scattering at the momentum of incident antiproton P = 287 MeV/c (c.m. momentum K = 144 MeV/c). The experimental data are taken from Ref. 5. The curves differ by the choice of the fitting parameter r. Solid curve is the calculation with $r_c(1^3S_1) = r_c(3^3S_1) = 0.50$ fm, $r(1^3P_2 = 0.72$ fm. Dashed curve is the same for $r_c(1^3S_1) = 0.47$ fm, $r_c(3^3S_1) = 0.60$ fm, $r_c(1^3P_2) = 0.65$ fm.

tion to the fact that with presence of large P-wave contribution into $p\bar{p}$-elastic scattering (besides the angular anisotropy it is seen clearly from the very rapid growth of the real-to-imaginary ratio for the forward elastic scattering amplitude from $\rho \simeq -1$ at the threshold to $\rho \simeq 0$ at $P \simeq 150$ MeV/c[3]) the polarization is small in both cases (for solid and dashed curves, which differ from each other only by the dominance of one of the triplet P-waves: for the solid curve $^3P_1$ is dominant, for the dashed - $^3P_1$). This effect (the smallness of the polarization with the presence of large P-wave contribution in the interaction) could be clarified to take into account the definition of the polarization[4]:

$$\vec{n} \cdot \vec{P} = P_y = \frac{Sp(f\vec{\sigma}f^+)}{Sp(ff^+)}\,\vec{n} = \frac{2Re\{(\alpha + \beta)\,\gamma^*\}}{|\alpha|^2 + |\beta|^2 + 2|\gamma|^2 + 2|\delta|^2 + 2|\varepsilon|^2} \tag{1}$$

where

$$\alpha = \frac{1}{4} \quad f_{^1S_0} + 3\,f_{^3S_1} + (f_{^3P_0} + 3\,f_{^1P_1} + 3\,f_{^3P_1} + 5\,f_{^3P_2})\,\mathrm{Cos}\theta \quad ,$$

$$\beta = \frac{1}{4} \quad f_{^3S_1} + f_{^1S_0} + (2\, f_{^3P_2} + f_{^3P_0} - 3\, f_{^1P_1})\ \mathrm{Cos}\theta \quad ,$$

$$\gamma = \frac{1}{4} \quad (\frac{5}{2}\, f_{^3P_2} - \frac{3}{2}\, f_{^3P_1} - f_{^3P_0})\ \mathrm{Sin}\theta ,$$

$$\delta = \frac{1}{4} (\ {}_{^3S_1} - f_{^1s_0}),$$

$$\varepsilon = \frac{3}{4} \quad (\frac{1}{2}\, f_{^3P_2} + \frac{1}{2}\, f_{^3P_1} - f_{^1P_1})\ \mathrm{Cos}\theta ;$$

$f_{^{2S+1}L_J}$ is an isospin averaged scattering amplitude with corresponding quantum numbers (for simplicity all formulae are written only for S- and P-waves; D-wave contribution into polarization in our calculations at energies in question is small); θ is angle in c.m.s.

Let us note that in our model the P-wave enhancement is caused by the existence of the near-threshold quasi-nuclear P-states, they exist close to the threshold in 4 of 6 triplet states (if the isospin is taken into account). Therefore we work in the situation when practically all triplet P-waves are enhanced equally and so the polarization could be small (due to the coefficient γ in the numeration in Ref. 1). In principle, if all triplet waves are equal, even being very big, the polarization would be zero (as for spinless amplitude).

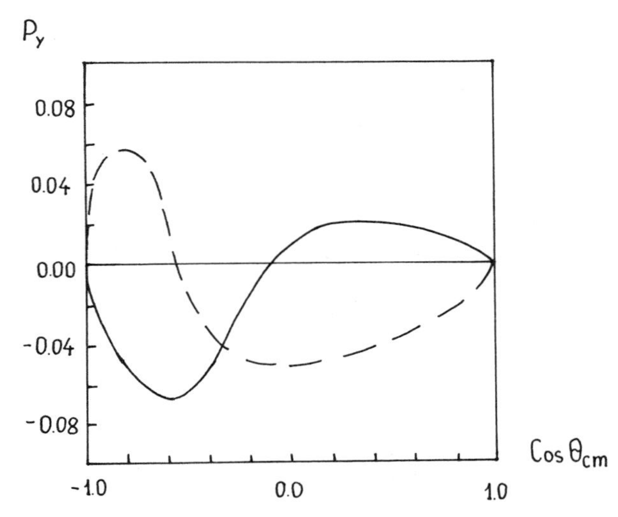

Fig. 2(a) Polarization in the $p\bar{p}$ elastic scattering at the c.m. momentum K = 144 MeV/c.. Solid and dashed curves correspond to the same choise of r_c as in Fig. 1.

The absence of rather large polarization is seen also and at smaller energies (Figs. 2(b) and 2(c)). But if the unusual behaviour is caused by the existence of the narrow isolated P- or D-state, as it is proposed[4], the visible polarization would be found in the experiment.

So, the measurements of polarization are very important, not only for the indication of spin characteristics of the $p\bar{p}$ elastic scattering amplitude, but also for the understanding of the unusual behaviour of ρ at small antiproton momenta.

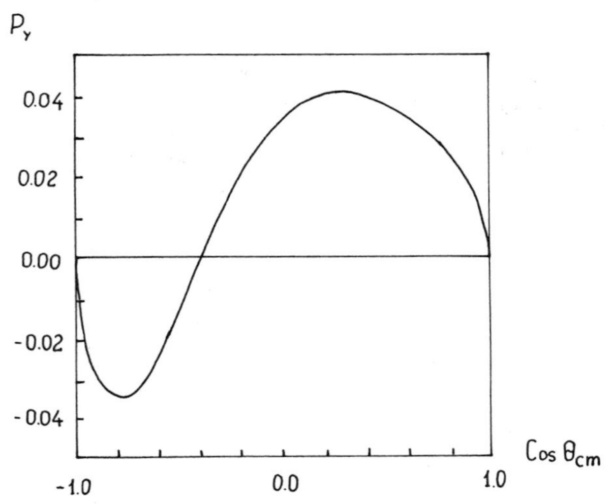

Fig. 2(b) The same as in Fig. 2(a) at K = 106 MeV/c. Curve corresponds to the choice of r_c as for solid curve in Fig. 1.

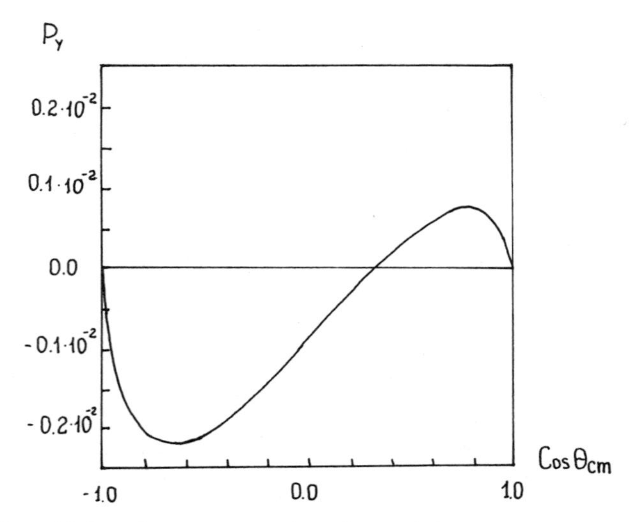

Fig. 2(c) The same as in Fig. 2(b) at K = 43 MeV/c.

REFERENCES

1. W. Brückner et al.; Phys.Lett. 166B (1986) 113;
Phys.Lett. 158B (1985) 180.
2. O.D. Dalkarov, K.V. Protasov and I.S. Shapiro; preprint 37, Moscow, FIAN, 1988.
3. O.D. Dalkarov and K.V. Protasov; JETP Lett. 44 (1986) 638.
4. M.L. Goldberger and K.M. Watson; Collision Theory, New York, 1964.
5. T. Ueda; Proc. IV LEAR Workshop, Villars-sur-Ollon, Switzerland, 6-13 Sept. 1987, p.247 (Ed. C. Amsler et al., Harwood Academic Press, New York).

NUCLEON ELECTROMAGNETIC FORM-FACTOR

IN THE TIME-LIKE REGION NEAR $N\bar{N}$ THRESHOLD

O.D. Dalkarov and K.V. Protasov *)

Lebedev Physical Institute
Leninsky Prospect 53
Moscow 117924, USSR

ABSTRACT

Nucleon electromagnetic form-factor in the time-like region near $N\bar{N}$ threshold is calculated. Its behaviour is shown to be dominated by the $N\bar{N}$ nuclear interaction in initial state. For this reason, an investigation of nucleon form-factor in the time-like region could provide useful information on the low energy $N\bar{N}$ parameters.

Study of nucleon-antinucleon annihilation channels (especially with definite quantum numbers) is an object of intense interest for understanding the nature of the low energy baryon-antibaryon interaction. Annihilation into lepton pair $p\bar{p} \to e^+e^-$ can proceed only from $p\bar{p}$ states with photon quantum numbers $J^{PC} = 1^{--}$ (for slow antinucleons it corresponds to S-states with isospin I = 0,1). Now there are some data[1,2] on this cross section behaviour at small relative momentum between $\bar{p}$ and p. In particular, in the experiment described in Ref. 2 the relative probability of $p\bar{p} \to e^+e^-$ annihilation for $\bar{p}$ incident momentum 300 MeV/c was measured, and the proton electromagnetic form-factor was found to be $G = 0.46^{+0.15}_{-0.09}$. The proton form-factor $G(q^2)$ is seen to be close to unity at the boundary of the physical region with approach from the side of time-like transferred momenta q^2 (i.e. for $-q^2 > 4M^2$): $G(-q^2 = 4M^2) \sim 1$ (q is the four-momentum and in the region in question $-q^2 = s$, where s is the $\bar{p}$ and p c.m.s. energy squared). If we try to approximate such value of G at $-q^2 > 4M^2$ using usual dipole formula, the value G will be less by the order of magnitude than one experimentally measured[2]. From our point of view large value of G is at least an evidence for a strong nuclear attraction between $\bar{p}$ and p.

---

*) Presented by K.V. Protasov

Antiproton-Nucleon and Antiproton-Nucleus Interactions
Edited by F. Bradamante *et al.*
Plenum Press, New York, 1990

Large value of the electromagnetic form-factor G in the known vector dominant models (VDM) can be obtained by taking into account besides light ρ, ω, ψ-mesons also heavy ρ', ψ'-mesons[3,4]. The coupling constants for these mesons with proton were in this case considered as free parameters. In our model the existence of the subthreshold resonances (i.e. bound $N\bar{N}$ quasi-nuclear states with masses close to 2M[5]) is an additional reason for increasing of the form-factor in the time-like region. This possibility was pointed out earlier[6].

Let us consider some common features of the nucleon-antinucleon annihilation into e^+e^--pair at low energy in $N\bar{N}$ system. This process (in the first order of α) is associated with the diagram shown in Fig. 1, where the dashed block denotes the amplitude of the initial state interaction, the black circle corresponds to the form-factor G_0 which is associated with the singularities far from $N\bar{N}$ threshold (for instance, with ρ, ω, ψ-poles in VDM). To take into account the effect of initial state interaction, we consider from the beginning the diagram without interaction in the initial state. To this diagram $\delta(\vec{r})$-potential is corresponded (since this diagram does not depend on $\vec{q}$-three-dimensional momentum). If this potential is denoted by $V(\vec{r}) = V\,\delta(\vec{r})$, the transition amplitude for the diagram on Fig. 1 can be written:

$$T \sim \int \Psi_{\nu\bar{\nu}}(\vec{r})\, V(\vec{r})\, \Psi_{e^+e^-}(\vec{r})\, d\vec{r} \tag{1}$$

Therefore the nucleon form-factor G has the form:

$$G = |\Psi(0)|\; G_0 \tag{2}$$

where G_0 is the form-factor corresponding to the diagram without initial state interaction. As for the concrete form $G_0 = G_0(s)$, it can in principle be approximated by the various models well described data in space-like region (dipole formula or VDM). Let us consider the factor $\Psi(0)$ which can be expressed by $\Psi(0) = 1,/f(-k)$, where f(k) is the Jost function for $N\bar{N}$ system. Jost function f(k) can be represented in the form[8]

$$f(k) = \tau(k)\, e^{i\,\delta(k)} \tag{3}$$

(Let us emphasize that this Jost function corresponds to S-wave since $\delta(\vec{r})$-potential extracts only this wave.) Function $\tau(k)$ is symmetrical on k, $\delta(k)$ being antisymmetrical (for this reason as usually for S-matrix we have $S(k) = f(k)/f(-k) = e^{2i\,\delta(k)}$). Phase $\delta(k)$ is complex a one since there are annihilation channels for $N\bar{N}$ system. Therefore

$$|\Psi_{N\bar{N}}(0)| = \frac{1}{|\tau(k)|}\, e^{-\mathrm{Im}\,\delta(k)} \tag{4}$$

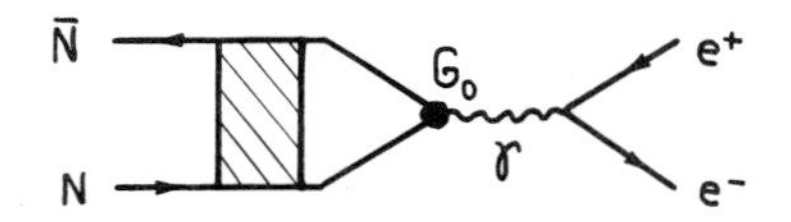

Fig. 1 The diagram for the $N\bar{N} \to e^+e^-$ annihilation with the initial state interaction.

Now we will consider only a small energy in the $N\bar{N}$ c.m. system. Let us use an expansion of $\delta(k)$ in the scattering length approximation, i.e. $\delta(k) = a\, k$ (we neglected the higher on k terms in this expansion). It is necessary to emphasize that expansion of $\tau(k)$ includes only k^2 terms since $\tau(k) = \tau(-k)$. Therefore an expansion of G in power series in k has the form:

$$G = C_1(1 - \mathrm{Im}\, a\, k + C_2 k^2 + \ldots) \qquad (5)$$

where C_1 and C_2 are the constants. No other term of the first order on k exist since all other factors are a function of s, so they give only even series in k. The value of Im a corresponds to the imaginary part of S-scattering length. The term with Im a has a principal significance: it occurs only in the case of initial state interaction. Neither in VDM (for instance[4]) nor in Regge-pole model[5] is it possible in principle, to get the linear behaviour in k.

Only LEAR experiments can now give a direct answer to the question about the presence of linear k-term in G behaviour. If a linear behaviour is observed in an experiment, it could be a strong indication of the necessity of taking into account an initial state interaction.

Moreover, the experiments with form-factor measurements could be used as an independent source of information about Im in addition to the investigation of the level shifts and widths of protonium. However, it is necessary to note that in the expansion of proton (neutron) form-factor in the power in k a slightly different value (not only Im a) will be included if the vertex is differed for various isospins in the above diagram. The correction for isospin dependence of $N\bar{N}$ interaction is rather trivial and it is not important for obtaining conclusion about the presence of a linear term.

For the numerical calculation of the $p\bar{p} \to e^+e^-$ annihilation cross section the coupled channel model (CCM) was used[7]. The channel 1 corresponds to $N\bar{N}$ system (interaction between $\bar{N}$ and N was described by the one boson exchange potential $V_{N\bar{N}}^{OBEP} = V_{11}$). Masses of proton and neutron were taken the same and were equal to M = 0.939 GeV. Channel 2 is annihilation one. The coupling between channels 1 and 2 is realized by short range Yukawa-type potential (V_{12}). All parameters for these channels are taken from Ref. 7 where the details of two-channel model were given (in the framework of this model all $p\bar{p}$ low energy data were described). The third channel corresponds to e^+e^- system. It is connected only with $N\bar{N}$ channel by transition potential. This transition potential (V_{13}) has a form of a Yukawa function, as mentioned above.

The electromagnetic proton form-factor is connected with $p\bar{p} \to e^+e^-$ cross section by the formula:

$$\sigma_{p\bar{p} \to e^+e^-} = \frac{\pi\, \alpha^2}{2\, M_K} |G|^2 \qquad (6)$$

(here assumed as usual $G_E(4M^2) = G_M(4M^2) = G$).

The details of numerical calculations in CCM (including the problem with δ-function potential) were discussed in Ref. 9.

The enhancement coefficients for $N\bar{N} \to e^+e^-$ reaction (I = 0, 1) are shown in Fig. 2. These coefficients are defined as

$$K = \sqrt{\frac{\sigma_{N\bar{N} \to e^+e^-}}{\sigma_{N\bar{N} \to e^+e^-}(V = 0)}} \tag{7}$$

where $\sigma_{N\bar{N} \to e^+e^-}$ is the cross section for $N\bar{N} \to e^+e^-$ reaction with initial state $N\bar{N}$ interaction, $\sigma_{N\bar{N} \to e^+e^-}(V = 0)$ is the same cross section but without one (both nuclear and annihilative into hadron channels interactions are switched off). The enhancement factor is shown as a function of $N\bar{N}$ c.m. relative momentum. The linear behaviour is seen to be revealed approximately up to $k \sim 100$ MeV/c in the enhancement factor, moreover the slope parameter is connected with the scattering lengths as it was shown above. The comparison with the calculation of the scattering lengths in the same CCM shows the following:

Im a $^3S_1$(I = 0) = 0.34 fm and Im a $^3S_1$(= 1) = 0.29 fm[7];

Im a $^3S_1$(I = 0) = 0.36 fm and Im a $^3S_1$(I = 1) = 0.33 fm

from slope parameters in Fig. 2, i.e. the approximation (5) is valid with 10% accuracy for $k \sim 100$ MeV/c.

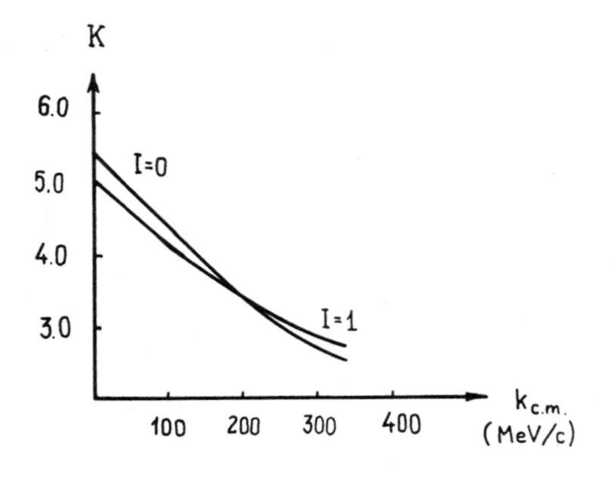

Fig. 2 The enhancement coefficient (see text) for different isospins as a function of c.m. momentum in $N\bar{N}$ system.

In Fig. 3(a,b) the energy dependence of proton and neutron form-factors correspondigly in time-like region are shown. Experimental points are taken from Ref. 1,2. The behaviour of proton electromagnetic form-factor G is seen from the Fig. 3(a) to be orrect (within the experimental accuracy), but its absolute value depends strongly on the input parameter G_0, i.e. on the model used in space-like region. Moreover, the main energetic behaviour of G is given by the initial state interaction, because input form-factor G_0 is practically constant (e.g. see dashed curve in Fig. 3(a)). As for neutron electromagnetic form-factor, it is interesting

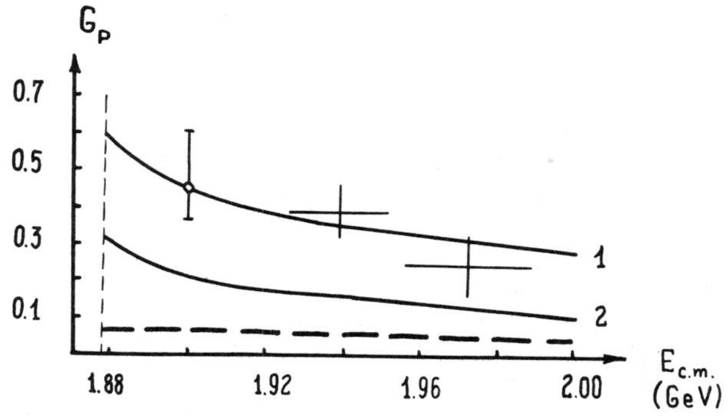

Fig. 3a The proton electromagnetic form-factor as a function of c.m. energy in $p\bar{p}$ system. Data are taken from Refs. 1 and 2. Solid curve 1 corresponds to the calculations from this work using G_0 calculated in VDM with coupling constants for ρ, δ, ψ-mesons[7] (the curve is normalized on experimental value at k = 300 MeV/c); solid curve 2 is the calculation of G with G_0 from dipole fit.

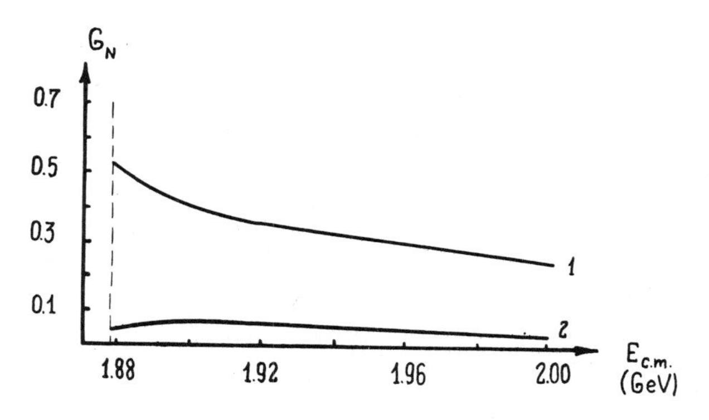

Fig. 3b The same as in Fig. 3a for neutron.

to note that its value in the case of isotopic independent input G_0 parameter is very small (curve 2 on Fig. 3(b)). If one of the isotopic G_0 form-factors is dominant, neutron form-factor will be close to the proton one (for instance, curves 1 in Figs. 3(a) and (b).)

The inverse $e^+e^- \to N\bar{N}$ reaction is also of interest. In principle, using this process one could measure a nucleon form-factor G just near $N\bar{N}$ threshold. A cross section for this reaction is connected with $N\bar{N} \to e^+e^-$ by the formula:

$$\sigma(e^+e^- \to N\bar{N}) = \left(\frac{k}{M}\right)^2 \sigma(N\bar{N} \to e^+e^-) \tag{8}$$

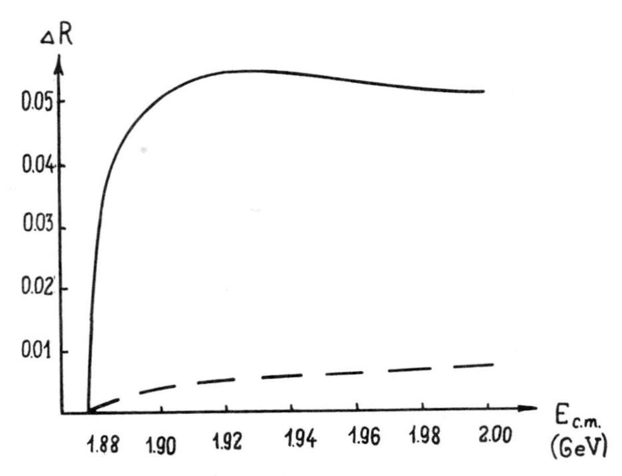

Fig. 4 The value $\Delta R = \sigma e^+e^- \to p\bar{p}/\sigma_{\mu\mu}$ as a function of c.m. energy. Solid curve corresponds to present calculation with form-factor G (solid curve 1 in Fig. 2a, dashed curve is the same without the interaction in initial state.

In Fig. 4 the dependence of the value $\Delta R = \sigma(e^+e^- \to p\bar{p})/\sigma_{\mu\mu}$ as a function of $p\bar{p}$ c.m. energy is shown ($\sigma_{\mu\mu}$ is the cross section for the $e^+e^- \to \mu^+\mu^-$ reaction).

REFRENCES

1. B. Delcourt et al.; Phys.Lett. 86B (1979) 395.
2. G. Bassompierre et al.; Phys.Lett. 68B (1977) 477.
3. P. Cesselli, M. Nigro and C. Voci; Proc. Workshop on Physics at LEAR with Low-Energy Cooled Antiprotons, Erice, Italy, 9-16 May 1982 (Ed. U. Gastaldi et al., Plenum Press, New York).
   F.M. Renard; Phys.Lett. 47B (1973) 361.
4. J.G. Körner and M. Kuroda; Phys.Rev. D16 (1977) 2165.
5. I.S. Shapiro; Nucl.Phys. A478 (1988) 665c.
6. O.D. Dalkarov, V.B. Mandelzweig and V.A. Khoze; Pis'ma v ZhETF 14 (1971) 131.
7. O.D. Dalkarov, K.V. Protasov and I.S. Shapiro; preprint 37, Moscow, FIAN, 1988.
8. V. de Alfaro and T. Regge; "Potential scattering", North-Holland Publishing Co., Amsterdam, 1965.
9. O.D. Dalkarov and K.V. Protasov; preprint 157, Moscow, FIAN, 1988.

MEASUREMENT OF SPIN PARAMETERS

A.D. Krisch

Randall Laboratory of Physics
The University of Michigan
Ann Arbor, MI 48109 U.S.A.

So far there has been little experimental work with polarized beams of antiprotons. However, the polarization experiments done with protons might serve as a useful guide in polarized antiproton projects. High energy polarized proton beams and polarized proton targets have allowed good measurements of spin effects in high energy elastic proton-proton scattering. Since I have written about this subject several times in recent years, I will only briefly review the earlier work and refer the interested reader to some rather detailed lectures[1] and publications[2].

During the period 1979 to 1986 we worked on the project of modifying the 30 GeV AGS to allow the acceleration of polarized protons. As shown in Fig. 1, this project required major changes in almost every part of the AGS. In early 1986 the Argonne-Brookhaven-Michigan-Rice-Yale team succeeded in accelerating a polarized proton beam to 22 GeV with over 40% polarization. The complex and difficult acceleration process is described in detail in a recent paper.[2]

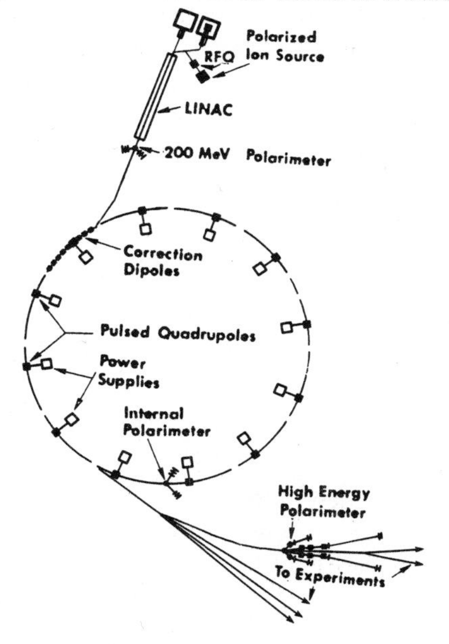

Fig. 1. AGS Layout

Antiproton–Nucleon and Antiproton–Nucleus Interactions
Edited by F. Bradamante *et al.*
Plenum Press, New York, 1990

The main goal of our group in initiating this accelerator improvement project was to extend to higher energy our surprising ZGS results[3] shown in Fig. 2.

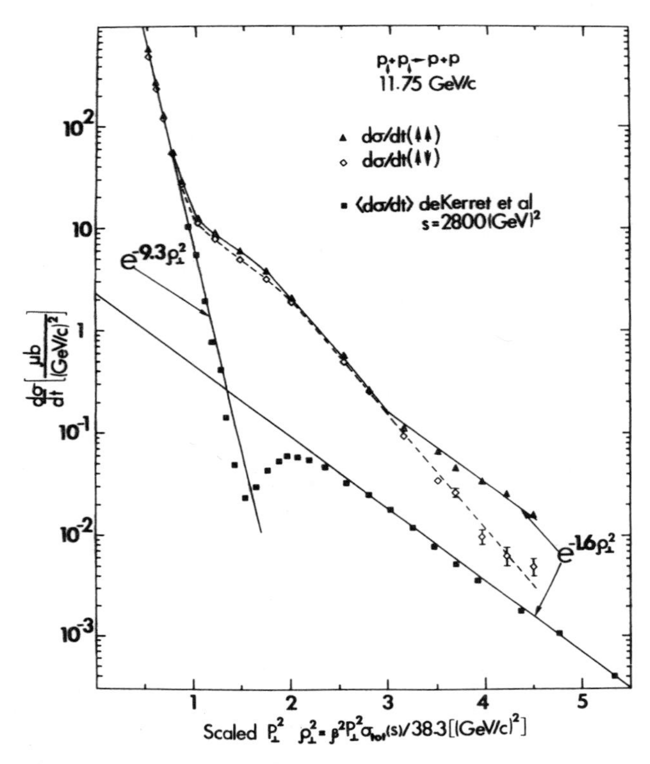

Fig. 2. Spin-parallel and Spin-Antiparallel proton-proton elastic cross-sections[3] plotted against scaled $P_\perp^2$ variable. The unpolarized p-p data of DeKerret et al.[4] are shown for comparison.

Notice that for small-$P_\perp^2$ glancing collisions the cross-section seems quite independent of both incident energy and incident spin state. This indicates that for all energies and all spin states, whenever two protons have a peripheral collision from the outer surfaces of each other, they see a characteristic size of about 1 fermi which causes the slope of exp (-9.3 $P_\perp^2$).

The spin-averaged data from the ISR at s = 2800 GeV^2 shows a sharp break followed by a large-$P_\perp^2$ hard-scattering component with the behavior exp (-1.6 $P_\perp^2$), which continues for several decades. This hard-scattering component is probably due to the direct collisions of the protons' constituents. The slope of 1.6 indicates that the effective size of the constituent-constituent interaction is about 1/3 fermi, which includes both the size of the constituents and the range of their interaction.

There is also a sharp break in the 12 GeV ZGS data; this break is followed by the same hard-scattering component with the same slope of 1.6. However, this break only occurs when the protons' spins are parallel. When the protons' spins are antiparallel the cross-section keeps dropping. At the maximum $P_\perp^2$ available at the ZGS, the ratio $d\sigma/dt$ (parallel): $d\sigma/dt$ (antiparallel) has reached a value of 4.

This large ratio of 4 causes a serious problem for the popular theory of strong interactions called perturbative QCD. Since this theory assumes that each interacting spin $\frac{1}{2}$ proton contains three spin $\frac{1}{2}$ quarks, it is most difficult to get a factor of 4 for the spin ratio. Perhaps the easiest way for an experimenter to understand this problem is to view each spin $\frac{1}{2}$ proton as a polarized target containing three spin $\frac{1}{2}$ quarks; the polarization of this polarized quark target is exactly $\frac{1}{3}$. Thus, it is impossible to get a proton-proton spin ratio of 4 even if the quark-quark spin-parallel cross-section is infinity and the quark-quark spin-antiparallel cross-section is zero.

During the early 1980's, while we were modifying the AGS to accelerate polarized protons, we decided to test our apparatus in a low priority experiment which measured A in $p + p \to p + p$ by scattering the AGS unpolarized proton beam from our polarized proton target.[5] Perturbative Quantum Chromodynamics, PQCD, predicts that A should be exactly zero and further states that this prediction should become more reliable as the incident energy and $P_\perp^2$ become larger. Our measurements of A at 28 GeV/c are plotted against $P_\perp^2$ in Fig. 3. Clearly the $A = 0$ prediction of PQCD is not supported by this data which indicates that A is growing at large $P_\perp^2$.

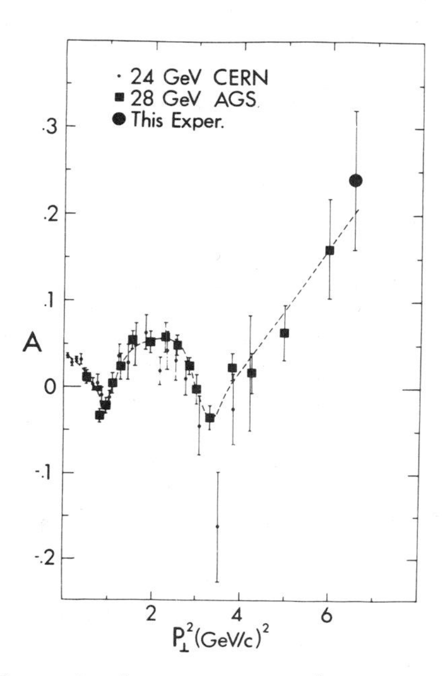

Fig. 3. The Analyzing Power is plotted against $P_\perp^2$ for $p + p \to p + p$ at 28 GeV/c.

All previous spin ratio measurements for high energy p-p elastic scattering are summarized in Fig. 4, which is a 3-dimensional plot of $d\sigma/dt$ (parallel): $d\sigma/dt$ (anti-parallel) against incident momentum and against $P_\perp^2$. This plot includes all of our AGS data through November 1987, which is interesting but is a bit sparse. At medium-$P_\perp^2$ the 13.3 and 16.5 GeV/c data is generally consistent with the 12 GeV/c ZGS data. Note that the large-$P_\perp^2$ fixed-angled 90°_{cm} points have exactly the same $P_\perp^2$ behavior as the fixed-energy 12 GeV/c points. This identical $P_\perp^2$ behavior indicates that these large spin effects are indeed high-$P_\perp^2$ hard-scattering effects and are not 90°_{cm} particle identity effects as suggested by Bethe and Weisskopf around 1978. The figure also shows that spin effects certainly do not seem to disappear with increasing P_{lab} and $P_\perp^2$ as suggested by perturbative QCD. The spin ratio in these violent collisions almost seems to be oscillating, an interesting possibility studied by Hendry, Brodsky, Tyurin, and others.

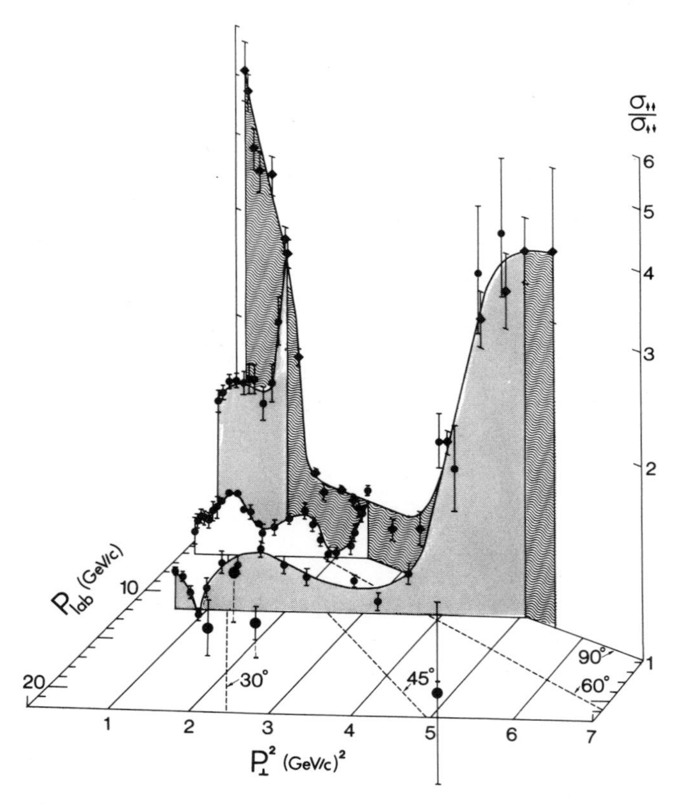

Fig. 4. 3-dimensional plot of $p + p \rightarrow p + p$ spin ratio

We had a very good AGS run from December 1987 to January 1988 with 18.5 GeV/c polarized protons.[6] We obtained rather detailed angular distributions for both A and A_{nn} which are shown in Fig. 5. Notice that A has a dip near $P_\perp^2 = 3$ (GeV/c)^2 which is quite similar to the dip seen in Fig. 3 at 24 and 28 GeV/c. Note that the errors in the new polarized beam data are generally smaller than the errors in Fig. 3.

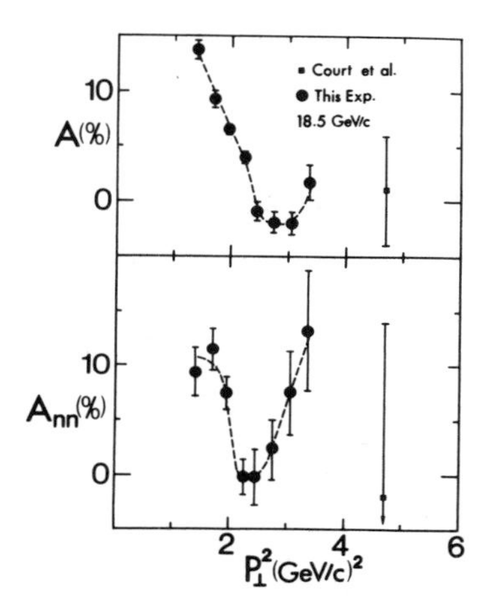

Fig. 5. The Analyzing Power, A, and the Spin-Spin Correlation Parameter, A_{nn}, are plotted against $P_\perp^2$ for $p + p \to p + p$ at 18.5 GeV/c.

The sharp dip in A_{nn} near $P_\perp^2 = 2.3$ (GeV/c)^2 at 18.5 GeV/c seems quite interesting. Notice from Fig. 4 that at 12 GeV/c the spin-spin effects have one sharp dip near $P_\perp^2 = 0.9$ (GeV/c)^2 where the diffraction peak ends, and a broad dip near $P_\perp^2 = 3.2$ (GeV/c)^2 where the hard-scattering component starts. It is not yet clear if the sharp dip at 18.5 GeV/c near $P_\perp^2 = 2.3$ (GeV/c)^2 corresponds to the small-$P_\perp^2$ dip moving up or the large-$P_\perp^2$ dip moving down. Our earlier $P_\perp^2 = 4.7$ (GeV/c)^2 measurement[7] with A_{nn} near zero had rather large errors and we must await our next run to study in detail the large $P_\perp^2$ behavior.

Finally, I will show Fig. 6 which is a plot of the relative pure initial spin state cross-sections:

$$\begin{aligned} \sigma_{\downarrow\downarrow} &\equiv d\sigma/dt(\uparrow\uparrow)/ < d\sigma/dt > = 1 + A_{nn} + 2A \\ \sigma_{\downarrow\downarrow} &\equiv d\sigma/dt(\downarrow\downarrow)/ < d\sigma/dt > = 1 + A_{nn} - 2A \\ \sigma_{\downarrow\uparrow} = \sigma_{\uparrow\downarrow} &\equiv d\sigma/dt(\uparrow\downarrow)/ < d\sigma/dt > = 1 - A_{nn} \end{aligned}$$

Near $P_\perp^2 = 1.4\ (GeV/c)^2$, $\sigma_{\uparrow\uparrow}$ is more than 50% larger than both $\sigma_{\downarrow\downarrow}$ and $\sigma_{\uparrow\downarrow}$. All three cross-sections come together near $P_\perp^2 = 2.5\ (GeV/c)^2$ and then appear to move apart again at larger $P_\perp^2$. We have also shown for comparison the spin-averaged p-p elastic cross-section[8] $< d\sigma/dt >$ at 19 GeV/c, which has the familiar break and dip

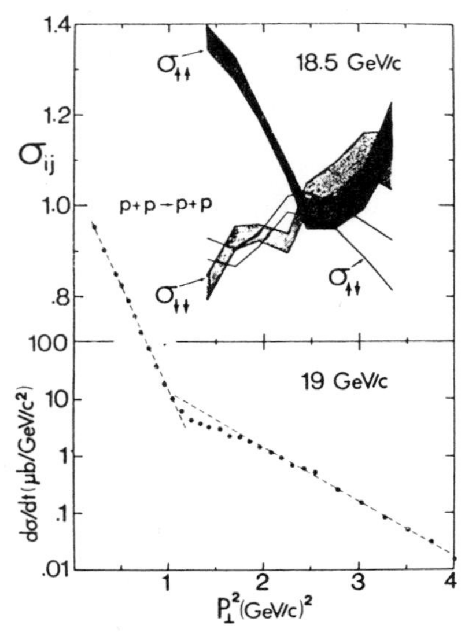

Fig. 6. The relative pure initial spin state cross-sections are plotted against $P_\perp^2$ for $p + p \rightarrow p + p$ at 18.5 GeV/c.

near $P_\perp^2 = 1\ (GeV/c)^2$ followed by the hard-scattering exponential. Note especially that $< d\sigma/dt >$ seems to drop smoothly near the sharp structure in both A and A_{nn} near $P_\perp^2 = 2.5\ (GeV/c)^2$. This suggests that the spin-averaged cross-section is insensitive to large and probably significant forces which appear quite clearly in the pure-spin cross-sections.

References

1. A.D. Krisch, Proc. Sixth Int. Symp. Polar. Phenom. in Nucl. Phys., Osaka, 1985, J. Phys. Soc. Jpn. 55 Suppl., p. 31 (1986)

   Proc. 1986 Serpukhov High Energy Spin Symposium, p.p. 41 and 272 (IHEP, 1987).
2. F.Z. Khiari et al., Phys. Rev., (January 1989) to be published.
3. D.G. Crabb et al., Phys. Rev. Lett., 41 , 1257 (1978).
4. H. DeKerret et al., Phys. Lett, 62B , 363 (1976) and 68B , 374 (1977).
5. P.R. Cameron et al., Phys. Rev. Rapid Comm., D32 , 2070 (1985).
6. D.G. Crabb et al., Phys. Rev. Lett. 60 , 2351 (1988).
7. G.R. Court et al., Phy. Rev. Lett., 57 , 507 (1986).
8. J.V. Allaby et al., Phys. Rev. Lett. , 23B , 67 (1968).

ANTIPROTON-PROTON CROSS SECTIONS AT SMALL MOMENTA

W. Brückner[1], B. Cujec[5], H. Döbbeling[2], K. Dworschak[1], F. Güttner[1]
H. Kneis[1], S. Majewski[2], M. Nomachi[2], S. Paul[1], B. Povh[1]
R.D. Ransome[4], T.-A. Shibata[3], M. Treichel[1] and Th. Walcher[3]

presented by T.-A. Shibata

INTRODUCTION

Since the startup of LEAR in 1983 we have undertaken a series of measurements of antiproton-proton cross sections for annihilation ($\bar{p}p \rightarrow$ mesons)[1], charge-exchange reaction ($\bar{p}p \rightarrow \bar{n}n$)[2], and elastic scattering ($\bar{p}p \rightarrow \bar{p}p$)[3,4]. We have focussed our efforts of the measurements on the beam momentum range between 160 and 600 MeV/c. Our detector is optimized to handle the antiproton beam at very small momenta[5]. The liquid hydrogen target is placed in the vacuum tank, and the vacuum is directly connected to the beam line. The incident beam is defined by thin scintillators also placed in the vacuum. With the apparatus of these unique designs we have explored the beam momentum region below 300 MeV/c with a good precision.

The main outcome of these measurements is that s- and p-wave components are identified in the cross sections, and a large p-wave contribution is found even at small momenta. This is quite contrary to nucleon-nucleon interaction where the p-wave contribution is known to be as small as 10% already at 300 MeV/c. In the next section we compare the annihilation cross section to the black disc value which gives the geometrical limit. In the third section the charge-exchange reaction is analysed with a simple potential which consists of one-pion-exchange and a real and an imaginary Woods-Saxon functions. In the fourth section we present the differential elastic cross section and its partial wave decomposition. The results are compared to the recent analyses with the scattering length - effective range expansion in which a consistent solution in the scattering data and the atomic data is found.

[1])Max-Planck-Institut für Kernphysik, Heidelberg, Germany
[2])Physikalisches Institut der Universität, Heidelberg, Germany
[3])Institut für Kernphysik der Universität, Mainz, Germany
[4])Rutgers University, Piscataway, NJ, USA
[5])Université Laval, Québec, Canada

Antiproton-Nucleon and Antiproton-Nucleus Interactions
Edited by F. Bradamante *et al.*
Plenum Press, New York, 1990

ANNIHILATION CROSS SECTION

The cross section for antiproton-proton annihilation from the present experiment is shown in Fig.1. The cross section above 400 MeV/c is already reported in ref.1. The cross section below 400 MeV/c is preliminary and contains a systematic error of 10% but it shows some interesting features.

In the following arguments we often refer to the interaction with a black disc. This is because it has a close resemblance to the antiproton-proton interaction at small momenta: all the incident waves are absorbed. The black disc value of the inelastic cross section is $\pi/k^2 \cdot (2\ell+1)$ for the partial wave ℓ. The curve in Fig.1 shows the black disc value for s-wave. Note that the black disc value presents the maximum of the inelastic cross section. Though the inelastic cross section consists of annihilation and charge-exchange cross sections, if the measured annihilation cross section alone already exceeds the s-wave black disc value we can conclude that there exists a p-wave contribution. Indeed we find in Fig.1 that annihilation cross section is substantially larger than the s-wave black disc value. This is a straightforward approach of identifying the p-wave contribution without assumptions.

In this connection it is worth making a comment on elastic and total cross sections. Namely, the comparison with the black disc value is less straightforward in these cross sections. As shown in Table 1 the black disc values do not give the maxima: elastic cross section becomes four times larger and total cross section becomes twice larger when elasticity η is 1.0 and the phase shift is 90°. Indeed so far measured elastic and total cross sections are smaller than these unitary limits of the s-wave at 200-300 MeV/c. Therefore we need another approach to examine the p-wave contribution in those cross sections.

Table 1. S-wave cross sections

	σ_{inel}	σ_{el}	σ_{tot}
Black disc (η=0)	π/k^2	π/k^2	$2\pi/k^2$
η=1.0, δ=90°	0	$4\pi/k^2$	$4\pi/k^2$

CHARGE-EXCHANGE CROSS SECTION

The differential cross section for charge-exchange reaction is measured at several momenta between 160 and 600 MeV/c. It shows in general a slope from forward to backward angles, indicating an admixture of the partial waves with $\ell \geq 1$. Furthermore, a shoulder following the peak at small angles is observed at the beam momenta higher than 400 MeV/c[6,7,2]. The example at 590 MeV/c is shown in Fig.2. Since various potential models predict the forward structure in very different ways the data provide a critical test of the theoretical models.

In order to study the origin of this structre we have analysed it with a simple potential expressed as[8]:

$$V = V_0 + iW + V_{OPE} \tag{1}$$

$$V_{OPE} = -(\tau_1 \cdot \tau_2)\,[(\sigma_1 \cdot \sigma_2)\, V_{ss} + S_{12} V_T] \tag{2}$$

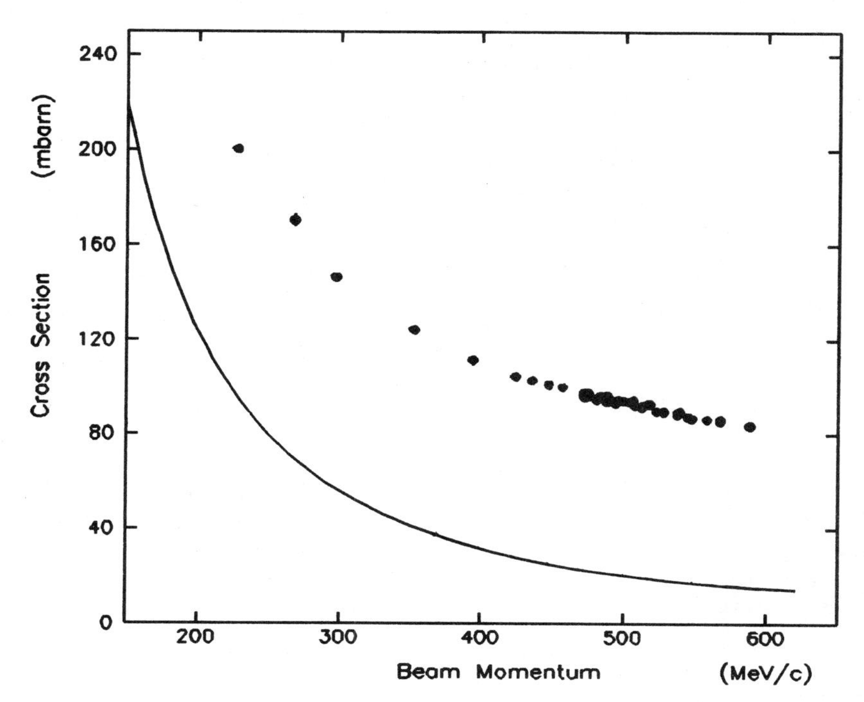

Fig. 1. Antiproton-proton annihilation cross section. The solid curve is the black disc value of s-wave inelastic cross section.

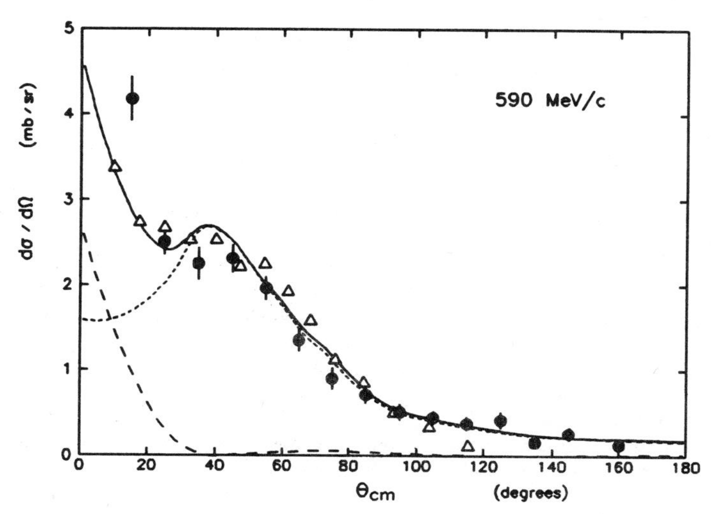

Fig. 2. The differential charge-exchange cross section at 590 MeV/c. The closed circles are from ref.2. The open triangles are from ref.7. The solid curve indicates the calculation with the potential in eq.1. The dashed curve shows the calculation when V_T is switched off. The dotted curve is when V_{SS} is switched off.

where V_0 and W are a real and an imaginary Woods-Saxon potential, V_{OPE} the one-pion-exchange potential, V_{ss} and V_T the radial functions of the spin-spin term and the tensor term, and S_{12} the tensor operator. The Woods-Saxon part is essentially the same as the annihilation potential in the Dover-Richard model[9]. The charge-exchange cross section is reproduced as shown with the solid curve in Fig.2, as well as annihialtion and elastic cross sections. The dotted curve in Fig.2 is the calculation when V_{ss} is switched off in eq.2 and the dashed curve is when V_T is swithed off. These observations suggest that the forward peak is due to the spin-spin term and the large angle part is due to the tensor term of one-pion-exchange. If another spin-spin term is added it would naturally affect the forward structure. So we understand that the above mentioned difference in the various potential model predictions arises from different spin terms in two-pion-exchange and in the annihilation potentials.

ELASTIC CROSS SECTION

The differential cross section for elastic scattering is measured at the beam momenta between 180 and 600 MeV/c[4]. The example at 287 MeV/c is shown in Fig.3. The data on proton-proton scattering are also shown for comparison. These two cross sections show a sharp contrast, and indicate a p-wave contribution in antiproton-proton interaction.

In order to discuss it in more quantitive way we have fitted the differential cross section with a formula:

$$d\sigma/d\Omega = 1/k^2 \left|\sum_\ell (2\ell+1)T_\ell P_\ell(\cos\theta)\right|^2 \qquad (3)$$

$$T_\ell = (\eta_\ell e^{2i\delta_\ell} - 1) / 2i \qquad (4)$$

To fix the parameter T_ℓ we use an additional experimental value: real-to-imaginary ratio of the forward elastic scattering amplitude[3] determined with the Coulomb-nuclear interference. Once T_ℓ is obtained, elastic cross section and inelastic (annihilation + charge exchange) cross section can be calculated as

$$\sigma_{el} = 4\pi/k^2 \sum_\ell (2\ell+1)|T_\ell|^2 \qquad (5)$$

$$\sigma_{inel} = \pi/k^2 \sum_\ell (2\ell+1)\ (1-\eta_\ell{}^2) \qquad (6)$$

The partial wave composition in inelastic and elastic cross section is shown in Fig.4. We find that the s-wave inelastic cross section is close to the maximum. There is a substantial p-wave component even at small beam momenta, which is cosistent with what we have discussed in annihilation cross section in the earlier section of this paper. In elastic cross section the mixture of the p-wave is also observed although the size is small.

The complex parameter T_ℓ is plotted in Argand diagram in Fig.5(a). It is interesting to compare it to the recent analyses with scattering length - effective range expansion[10,11]. The scattering length and the effective range for s-wave and the scattering volume and the effective range for p-wave are fundamental parameters which can be determined without assuming potentials etc. In ref.10 the data on the $\bar{p}p$ scattering, $\bar{n}p$ reaction as well as antiprotonic hydrogen are fitted. The Argand diagram from this analysis is shown in Fig.5(b). The main result is that there exists a consistent solution in the fit to the scattering data and to the atomic data. The real part of T_0 (s-wave) is negative or repulsive. The role of the p-wave in the oscillatory behavior of the real-to-imaginary ratio of the forward elastic scattering amplitude[3] is also discussed in refs.10,11.

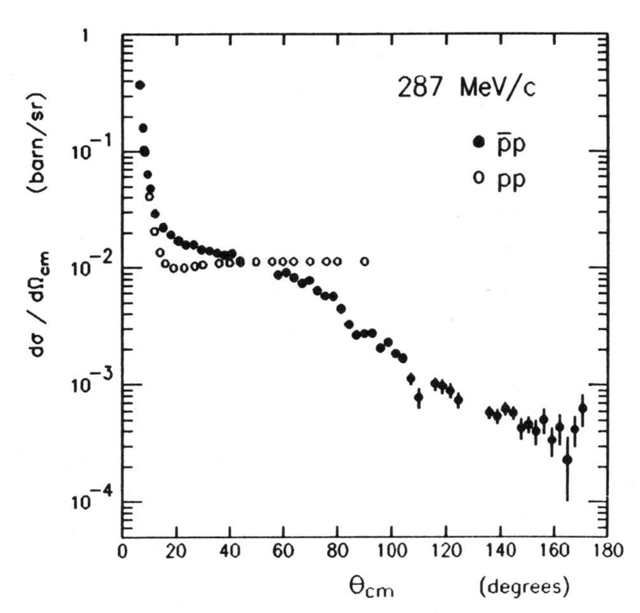

Fig. 3. Comparison of the differential cross sections for $\bar{p}p$ and pp elastic scattering.

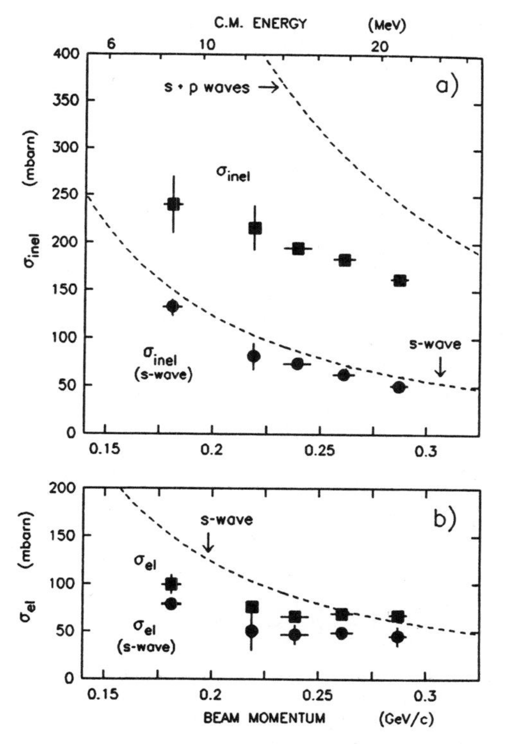

Fig. 4. (a) Inelastic and (b) elastic cross sections are shown with black squares. The black circles indicate the s-wave components. The black disc values are also shown with the dotted curves for comparison.

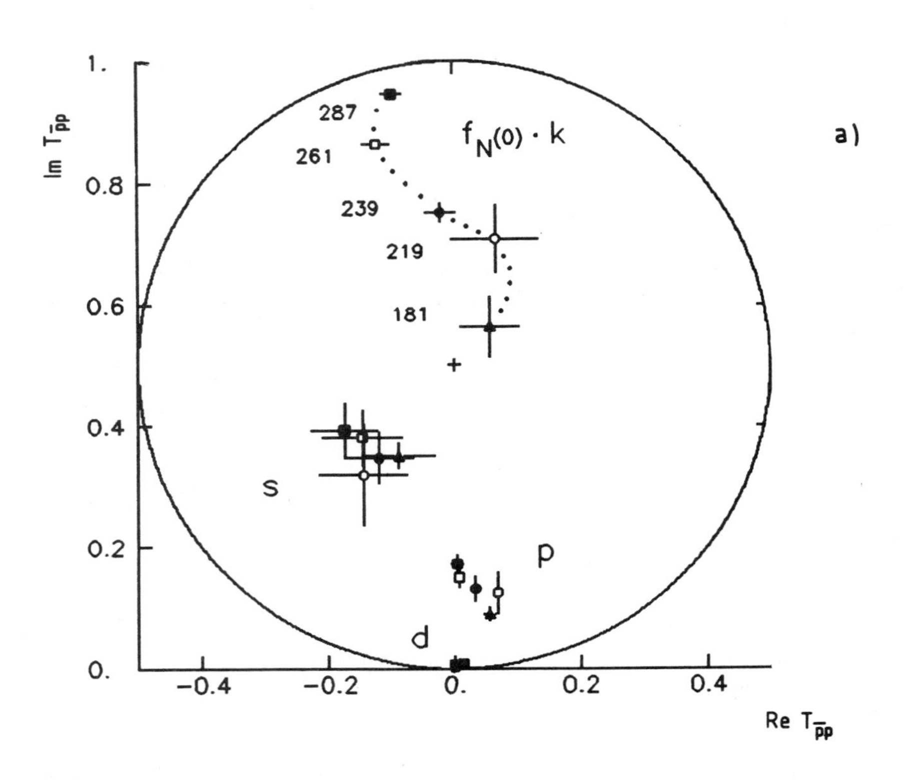

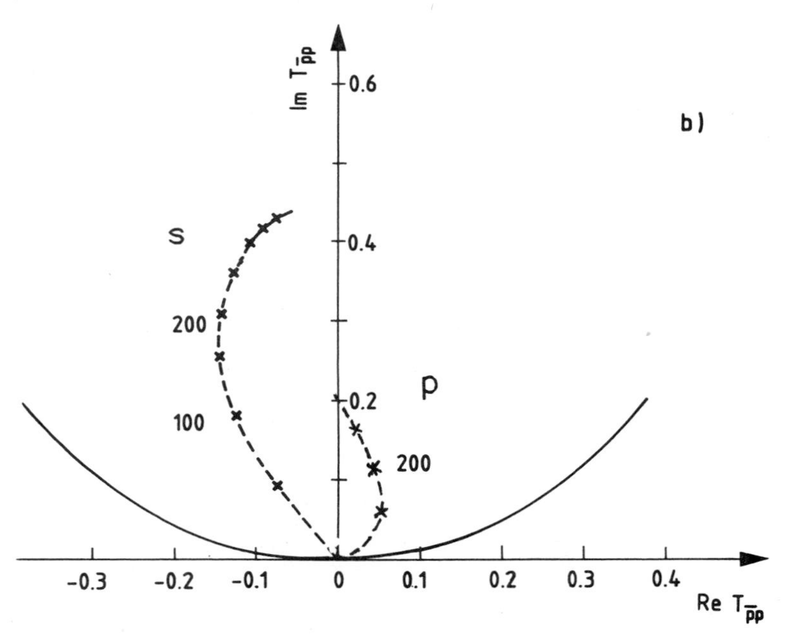

Fig. 5. (a) Partial wave amplitudes T_ℓ in Argand diagram. The forward elastic scattering amplitude $f_N(0) \cdot k = T_0 + 3T_1 + \ldots$ is also plotted. The numbers in the figure indicate the beam momenta. (b) Patial wave amplitudes T_ℓ determined with the scattering length - effective range expansion using the data on the scatterings and the antiprotonic hydrogen.

Finally, we have studied the range of annihilation and the range of elastic scattering and charge-exchange reaction using the potentials. The result is that the annihilation is mostly determined around 1 fm while charge-exchange reaction is sensitive to the potential up to 4 fm. The details are described in refs.12,8.

CONCLUSIONS

The measurements of antiproton-proton cross sections for annihilation, charge-exchange reaction and elastic scattering have been performed in the beam momentum range between 160 and 600 MeV/c. A p-wave contribution is identified in the cross sections even at small beam momenta, which is quite contrary to the nucleon-nucleon interaction.

The structure in the differential charge-exchange cross section is studied with a simple potential model. The differential elastic cross section is expanded with the partial wave amplutides. The Argand diagram from this analysis is compared to that from the scattering length - effective range expansion. A consistent solution exists in the scattering data and the atomic data.

REFERENCE

1. W. Brückner et al., Phys. Lett. 169B (1986) 302.
2. W. Brückner et al., Phys. Lett. 197B (1987) 463.
3. W. Brückner et al., Phys. Lett. 158B (1985) 180.
4. W. Brückner et al., Phys. Lett. 166B (1986) 113.
5. W. Brückner et al., Nucl. Instrum. Mothods A269 (1988) 527.
6. M. Bogdanski et al., Phys. Lett. 62B (1976) 117.
7. K. Nakamura et al., Phys. Rev. Lett. 53 (1984) 885.
8. T.-A. Shibata, Phys. Lett. 189B (1987) 232.
9. C.B. Dover and J.M. Richard, Phys. Rev. C21 (1980) 1466.
10. J. Mahalanabis, H.J. Pirner and T.-A. Shibata, Nucl. Phys. A485 (1988) 546.
11. B.O. Kerbikov and Yu.A. Simonov, ITEP-38 (1986), I.L. Grach, B.O. Kerbikov and Yu.A. Simonov, ITEP preprint 210 (1987).
12. B. Povh and Th. Walcher, Comm. Nucl. Part. Phys. 16 (1986) 85.

COMPARISON OF $d\sigma/d\Omega$ AND A_{on} RESULTS IN $\bar{p}p$ ELASTIC SCATTERING WITH MODEL PREDICTIONS

R.A.KUNNE, S.DEGLI-AGOSTI, E.HEER, R.HESS,
C.LECHANOINE-LELUC, Y.ONEL, D.RAPIN,
DPNC, University of Geneva, Geneva, Switzerland

C.I.BEARD, D.V.BUGG, J.R.HALL,
Queen Mary College, London, Great Britain

A.S.CLOUGH, R.L.SHYPIT,
University of Surrey, Surrey, Great Britain

R.BIRSA, F.BRADAMANTE, S.DALLA TORRE-COLAUTTI, A.MARTIN, A.PENZO, P.SCHIAVON,
F.TESSAROTTO, A.VILLARI,
INFN, Trieste and University of Trieste, Trieste, Italy

K.BOS, J.C.KLUYVER, L.LINSSEN,
NIKHEF-H, Amsterdam, The Netherlands

(Presented by R.A.Kunne)

ABSTRACT

Recent measurements by the SING collaboration (experiment PS172 at LEAR) of $d\sigma/d\Omega$ and A_{on} in $\bar{p}p$ elastic scattering between 500 and 1550 MeV/c may be used to critically examine $\bar{N}N$ potential models. In this paper four of such models are compared with the experimental data. All models fall short in explaining the details of the asymmetry distributions, although the general trend is predicted more or less correctly.

THE PS172 TWOBODY EXPERIMENT AT LEAR

The purpose of the twobody experiment of the SING collaboration was the measurement of the differential cross section $d\sigma/d\Omega$ and asymmetry A_{on} of the three twobody reactions $\bar{p}p \rightarrow \bar{p}p$, $\bar{p}p \rightarrow \pi^+\pi^-$ and $\bar{p}p \rightarrow K^+K^-$ at LEAR energies.

Antiproton-Nucleon and Antiproton-Nucleus Interactions
Edited by F. Bradamante *et al.*
Plenum Press, New York, 1990

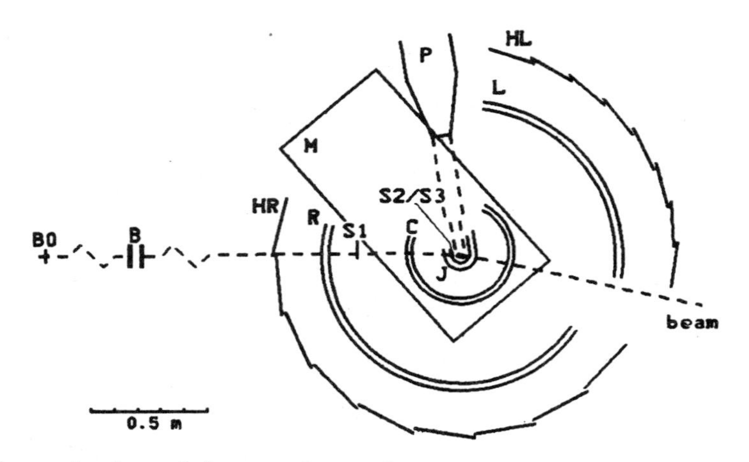

Fig. 1. Schematic view of the experimental setup.
B0,S1,S2,S3: beam defining counters; B,J,C,L,R: multiple wire proportional chambers; HL,HR: hodoscope counters; M: polarized target magnet; P: cryostat with polarized target; beam: incident and throughgoing beam track;

The setup, shown in figure 1, was designed primarily the cover the full angular range of the two meson reactions. The elastic reaction was covered only for those angles where both proton and antiproton had sufficient energy to leave target and setup.

The target was placed between the poleshoes of the magnet. Three sets of cylindrical MWPCs surrounded the target and permitted reconstruction of and distinction between the tracks of the outgoing protons, pions and kaons. In addition, two beam chambers were placed in the incoming beam region. The beam was defined by a coincidence between three or four of the beam counters. Two arrays of eight and nine hodoscope counters respectively, surrounded the setup and triggered on twobody reactions. Below and above the target were two counters, that intercepted and vetoed pions originating from multipion annihilation events.

Elastic scattering data were collected at 15 different momenta. Per momenta 1.5 to 2.0×10^6 events were registered. After reconstruction, using the matrix method described in Aprile et al. (1981), typically 2×10^5 events having $\chi^2/DF < 8$ were written to DST for each momentum. Two sets of cuts defined the samples used for the calculation of the differential cross sections and asymmetry, respectively. For the asymmetry sample a vertex cut of $|v_Z| < 12$ mm assured that the events were lying well inside the 30 mm long target. For the differential cross section sample this vertex cut was applied to the data, as well as a cut on the azimuthal angle at 6°. This second cut assured that the events were lying in the fully efficient part of all chambers. A special counter, S3 in figure 1, that was smaller in diameter than the target and placed just in front of it, selected those events where the incoming beam traversed the central part of the target. The two sets of cuts reduced the number of events to 80K per momentum for the A_{on} measurement and to 20K per momentum for the $d\sigma/d\Omega$ measurement.

Several corrections were applied to the event numbers. First of all the background from events on protons bound in C and O nuclei was subtracted. This background was calculated using runs with a dummy target, containing only bound protons. These data were analyzed in the same way as the polarized target data. The background subtraction in both samples amounted to about 15%. In the case of the differential cross section measurement further corrections for chamber inefficiency (20 – 40%), absorption of secondaries (10 – 15%), random triggers (1 – 3%), random vetoes (0 – 1%), bad beam (1 – 3%), attenuation of beam before fiducial volume (4 – 8%) and effective target length (0 – 1%) were applied. The $d\sigma/d\Omega$ data were normalized using the value of the forward differential cross section as calculated from σ_{tot} and ρ measurements via the optical theorem (Coupland et al, 1977).

The systematic errors on the asymmetry and differential cross section measurements were estimated to be 5% and 10%, respectively. In the first case the dominant error is the value of the target polarization, in the second case the uncertainty in the position of S3 and the density of the target. A more extensive description of the setup, analysis and results of the experiment is given in Kunne (1988). The asymmetry data were published earlier (Kunne et al. 1988).

THE POTENTIAL MODELS

NN – potential models, based on meson exchanges to describe the long and medium range potential and complemented with a phenomenological short range potential have been very successful in explaining the available NN scattering data, which form at certain energies a complete set. Fundamentally, $\bar{N}N$ and NN interactions have the same Feynman diagrams. Considering only meson exchange diagrams the $\bar{N}N$ and NN are related by a G – parity transformation. The boson exchanges define a potential V(NN) which can be transformed to a V($\bar{N}N$) potential. $\bar{N}N$ scattering is therefore a completely new territory where the NN potentials can be tested indirectly.

Models based on transformed NN potentials are for example the Dover – Richard model (Dover and Richard, 1980 and 1982) the Paris model (Côté et al., 1982) the Bonn model (Hippchen et al., 1987) and the Nijmegen model (Timmers et al., 1984) The first two are both based on the Paris NN potential, the last two are based on their own NN precursors. After the G – parity transformation to each of the potentials a phenomenological short range part was added with 4 to 15 free parameters (depending on the model). The free parameters of each model were adjusted to fit the available $\bar{N}N$ data, i.e. differential and total cross sections. All four models are non – relativistic and claim validity only below 900 MeV/c.

The models are capable to explain the gross features of the data. Of course it is not surprising, that the shape of the differential cross section is reproduced by the models, at least in the forward angular region, as these models are adjusted to fit these data from previous experiments. But on top of this, all models predict correctly the general shape of A_{on}, two maxima and a central minimum, more or less at the correct place. However, most models predict large negative values for the central minimum, which are not observed (see the example of figure 2).

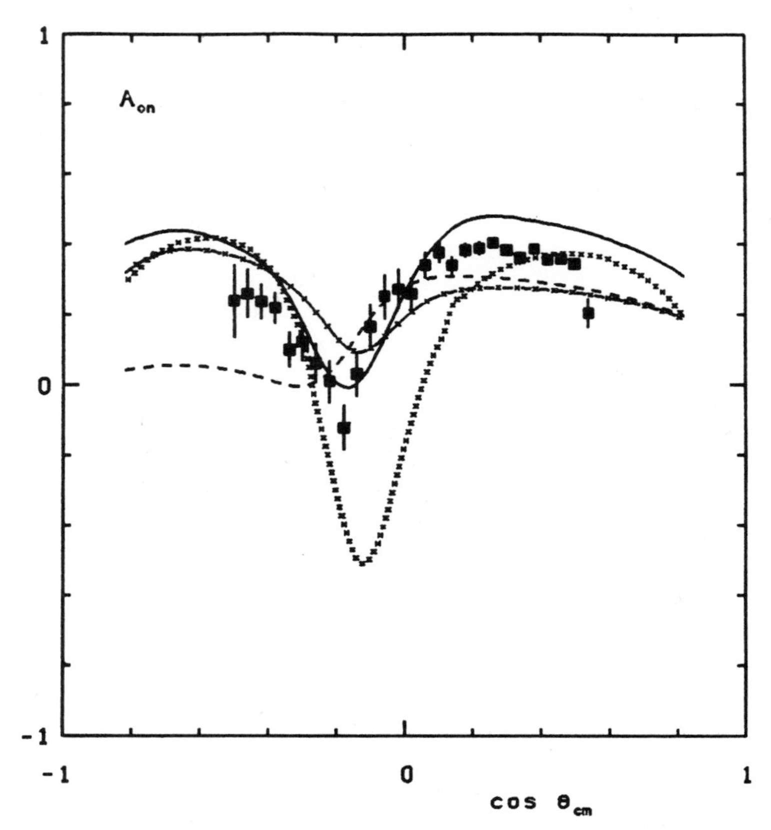

Fig. 2. Model predictions for the asymmetry at 679 MeV/c.
The data points are those of this experiment. The curves are the prediction of the Paris model (full curve), Dover – Richard model II (dashed curve), Nymegen potential (crossed curve) and Bonn potential (crossed – dashed curve).

THE PARIS MODEL

The Paris model seems to be the best of the four models described, although all models may be able to adjust their parameters in such a way that the asymmetry data are better fitted. The starting point of the model is the Paris NN potential model (Lacombe 1975, 1980 and 1981) The potential has the form $V(\bar{N}N) = U(\bar{N}N) - i\ W(\bar{N}N)$. Here $U(\bar{N}N)$ is the G – parity transformed medium and long range parts of the Paris NN potential complemented by a quadratic cut – off function below 0.9 fm. The absorptive part $W(\bar{N}N)$ is of short range and isospin dependent and derived from two meson annihilation diagrams. It is parametrized to be linear dependent on kinetic energy T. It contains twelve parameters which are fitted to 915 differential elastic cross sections and total and charge exchange cross sections in the momentum range 200 MeV/c $< p <$ 900 MeV/c measured between 1968 and 1981. The energy dependence built into the absorp-

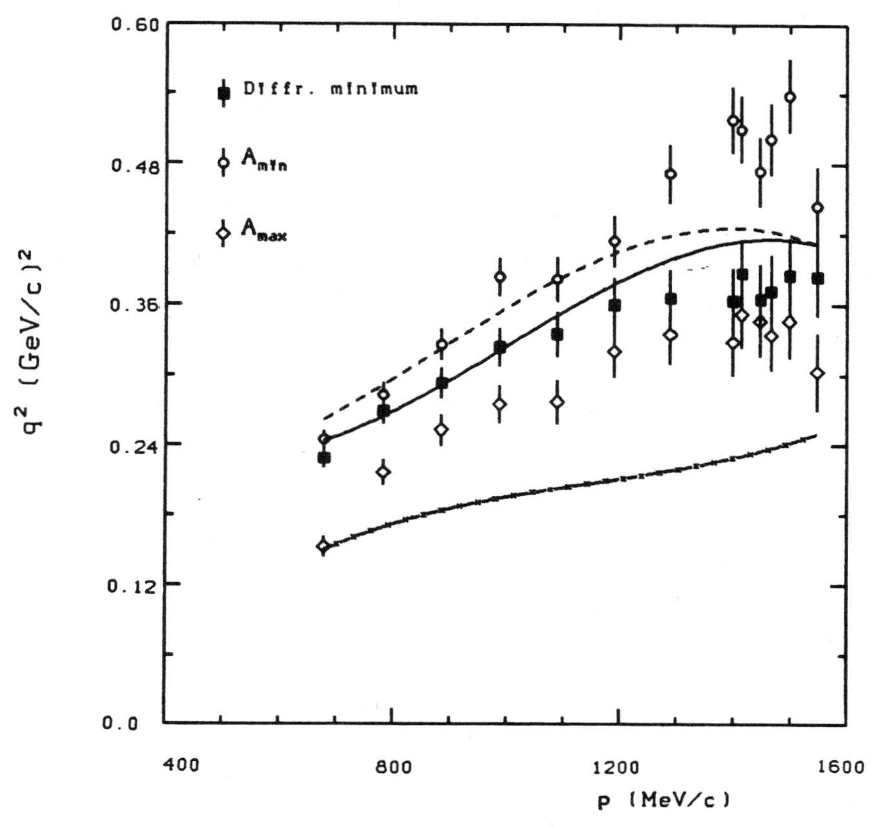

Fig. 3. Position of the extrema in the differential cross section and asymmetry data compared with the predictions of the Paris model. Full, dashed and crossed – dashed curve are the predictions for the position of the differential minimum, the central minimum in A_{on} and the maximum in A_{on}.

tive part of the model makes extrapolation to momenta outside the fitted momentum range feasible. In fact the model predicts the asymmetry amazingly well over the whole range of our data.

In figure 3 the position of the first differential minimum, and the first maximum and central minimum in the A_{on} data are plotted as a function of energy, together with the corresponding predictions from the Paris potential. The diffraction minimum and A_{min} are, at least below 1 GeV/c, predicted in the right place. The first maximum A_{max} occurs in the data at higher values of the squared momentum transfer than is foreseen. The serious disagreement between model predictions and experimental data is shown in figure 4. In this figure the value of the central minimum in A_{on} is plotted as a function of the momentum. The measured values show a preference for a slowly rising, positive asymmetry at higher momenta, while the model favours an oscillating, negative A_{on}.

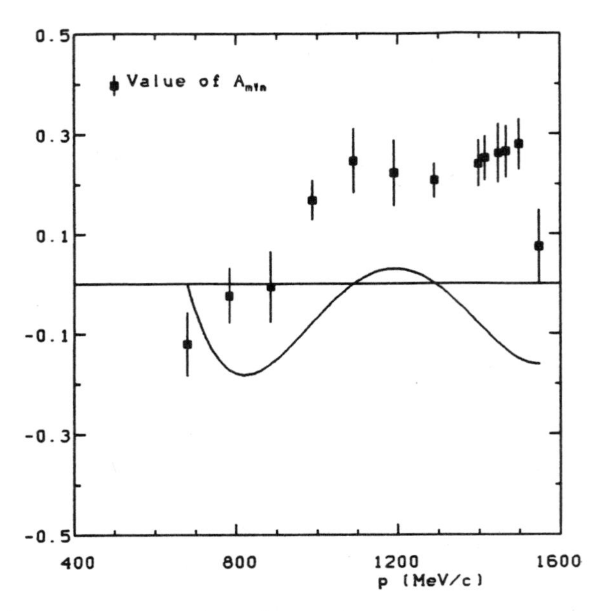

Fig. 4. Value of the central minimum in the asymmetry data compared with the predictions of the Paris model (full line).

REFERENCES

Aprile, E. et al. (1981) A fast online event reconstruction procedure based on a linearized least − squares fit. CERN Yellow Report 81 − 07, pp.124 − 131.

Côté, J. et al., Phys. Rev. Lett. 48 (1982) 1319.

Coupland, M. et al., Phys. Lett 71B (1977) 460.

Dover, C.B., Richard, J.M., Phys. Rev. C21 (1980) 1466.

Dover, C.B., Richard, J.M., Phys. Rev. C25 (1982) 1952.

Hippchen, T. et al. (1987), Submitted to Phys. Rev. C.

Kunne, R.A. (1988), Ph.D. thesis, University of Amsterdam.

Kunne, R.A. et al., Phys. Lett. 206B (1988) 557.

Lacombe, M. et al., Phys. Rev. D12 (1975) 1495.

Lacombe, M. et al., Phys. Rev. C21 (1980) 861.

Lacombe, M. et al., Phys. Rev. C23 (1981) 2405.

Timmers, P.H. et al., Phys. Rev. D29 (1984) 1928.

MEASUREMENT OF SPIN DEPENDENT OBSERVABLES IN THE $\bar{p}-p$ ELASTIC SCATTERING FROM 450 TO 700 MeV/c

R.Bertini[1,2], M.Costa[1], F.Perrot[1],
H.Catz[2], A.Chaumeaux[2], J.-C.Faivre[2], E.Vercellin[2,7],
J.Arvieux[3], J.Yonnet[3],
B.van den Brandt[4], J.A.Konter[4], D.R.Gill[4,8],
S.Mango[4], G.D.Wait[4,8],
E.Boschitz[5], B.Gyles[5], W.List[5], C.Otterman[5],
R.Tacik[5], M.Wessler[5],
E.Descroix[6], J.-Y.Grossiord[6], A.Guichard[6],

1. CERN-EP, CH12-11, Geneva 23, Switzerland
2. DPhNME, CEN Saclay, F91191 Gif/Yvette cedex, France
3. L.N.S., CEN Saclay, F91191 Gif/Yvette cedex, France
4. P.S.I. (formerly S.I.N.), CH 5234 Villigen, Switzerland
5. Univ. und Kernforschungzentrum, D7500 Karlsruhe, R.F.A.
6. I.P.N. Lyon, F69622 Villeurbanne cedex, France
7. present adr.: Istituto Fisica, I10125 Torino, Italy
8. present adr.: TRIUMF, Vancouver, B.C. Canada V6T2A3

(presented by F.Perrot)

INTRODUCTION

We report here on the experiment PS198 at LEAR which was completed in June 1988. The aim of that experiment was the measurement of spin observables in $\bar{p}-p$ and $\bar{p}-d$ elastic scattering in order to study the $\bar{N}-N$ interaction in the two isospin states $I=1$ and $I=0$. It was foreseen to measure the full angular distribution of the differential cross section and of the analyzing power in $\bar{p}-p$ elastic scattering at few incident momenta between 450 and 750 MeV/c.

In that energy domain, data have already been published on $\bar{p}-p$ total cross sections, as well as on the angular distributions of differential cross sections in $\bar{p}-p$ elastic scattering [1-3]. However, even several years after LEAR came into operation, data on spin dependent parameters are still incomplete. Measurements of the $\bar{p}-p$ analyzing power have already been performed but in a more restricted angular

Antiproton-Nucleon and Antiproton-Nucleus Interactions
Edited by F. Bradamante *et al.*
Plenum Press, New York, 1990

range and at different incident momenta [4,5]. Complete angular distributions of the $\bar{p}-p$ analyzing power are still missing, as well as data on spin rotation parameters. Some experimental facts can explain this situation: $\bar{p}$ polarized beams are not yet available, and the $\bar{p}$ polarization cannot be easily measured because of the low value of the $\bar{p}-C$ analyzing power at those energies [6]. Therefore spin dependent measurements are restricted to scattering on polarized target, and to double scattering experiments where the outgoing proton, rather than the antiproton, is rescattered on a carbon analyser. In this perspective, the present experiment combined the use of a polarized target, a spin rotation device (the field of the spectrometer), and a proton recoil polarimeter. This set-up allows the measurement of the $\bar{p}-p$ analyzing power A_y, as well as the depolarization parameter D, and the spin rotation parameters. Unfortunately because of the deadline for the floorspace avaibility in the experimental area, only the analyzing power could be measured.

Let us now discuss our theoretical understanding of the $\bar{N}-N$ interaction. The $\bar{N}-N$ system is more complicated than the well known $N-N$ system: there is no generalized Pauli principle in the $\bar{N}-N$ scattering that excludes, like in $N-N$ scattering for each isospin some partial waves. Moreover the phase shifts become complex due to the presence of the annihilation. These two features each double the number of the required parameters needed to define the amplitudes. All the theoretical descriptions of $\bar{N}-N$ scattering, are essentially based on potential models. In these models the real part of the potential is obtained by the G-parity tranformation of the $N-N$ potentials just reversing the sign of the odd G-parity meson exchange contributions. Cancellations between the different meson exchange contributions do not occur here like in the $N-N$ case [7,8]. Because of that, the real part of the potential is more attractive. The imaginary part of the potential is expected to be large because the annihilation cross section is very important. Different approaches have been used to describe the annihilation. The calculations performed in the framework of these potential models [9-12] fit rather correctly to the data of spin integrated cross sections, but differ on the predictions of spin dependent parameters.

Therefore the measurement of the $\bar{p}-p$ analyzing power A_y in a wide angular range can provide a good check of the theoretical inputs of the existing models. The present experiment was designed to measure full angular distributions of this parameter by scattering antiprotons on a polarized target, and detecting a single outgoing particle in the high resolution spectrometer SPES II.

EXPERIMENTAL SET-UP

The experimental set up is shown in Fig.1. The antiproton beam of LEAR, with an intensity of 5.10^5 to 5.10^6 $\bar{p}$/sec., was scattered on a polarized proton target. The intensity of the beam was monitored with counter (1), consisting of a thin circular scintillator F associated to an antihalo scintillator FH with a circular hole of 12mm. Typical counting rates of FH were less than 2% of F. An additional relative monitoring was provided with counter (3) (Fig.1) consisting of two scintillator slabs put in view of the target at 0^0, out of the scattering plane.

The target was made of a solid slab of pentanol 5mm thick, 18mm high and 18mm wide. It was polarized in a 25 kGauss magnetic field produced by a superconducting coil magnet. The polarization was held in the frozen spin mode, in the vertical direction at a temperature of 70 mK. The magnetic field was lowered to

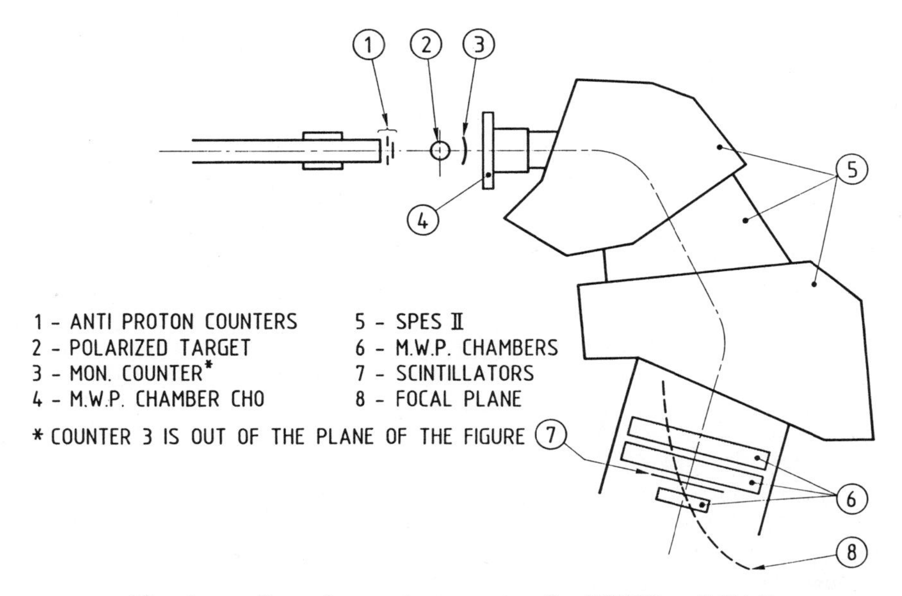

Fig. 1. Experimental apparatus for PS198 at LEAR

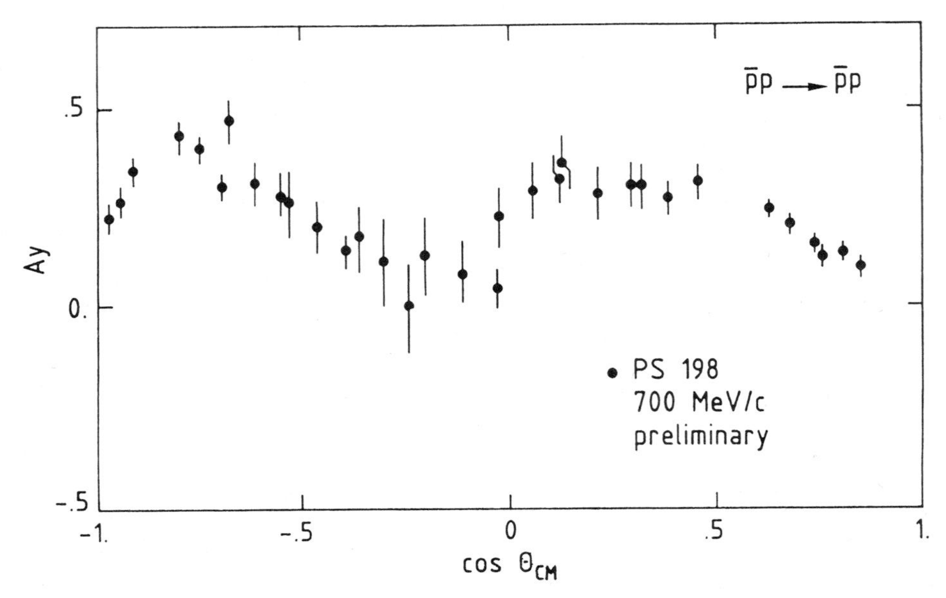

Fig. 2 Preliminary results of the analyzing power $A_y(\theta)$ in $\bar{p}-p$ elastic scattering at 700 MeV/c

7 kGauss in order to decrease the influence of the field on the trajectories of the incoming and outgoing particles. Typical value of the polarization was about 70%, with a relaxation time of about 150 hours.

At each energy, the differential cross section was successively measured for the 2 opposite states (up and down) of the target polarization. A first series of measurements was done detecting the $\bar{p}$, by moving the spectrometer between 10 and 45°lab. (covering the CM angles with $\cos\theta \geq 0$) and another series detecting the proton in the same laboratory angular range ($\cos\theta \leq 0$). In this way, always the most energetic particle was detected, thus limitating straggling and multiple scattering effects. The antiprotons (protons) scattered at forward (backward) angles, were momentum analyzed in the high resolution spectrometer SPES II, and detected in the 4 MWPC's (4 and 6 in Fig.1), all of them with X and Y planes, and with scintillator S (7 in Fig.1). The time of flight between F and S was measured in order to select the required particle $\bar{p}$ or p, among the other products of the interaction, mainly pions. Using the reconstructed tracks of the particles and the transfer matrix of SPES II, the missing mass was calculated and used to select the $\bar{p}-p$ elastic channel from any other contribution. Around $\theta_{lab}=45^0$, where the energy of the scattered particle was minimal, angular and energy straggling considerably deteriorated the resolution on the missing mass and therefore the peak over background ratio. In order to study the possible contaminations from quasi elastic scattering and other background in this angular domain, a recoil counter was added at 90° with the spectrometer, and rotated with it. It consisted of a scintillating counter and a MWPC, and was efficient to detect the recoil particle for the angular range $70^0 \leq \theta_{CM} \leq 110^0$. The coplanarity of the scattered and recoil particles was checked with this device. It was found that the contribution of the quasi elastic scattering to our data was negligible.

RESULTS

In the $\bar{p}-p$ elastic scattering, data were taken at 3 momenta of the incident antiproton beam, 450, 550 and 700 MeV/c. The analysis of the data is still in progress. Preliminary results of the angular distribution of the analyzing power at 700 MeV/c are available. They represent the first data that cover the whole angular domain, and they will be sensitive to the contributions of all the different helicity amplitudes. We have increased the error bars of these preliminary data because some corrections are still to be applied. For instance the eventual variation of the efficiency of the chambers during the experiment was not yet taken into account. For part of the data, the values of the target polarization are still preliminary. However we expect that the data will not change by more than a few percent. It is therefore reasonable to compare these data with the theoretical predictions. Of all the theoretical calculations shown in the litterature, we have retained only the two that are in reasonable agreement with our data [11,13]. It should be stressed that in these calculations there is a very different treatment of the annihilation. Of them, the Paris group predictions [11] show the closer agreement. A more complete discussion will be presented when the final data will be available, treating also the energy dependence of $A_y(\bar{p}-p)$. The analyzing power in the $\bar{p}-d$ elastic scattering at 700 MeV/c has also been measured. Combining these data with the $\bar{p}-p$ elastic scattering data and with the few points in $\bar{p}-p$ and $\bar{p}-n$ quasi elastic scattering also measured at the same energy we expect to extract isospin dependence of the amplitudes.

REFERENCES

1. A.S.Clough et al., Phys. Lett. 146B (1984) 299
2. C.I.Beard et al., 3rd LEAR Workshop, U.Gastaldi, R.Klapisch, J.-M.Richard and J.Tran Thanh Van, eds., Frontieres, Gif-sur-Yvette (1985) p225 et 229
3. W.Bruckner et al., Phys. Lett. B158 (1985) 180; Phys. Lett. 166B (1986) 113; Phys. Lett. 169B (1986) 302
4. R.A.Kunne et al., Phys. Lett. B206 (1988) 557
5. M.Kimura et al., Nuovo Cimento A71 (1982) 438
6. A.Martin et al., IV LEAR Workshop, Nucl. Science Res. Conf. Series Vol.14 (1987) 685, Harwood Academic Publishers
7. W.W.Buck et al., Ann. Phys. (N.Y.) 121 (1979) 47
8. C.B.Dover et al., Ann. Phys. (N.Y.) 121 (1979) 70
9. A.Delville et al., Am. J. Phys. 46 (1978) 907
10. R.A.Bryan et al., Nucl. Phys. B5 (1968) 201
11. J.Cote et al., Phys. Rev. Lett. 48 (1982) 1319
12. P.H.Timmers et al., Phys. Rev. D29 (1984) 1928
13. C.B.Dover and J.M.Richard, Phys. Rev. C25 (1982) 1952

STUDY OF THE SPIN STRUCTURE OF THE $\bar{p}p \rightarrow \bar{n}n$ CHANNEL AT LEAR

M.P. Macciotta, A. Masoni, G. Puddu and S. Serci

Dipartimento di Scienze Fisiche dell'Università and
Sezione INFN, Cagliari, Italy

A. Ahmidouch, E. Heer, R. Hess, R. Kunne,
C. Lechanoine-Leluc and D. Rapin

Département de physique nucléaire et corpusculaire
Université de Genève, Geneva, Switzerland

R. Birsa, F. Bradamante, S. Dalla Torre-Colautti,
M. Giorgi, A. Martin, A. Penzo, P. Schiavon, F.Tessarotto,
A. Villari and A.M. Zanetti

Dipartimento di Fisica dell'Università and Sezione INFN
Trieste, Italy

F. Iazzi, B. Minetti and M. Agnello

Dipartimento di Fisica del Politecnico and Sezione INFN
Turin, Italy

T. Bressani, E. Chiavassa, N. De Marco, M. Gallio,
A. Musso and A. Piccotti

Dipartimento di Fisica dell'Università and Sezione INFN
Turin, Italy

R. Bertini, H. Catz, J.-C. Faivre and F. Perrot

DPhNME, CEN Saclay, Gif/Yvette, France

J. Arvieux

L.N.S., CEN Saclay, Gif/Yvette, France

(Presented by D. Rapin)

Antiproton-Nucleon and Antiproton-Nucleus Interactions
Edited by F. Bradamante *et al.*
Plenum Press, New York, 1990

ABSTRACT

Experiment PS 199 will measure the differential cross-section, the polarisation parameter P and the spin transfer D of the charge exchange reaction $\bar{p}p \to \bar{n}n$ from 500 to 1500 MeV/c.

Measurement techniques and future possibilities offered by the present apparatus are presented.

INTRODUCTION

Following the notation of[1], the scattering matrix of the $\bar{p}p \to \bar{n}n$ reaction can be parametrized by five nuclear amplitudes.

$$M = \tfrac{1}{2}[(a+b)+(a-b)\vec{\sigma}_1.\hat{n}\vec{\sigma}_2.\hat{n}+(c+d)\vec{\sigma}_1.\hat{m}\vec{\sigma}_2.\hat{m}+(c-d)\vec{\sigma}_1.\hat{\ell}\vec{\sigma}_2\hat{\ell}+e(\vec{\sigma}_1.\hat{n}+\vec{\sigma}_2.\hat{n})]$$

where $\vec{\sigma}_1$ and $\vec{\sigma}_2$ are the Pauli spin matrices operating on the incident and target particles respectively; $\vec{n}$, $\vec{m}$, and $\vec{\ell}$ are three directions in the c.m. frame which are defined by $\vec{n} = \vec{k}_1 \times \vec{k}_3$, $\vec{m} = \vec{k}_1 - \vec{k}_3$, $\vec{\ell} = \vec{k}_1 + \vec{k}_3$; here, $\vec{k}_1$ and $\vec{k}_3$ are the momenta of the incomping $\bar{p}$ and of the scattered $\bar{n}$ respectively.

An extensive list of relations between observables and amplitudes can be found in[1]. A few of them are listed below.

$$\frac{d\sigma}{d\Omega} = \frac{1}{2}\left(|a|^2 + |b|^2 + |c|^2 + |d|^2 + |e|^2 \right)$$

$$\frac{d\sigma}{d\Omega} . P = \mathrm{Re}(a^* e);\ \frac{d\sigma}{d\Omega}(1-D) = |c|^2 + |d|^2;\ \frac{d\sigma}{d\Omega}(1-D_t) = |b|^2 + |d|^2$$

Measurements of spin observables provide information on relative phases between amplitudes and on their relative contributions to the interaction.

Considerable insight can be gained by studying different isospin channels. Each amplitude can be written in terms of isosinglet T° and isotriplet T^1:

$$T(\bar{p}p \to \bar{p}p) = T(\bar{n}n \to \bar{n}n) = \tfrac{1}{2}(T^0 + T^1),$$

$$T(\bar{p}p \to \bar{p}p) = T(\bar{n}p \to \bar{n}p) = T^1$$

$$T(\bar{p}p \to \bar{n}n) = T(\bar{n}n \to \bar{p}p) = \tfrac{1}{2}(T^0 - T^1),$$

The present goal of PS 199 is the study of the charge exchange channel $\bar{p}p \to \bar{n}n$, in particular the measurement of the cross section $d\sigma/d\Omega$ and of the spin parameters P (= A_{on}) and D (= D_{nn}). Predictions of the Dover-Richard model are shown on fig. 1 for those parameters[2]. The members of PS 199 collaboration come from 3 previous LEAR experiments. As shown in table 1, the present program is the natural continuation of previous works.

THE MEASUREMENT AND THE EXPERIMENTAL SET-UP

The experimental set-up will be used to measure $d\sigma/dt$ and P is shown in figure 2.

The polarized proton target is completely surrounded by thin scintillator counters to reject events with charges secondaries or γ_s outside the azimuthal acceptance.

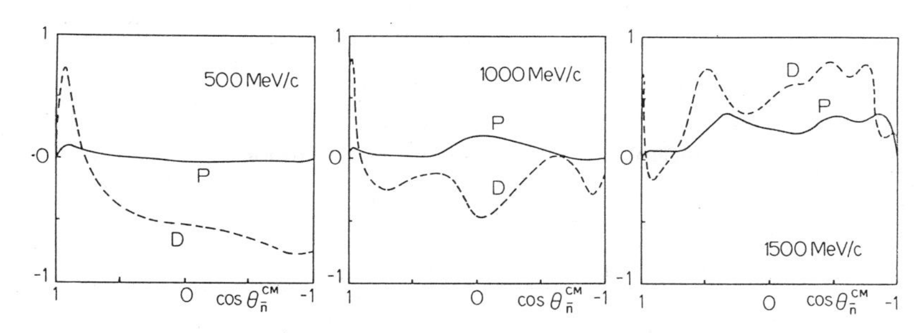

Fig. 1. Predictions for P and D spin parameters from Dover-Richard model (Ref. 2).

Table 1. Genealogy of PS 199

LEAR Experiment	Channel	Observables
PS 172 "SING"	$\bar{p}p$ small angle $\bar{p}C$ " " $\bar{p}p$ $\bar{p}p\uparrow \to \bar{p}p$ $\to \pi^+\pi^-$ $\to k^+k^-$	$d\sigma/d\Omega$; ρ Ac analyzing power σ_{tot} } $d\sigma/d\Omega$, P
PS 178 "ANTIN"	$\bar{p}p \to \bar{n}n$ $\bar{p}A \to \bar{n}A'$ (0°) $\bar{n}p \to n\bar{p}$	} cross-sections at low energy
PS 198 "SPES II"	$\bar{p}p\uparrow \to \bar{p}p$ $\bar{p}n\uparrow \to \bar{p}n$ (deuterium)	} $d\sigma/d\Omega$, P

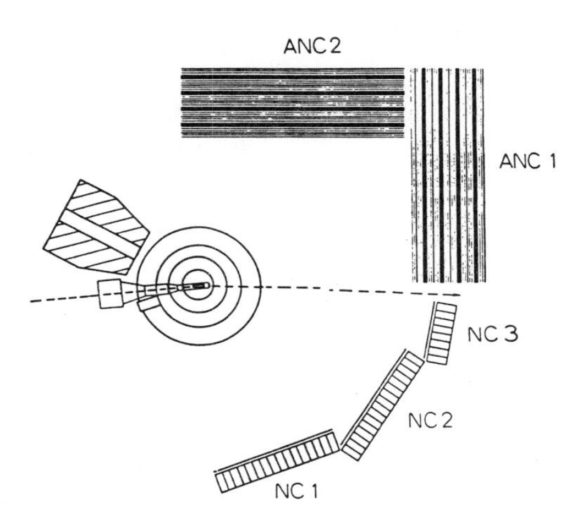

Fig. 2. Experimental apparatus for the measurements of $d\sigma/d\Omega$ and P.

The neutrons will be detected in three hodoscopes (NC1, NC2 and NC3) of vertical scintillator counters, each bar being 8cm x 20cm x 130cm.

The $\bar{n}$ detector consists of two identical modules (ANC1 and ANC2). Each module is a sandwich of iron slabs (converter), planes of limited streamer tubes (to reconstruct the trajectories of the annihilation products) and hodoscopes of scintillation counters for triggering purposes. Both $\bar{n}$ and n detectors will ensure full angular coverage. TOF measurements by these scintillation counters as well as by the neutron counters allow to reject γ_s from annihilation in the target region.

Charge-exchange events are identified by angular correlation and coplanarity. The main source of background is due to $\bar{p}$ charge exchange on bound protons in the polarized target and is expected to be less than 5%. The spectra of the background events will be determined with dummy target measurements.

For absolute normalization, measurements with a liquid hydrogen target are also foreseen.

POLARIZATION TRANSFER

In order to obtain the D parameter, one has to measure the polarization of the outgoing neutron. The np elastic reaction, having a large analyzing power, is adequate for this purpose.

The existing detectors will be rearranged as in fig. 3. The polarization of the neutrons produced in the charge-exchange scattering will be analyzed by measuring the azimuthal distribution in elastic scattering off the free protons present in the scintillator counters of hodoscope NC3 and by detecting the rescattered neutrons in NC1 or NC2. Elastic np scattering in NC3 will be identified by pulse height analysis.

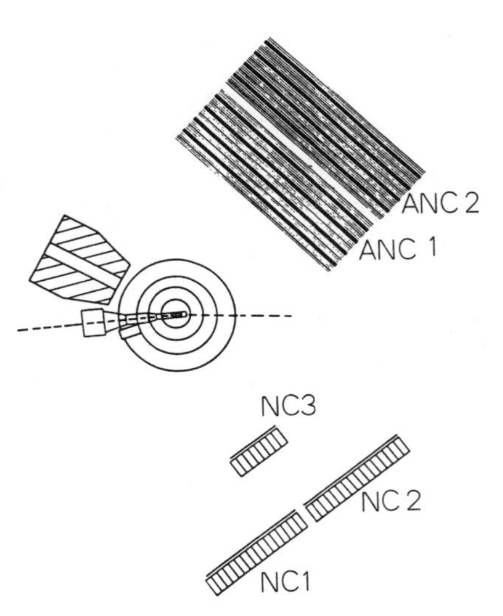

Fig. 3. Experimental apparatus for the measurement of D

Such method has been recently used with the PSI (formerly SIN) polarized neutron beam[3].

OTHER POSSOBILITIES

Predictions from models[4] as well as experimental results[5] indicate non-negligible analyzing power for the $\bar{n}p$ (= $\bar{p}n$)channel. This would allow the measurement of the polarization of the anti-neutrons in a way similar to neutrons. This would give access to D_T spin transfer.

By adding an horizontal field to the frozen spin target, one could measure A, R. A_T and R_T spin transfers.

Replacing the polarized proton target by a polarized deuterium target, the quasi-elastic scattering (with spectator proton) allows the measurement of the $\bar{p}n$ channel).

REFERENCES

1. J. BYSTRICKY ET AL., J. PHYS. (FRANCE) 39, 1 (1978)
2. C.B. Dover and J.M. Richard, Phys. Rev. C21, 1466 (1980)
3. R. Binz et al., SIN Newsletter 19, 48–50 (1986)
4. B. Loiseau & J.M. Richard, private communications
5. A. Martin et al., CERN EP 88, 35

TOTAL AND DIFFERENTIAL CROSS SECTIONS FOR THE REACTION $\bar{p}p \rightarrow \bar{\Lambda}\Lambda$

David W. Hertzog
University of Illinois at Urbana-Champaign
Department of Physics
Urbana, IL 61801

representing the PS185 collaboration:

P.D. Barnes[1], P. Birien[3], B.E. Bonner[6], W. Breunlich[8], G. Diebold[1], W. Dutty[3], R.A. Eisenstein[4], G. Ericsson[7], W. Eyrich[2], R. von Frankenberg[5], G. Franklin[1], J. Franz[3], N. Hamann[3], D. Hertzog[4], A. Hofmann[2], T. Johansson[7], K. Kilian[5], N. Nägele[8], W. Oelert[5], S. Ohlsson[7], B. Quinn[1], E. Rössle[3], H. Schledermann[3], H. Schmitt[3], G. Sehl[5], J. Seydoux[1], and F. Stinzing[2]

[1]Carnegie-Mellon Univ. - [2]Erlangen-Nürenberg Univ. - [3]Freiburg Univ.
[4]Univ. of Illinois at Urbana-Champaign - [5]Jülich IKP-KFA - [6]Rice Univ.
[7]Uppsala Univ.- [8]Vienna IMEP- ÖAW

INTRODUCTION AND MOTIVATION

One of the methods to explore the dynamics of proton-antiproton annihilations is to study reactions in which associated strangeness production occurs. The PS185 collaboration at LEAR has been involved in a systematic study of the process $\bar{p}p \rightarrow \bar{Y}Y$ where Y is a Λ or a Σ hyperon. A rather complete data set is beginning to emerge from experimental runs performed over the last four years. Data on the total and differential cross sections for the reaction $\bar{p}p \rightarrow \bar{\Lambda}\Lambda$ will be presented below.

Physics interest in the basic properties and interactions of hyperons has always been high due to the great degree of similarity, yet somewhat exotic differences, between these particles and the ordinary and thoroughly explored nucleons. Within the context of broken SU(6), static properties of the baryon octet such as masses and magnetic moments are rather successfully predicted (within 10%). Thus for hyperons, we might expect largely similar dynamical behavior as well. Exceptions can then be ascribed to the effects of the embedded strange quark(s). In a reaction such as $\bar{p}p \rightarrow \bar{\Lambda}\Lambda$, the $\bar{p}p$ initial state interactions (distortions) can be well fit from precise elastic scattering data. If it is assumed that the outgoing $\bar{\Lambda}\Lambda$ interaction is similar to that of the incoming $\bar{p}p$, then the main unknown in a theoretical description is the fundamental associated strangeness production.

The strangeness production can be thought of in two complimentary ways. From a quark point of view, a $\bar{u}u$ quark pair is annihilated in the entrance channel and a $\bar{s}s$ quark pair is created in the exit channel. This view is shown in Fig 1a. The $\bar{s}s$ creation is mediated by either a "vacuum" or "gluon" interaction. Two models used to describe such interactions are the $^3P_0$ and $^3S_1$ models, respectively. If the reaction is

Antiproton–Nucleon and Antiproton–Nucleus Interactions
Edited by F. Bradamante *et al.*
Plenum Press, New York, 1990

driven by only one of these interactions, then the differential cross section would show the characteristic S- or P-wave shape. An alternate description of the dynamics of the reaction is based on meson exchange as shown in Fig 1b. Here K , K^*, or K_2^* bosons are exchanged between the antiproton and proton in the initial state to form the outgoing lambda and antilambda. The relative importance and inclusion of the various mesons is one of the parameters of the models. For example, the heavy K_2^* governs interactions at very short range, while K exchange describes more of a long-range interaction. A common ingredient to both the "quark-gluon" and the "meson exchange" descriptions is the importance of the initial and final state distortions which makes interpretation of the results somewhat more difficult.

At threshold, the $\bar{\Lambda}\Lambda$ relative velocity approaches zero and there it is expected that only one partial wave will dominate. The total cross section for the Lth partial wave is expected to rise in proportion to $\varepsilon^{L+1/2}$, where $\varepsilon = \sqrt{s} - 2m_\Lambda$ is the $\bar{\Lambda}\Lambda$ kinetic energy in the COM and L is the orbital angular momentum quantum number. Hence its shape can be used to indicate which waves are important to the reaction at those energies. The shape of the differential cross section can be used in a similar fashion.

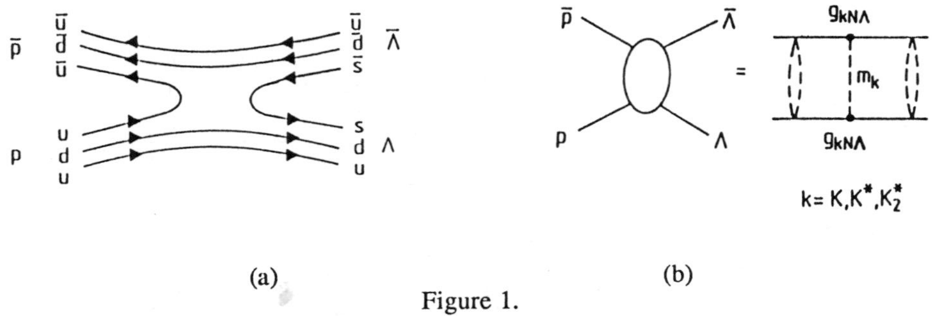

(a) (b)

Figure 1.

"Quark-gluon" and "meson exchange" description of the reaction $\bar{p}p \rightarrow \bar{\Lambda}\Lambda$.

EXPERIMENTAL SIGNATURE

The experiment has been performed at the Low-Energy Antiproton Ring (LEAR) at CERN where high-intensity, cooled beams of antiprotons ($\bar{p}$) are available at extracted momenta of (0.1 to 2.0) GeV/c. The reaction $\bar{p}p \rightarrow \bar{\Lambda}\Lambda$ has a threshold at 1.435 GeV/c. The proper lifetime, $c\tau$, of a Λ is 7.9 cm and hence Λ's produced in this reaction recoil downstream approximately 3 to 10 cm before they decay. The branching ratio for $\Lambda \rightarrow p\pi^-$ is 64%. At the highest LEAR momentum the outgoing decay proton is constrained to a lab angle of less than 42°. The pion, while Lorentz boosted downstream, can still be found at any laboratory angle. Thus the detector which is shown in Fig. 2 has a 4π center-of-mass acceptance only for the protons. In practice, the total acceptance is about 60% due to those events where the outgoing pions are at very steep lab angles or from events where the lambdas decay either before or after the tracking chamber fiducial volume.

Although the total $\bar{p}$ cross section is large (about 100 mb), data can be accumulated at high beam intensities (10^6 incident $\bar{p}$/sec) by use of a "charged-neutral-

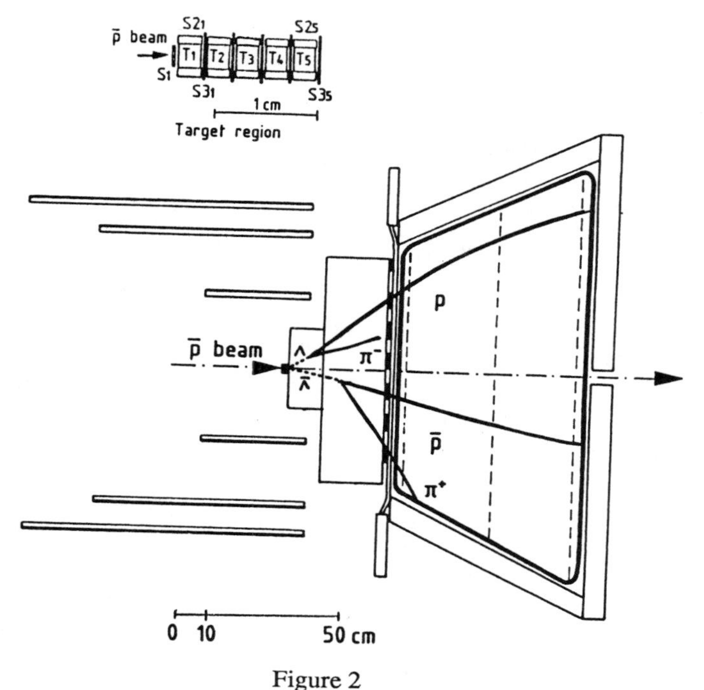

Figure 2

Plan view of the detector.

charged" trigger. The process is $\bar{p}p \rightarrow \bar{\Lambda}\Lambda \rightarrow \bar{p}\pi^{+}p\pi^{-}$. An incoming $\bar{p}$ is detected by a multielement scintillator telescope and enters a target system consisting of 4 individual cells containing 2.5 mm long cylinders of CH_2 and each surrounded by thin veto scintillators. A fifth cell containing ^{12}C is employed to determine the number of $\bar{p}p \rightarrow \bar{\Lambda}\Lambda$ background" events from ^{12}C which survive event reconstruction kinematic cuts (typically ~6%). The use of this cellular structure allows subdivision of the cross section near threshold into sub-MeV bins. A decay volume is sandwiched between the target and a scintillator hodoscope (Fig. 2). This completes the online trigger. The decay volume is filled with both MWPC planes and drift chamber planes in order to record the charged-particle tracks whose pattern forms a distinctive double "Vee" signature of the events. Finally, a low-field solenoidal magnet containing 3 additional drift chamber planes is utilized in order to distinguish the charge of the penetrating baryons and thus tag the respective lambda and antilambda. On the upstream side of the detector, 3 concentric boxes containing streamer tube planes have been recently added in order to detect tracks of high-angle decay pions. The beam and the target are unpolarized.

THE DATA SETS

The first experiments were performed in the spring of 1984 and subsequent runs have occurred in the following years. Most of the completed analysis to date is on the channel $\bar{p}p \rightarrow \bar{\Lambda}\Lambda$; however, an analysis of the $\bar{\Lambda}\Sigma + \Lambda\bar{\Sigma}$ final state for one set of data above this threshold (1653 MeV/c) is also nearly complete[1]. In the summer of 1986, we performed a dedicated "scan" in the momentum range 1350 to 1550 MeV/c in an attempt to look for the $\xi(2230)$ resonance as a bump in the total cross section of the channel $\bar{p}p \rightarrow K_s K_s$. The results of this search will be reported in the next few months, although preliminary results on the total and differential cross sections were shown at the recent Mainz conference[2]. While that experiment was optimized for the

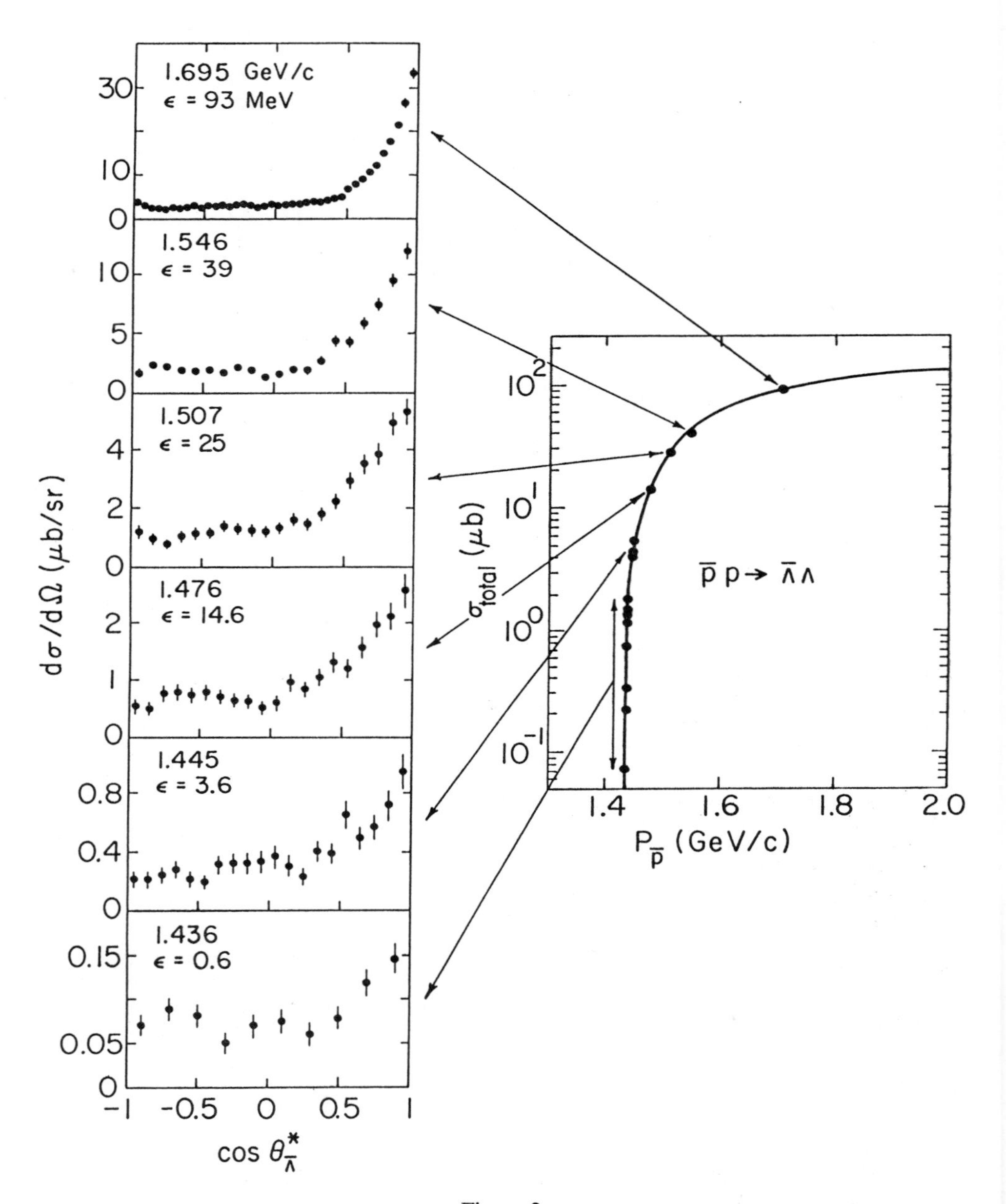

Figure 3

Compilation of total and differential cross section results for $\bar{p}p \rightarrow \bar{\Lambda}\Lambda$.

K_sK_s final state and the momentum choices bracketed the center-of-mass energy of 2230 MeV (1435 MeV/c, which is interestingly enough also the $\bar{\Lambda}\Lambda$ threshold), about half of the data was just above the $\bar{\Lambda}\Lambda$ threshold. Hence, in addition to what is presented here, higher statistics near-threshold data on $\bar{p}p \rightarrow \bar{\Lambda}\Lambda$ can be expected soon[3]. In the past year, we have accumulated data at 1900 and 1910 MeV/c incident antiproton momentum. Beside being the highest extracted antiproton momentum to date, these data were accumulated at the thresholds of the complimentary $\bar{\Sigma}\Sigma$ channels, where interesting cusp effects might be expected. The data presented here are exclusively on the the reaction $\bar{p}p \rightarrow \bar{\Lambda}\Lambda$ and are summarized in Table I.

TABLE I

Summary of the analyzed $\bar{p}p \rightarrow \bar{\Lambda}\Lambda$ data.

$\bar{p}$ Momentum (MeV/c)	ε (MeV)	σ_{total} (μb)	Ref.
1435.9	0.29	0.84±0.20	4
1436.9	0.64	1.44±0.32	4
1445.3	3.6	4.86±0.42	4
1476.5	14.6	13.8±0.5	5
1507.5	25.5	26.6±0.7	5
1546.2	39.2	44.6±1.5	6
1695.0	92.9	85±8	7

The near-threshold experiments were performed in order to observe the reaction in its most simple form, while the higher cross section data set was targeted toward a determination of spin parameters such as polarization, spin correlation coefficients and CP violation checks. These results were presented at this school by Hamann[8] and are contained in his report. Fig. 3 contains a summary of our total and differential cross section data. At the highest antiproton momentum, the differential cross section data reveals an anisotropy with a strong forward peaking of the antilambda in the center-of-mass system. This effect is observed to decrease in magnitude as the momentum is lowered toward threshold, *but not to disappear* even at the lowest momentum values. Clearly, even at 600 keV excess energy above threshold, there remains a strong mix of both S and P waves in the interaction. This fact has led to several theoretical explanations; a popular one appears to be that the S wave is simply *suppressed* by initial and final state interactions of the respective particle and antiparticles. S-wave implies a close proximity of the two particles and thus annihilations there remove events from the otherwise observed cross sections. It should be pointed out that an alternate approach discussed at this school by Shapiro is based on P-wave *enhancement* due to near-threshold resonances.

WHAT HAVE WE LEARNED?

The data reported here constitute only a subset of the observables we have measured on the reaction $\bar{p}p \rightarrow \bar{\Lambda}\Lambda$. With the inclusion of the excellent polarization and the preliminary spin correlation coefficient results[8], there is now quite a complete data set on hyperon-antihyperon production from antiproton-proton annihilations. Many theoretical models have been used to describe these data[9]. While it would be impractical to highlight the various features of each these models here, it should be pointed out that from almost any of the models reasonable "fits" exist to some subset of the data. Although the agreement with the data is often achieved by adjusting quite different physical processes in the models, there are some general or generally agreed upon conclusions from the works as a whole based on the earliest data that have been published.

For example, since both the entrance and exit channel of the reaction $\bar{p}p \rightarrow \bar{\Lambda}\Lambda$ have strong absorption potentials it is necessary to use distorted waves (DWBA). The strength of the absorption can be tuned to fit the total cross section. As is known from other nucleon-antinucleon systems, it is found to have a large effect. The result of this is that the fundamental exchange potential, whether described by meson or gluon exchange, is rather unimportant in detail. The shapes of the observables are thus driven by the initial- and final-state interactions and a "simple" interpretation is not immediately possible. Nevertheless, the data set is continuing to mature with a precision that now includes the energy dependence of the observables and soon, the inclusion of new channels which bring the isospin dependence of the reactions into the picture, such as $\bar{p}p \rightarrow \bar{\Lambda}\Sigma + cc$ and $\bar{p}p \rightarrow \bar{\Sigma}\Sigma$. The theoretical community, which has been actively testing a variety of models on the data presented to date, can expect a very rich and even more complete picture from the experimental side on hyperon-antihyperon formation from antiproton-proton annihilations. When this data set is "fit" as a whole, we hope to have learned something quite fundamental about the dynamics of strangeness production in these systems.

This collaborative research is supported in part from the Austrian Science Foundation, the German Bundesministerium für Forschung und Technologie, the Swedish Natural Science Research Council, the United States Department of Energy and the United States National Science Foundation grant NSF PHY 86-10493.

REFERENCES

1. Analysis in progress by S. Ohlsson, Univ. of Uppsala.
2. Analysis in progress by H. Schledermann, Univ. of Freiburg and J. Seydoux, Carnegie-Mellon Univ. Preliminary results will appear in H. Schmitt et al, proceedings of the IXth European Symposium on Antiproton-Proton Interactions and Fundamental Symmetries, Mainz, Sept. 1988, in press.
3. Analysis in progress by F. Stinzing, Univ. of Erlangen-Nürenberg.
4. R. von Frankenberg, et al., *Threshold Measurement of the Reaction* $\bar{p}p \rightarrow \bar{\Lambda}\Lambda$ at LEAR, Proc. 4th LEAR Workshop, Villars-sur-Ollon, Sept. 1987, edited by C. Amsler, G. Backenstoss, R. Klapisch, C. Leluc, D. Simon, and L. Tauscher., and R. Von Frankenberg, Ph.D. dissertation, Univ. of Erlangen-Nürenberg (1987) unpublished.
5. P.D. Barnes et al., Phys. Lett., **189B** 249 (1987), and C. Maher, Ph.D. dissertation, Carnegie-Mellon University (1986) unpublished.
6. Ph.D. dissertation, W. Dutty, Univ. of Freiburg (1988) unpublished.
7. Preliminary results, G. Sehl, Jülich IKP-KFA.
8. N. Hamann, these proceedings.
9. For example, F. Tabakin and R. A. Eisenstein, Phys. Rev. C 31, 1857 (1985); S. Furui and A. Faessler, Nucl. Phys. A 468, 669 (1987); M. Kohno and W. Weise, Phys. Lett. 206B, 584 (1988); O. D. Dalkarov, K. V. Protasov and I. S. Shapiro, preprint 37, Moscow, FIAN, 1988.

SPIN OBSERVABLES IN THE REACTION $\bar{p}p \rightarrow \bar{\Lambda}\Lambda$

Nikolaus H. Hamann

University of Freiburg, 7800 Freiburg, Fed. Rep. Germany

representing the PS185 Collaboration:

P.D. Barnes[1], P. Birien[3], B.E. Bonner[6], W. Breunlich[8], G. Diebold[1], W. Dutty[3], R.A. Eisenstein[4], G. Ericsson[7], W. Eyrich[2], R. von Frankenberg[5], G. Franklin[1], J. Franz[3], N. Hamann[3], D. Hertzog[4], A. Hofmann[2], T. Johansson[7], K. Kilian[5], N. Nägele[8], W. Oelert[5], S. Ohlsson[7], B. Quinn[1], E. Rössle[3], H. Schledermann[3], H. Schmitt[3], G. Sehl[5], J. Seydoux[1] and F. Stinzing[2]

[1] Carnegie – Mellon University, Pittsburgh, PA 15213, USA.
[2] University of Erlangen – Nürnberg, 8520 Erlangen, Fed. Rep. Germany.
[3] University of Freiburg, 7800 Freiburg, Fed. Rep. Germany.
[4] University of Illinois, Urbana – Champaign, IL 61820, USA.
[5] Inst. Kernphysik KFA Jülich, 5170 Jülich, Fed. Rep. Germany.
[6] Rice University, Houston, TX 77251, USA.
[7] University of Uppsala, 75121 Uppsala, Sweden.
[8] Inst. Mittelenergiephysik ÖAW, 1090 Vienna, Austria.

Abstract
The reaction $\bar{p}p \rightarrow \bar{\Lambda}\Lambda$ has been measured from threshold up to 1.7 GeV/c incident $\bar{p}$ momentum. Large polarizations of Λ and $\bar{\Lambda}$ have been observed even at the lowest energies, and they appear to have a characteristic dependence on the momentum transfer involved. The measured $\Lambda - \bar{\Lambda}$ spin correlations are compatible with the particles being created in a spin-one state. As tests of the fundamental symmetries CPT and CP, mean lives and decay asymmetry parameters, respectively, are compared from Λ and $\bar{\Lambda}$ decay distributions.

INTRODUCTION AND MOTIVATION

While exclusive processes such as $\bar{p}p \rightarrow \bar{\Lambda}\Lambda$, $\bar{p}p \rightarrow \bar{\Lambda}\Sigma^0 + \Lambda\bar{\Sigma}^0$ and $\bar{p}p \rightarrow \bar{\Sigma}\Sigma$ represent only a small fraction of the total $\bar{p}p$ cross section, they provide a laboratory for detailed studies of specific features of the reaction and in particular the spin dynamics. This paper is restricted to the reaction $\bar{p}p \rightarrow \bar{\Lambda}\Lambda$ which is part of an ongoing programme in the framework of experiment PS185 at LEAR. With an unpolarized beam incident on an unpolarized target, we measure in our experiment a complete set of observables: the production cross section $\sigma(\bar{p}p \rightarrow \bar{\Lambda}\Lambda)$, the differential cross section $d\sigma/d\Omega$, the Λ and $\bar{\Lambda}$ polarizations P, and the $\Lambda - \bar{\Lambda}$ spin-correlation coefficients C_{ij}. The cross sections are treated in an accompanying paper [1], whereas the spin observables are the subject of this one.

Antiproton-Nucleon and Antiproton-Nucleus Interactions
Edited by F. Bradamante *et al.*
Plenum Press, New York, 1990

The selectivity of $\bar{p}p \to \bar{\Lambda}\Lambda$ is mainly due to the underlying process of associated strangeness production. This typically involves large momentum transfers of the order of 3 fm^{-1}. The production cross section is only a fraction 10^{-3} to 10^{-4} of the total $\bar{p}p$ cross section. With the kinematical threshold at 1.435 GeV/c beam momentum, the range of LEAR momenta used so far translates into a range of about 1 to 100 MeV kinetic energy in the $\bar{\Lambda}\Lambda$ system. Near the threshold it is expected that only a few partial waves contribute to the production process.

In the static quark model, the Λ hyperon is composed of single u, d and s quarks in a relative S state. The ud quark pair is coupled to spin and isospin zero, so that the spin vector of the s quark is that of the Λ hyperon itself. The simplest quark-line diagram associated with $\bar{p}p \to \bar{\Lambda}\Lambda$ is shown in figure 1. The reaction dynamics is determined by the $\bar{u}u$ annihilation and $\bar{s}s$ creation, while the spin- and isospin-zero ud and $\bar{u}\bar{d}$ quark pairs are spectators. This makes the process an attractive tool to study quark–gluon dynamics, and in particular the quantum numbers of the $\bar{s}s$ vertex. If the $\bar{\Lambda}\Lambda$ pairs are produced with S waves, and if the final state is a spin triplet, then the $\bar{s}s$ vertex has the "gluon quantum numbers" $J^P = 1^-$. In the case of P-wave production, and a final state spin triplet, the vertex has the "vacuum quantum numbers" $J^P = 0^+$.

It is well known from experiments that hyperons produced in high-energy reactions emerge polarized [2]. Pronounced Λ polarizations of the order of $P \approx -0.5$ have been measured in a variety of processes, such as $\gamma p \to K^+\Lambda$, $\pi^-p \to K^0\Lambda$, $K^-p \to \Lambda + X$, and $p + A \to \Lambda + X$. This has been observed at transverse momenta of the order of 1 GeV/c, and for a given reaction it seems to be roughly independent of the centre-of-mass energy. Several ideas in the framework of the static quark model have been put forward to explain the observed hyperon polarization. For the case of inclusive Λ production with proton beams, one assumes that s quarks are produced polarized and then recombine with the incident baryon fragment ("spectator diquark") to form polarized Λ hyperons. The suggested mechanisms for polarizing the s quarks invoke string breaking [3] or Thomas precession [4]. In a recent experiment [5] with polarized proton beams at BNL, the spin transfer and the analyzing power have been measured for $p + Be \to \Lambda + X$. The Λ polarization appears to be independent of the incident proton polarization. This result is consistent with the idea that the spin of the Λ hyperon is essentially carried by its s quark. In view of our $\bar{p}p \to \bar{\Lambda}\Lambda$ experiments at LEAR, the results from BNL may then also support the simple valence quark picture indicated in figure 1.

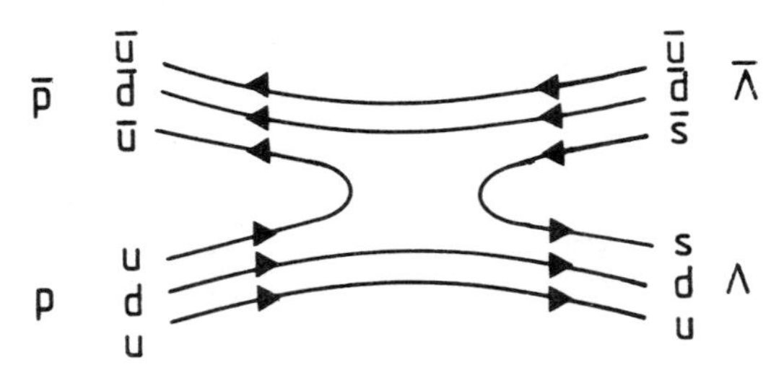

Fig. 1. Quark-line diagram of the reaction $\bar{p}p \to \bar{\Lambda}\Lambda$.

APPARATUS AND EVENT SIGNATURE

The PS185 apparatus is a non-magnetic forward decay spectrometer with large centre-of-mass acceptance. The delayed decays $\bar{\Lambda}\Lambda \to \bar{p}\pi^+ p\pi^-$ are recorded in a stack of proportional chambers and drift chambers. This chamber stack is sandwiched between devices entering the "charged–neutral–charged" on-line trigger scheme: a CH_2 target system triggering on the $\bar{p}$ beam and vetoing the production of charged particles, and a scintillator hodoscope triggering on the detection of charged particles from delayed decays. A 0.1 T solenoid with three drift chambers serves to identify Λ and $\bar{\Lambda}$ by means of charge distinction. The detector is described in more detail elsewhere [6].

The events are reconstructed off-line from their tracks recorded in the chamber stack. Those events which exhibit a distinctive "2 V^0" signature are kinematically fitted to the hypothesis $\bar{p}p \rightarrow \bar{\Lambda}\Lambda \rightarrow \bar{p}\pi^+p\pi^-$. With the determined momenta of the hyperons, the spatial distribution of their decay vertices can be transformed into lifetime distributions. Figure 2 shows Λ and $\bar{\Lambda}$ lifetime distributions for 4063 reconstructed events at 1.546 GeV/c incident $\bar{p}$ momentum [7]. The fitted mean lives of Λ and $\bar{\Lambda}$ agree within errors with each other and also with the world average [8]. From our data we determined the ratio

$$R = 2(\tau - \bar{\tau})/(\tau + \bar{\tau}) = 0.02 \pm 0.05 ,$$

which is consistent with zero as required from CPT invariance. However, our error is smaller than that reported from previous work [8].

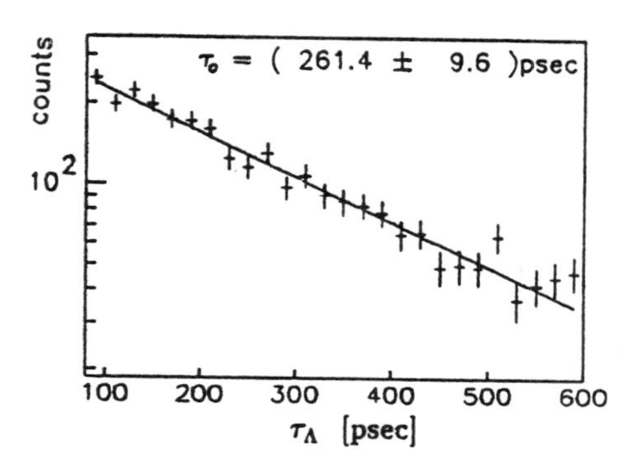

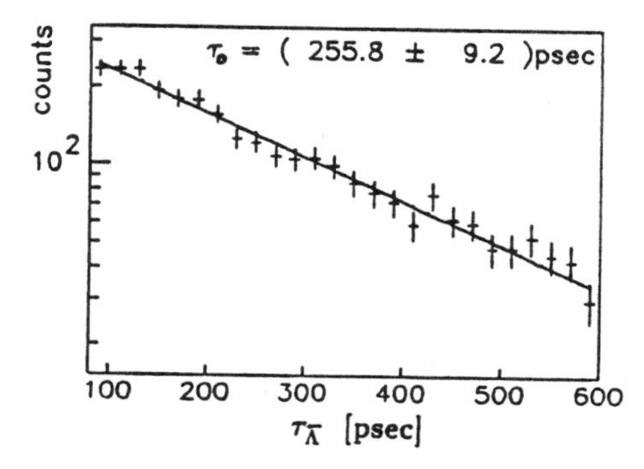

Fig. 2. Λ and $\bar{\Lambda}$ lifetime distributions from $\bar{p}p \rightarrow \bar{\Lambda}\Lambda$ at 1.546 GeV/c incident $\bar{p}$ momentum.

Λ AND $\bar{\Lambda}$ POLARIZATION

Owing to parity conservation in $\bar{p}p \rightarrow \bar{\Lambda}\Lambda$, the hyperons can only be polarized transverse to the production plane. The decays $\Lambda \rightarrow p\pi^-$ and $\bar{\Lambda} \rightarrow \bar{p}\pi^+$ are parity-violating weak decays which are characterized by mixtures of S- and P-wave amplitudes. For a sample of Λ's with polarization P, the angular distribution of the decay protons in the Λ rest frame is given by

$$dN/d(\cos\theta_p) = N(1 + \alpha P\cos\theta_p) ,$$

where θ_p is measured between the normal of the production plane and the proton momentum vector, and $\alpha = 0.642 \pm 0.013$ is the $\Lambda \rightarrow p\pi^-$ decay asymmetry parameter. Parity violation in the decays thus manifests itself in an up–down asymmetry of the decay angular distribution with respect to the production plane. The degree of asymmetry in these "self-analyzing" decays is determined by αP.

For the evaluation of αP the "method of weighted sums" has been adopted [7]. It only requires a symmetry condition of the detector acceptance function, $\eta(\theta_p) = \eta(180° - \theta_p)$, and no corrections are needed. The condition is well fulfilled in our case. Simulations showed that the method did not bias the polarizations extracted from the real data. The product αP has been evaluated for Λ and $\bar{\Lambda}$ separately. However, the polarization distributions can be combined because charge-conjugation invariance in $\bar{p}p \rightarrow \bar{\Lambda}\Lambda$ requires the polarizations of the outgoing particles to be equal. In addition, one has $\alpha = -\bar{\alpha}$ if CP conservation is assumed for the decays $\Lambda \rightarrow p\pi^-$ and $\bar{\Lambda} \rightarrow \bar{p}\pi^+$.

Figure 3 displays a compilation of $\bar{\Lambda}\Lambda$ polarization data obtained with our experiment at incident $\bar{p}$ momenta of 1.445 GeV/c (848 events), 1.477 GeV/c (1185 events), 1.508 GeV/c (1845 events), 1.546 GeV/c (4063 events) and 1.695 GeV/c (11427 events). The polarizations are shown as a function of the four-momentum transfer squared,

$$t = (p_{\bar{p}} - p_{\bar{\Lambda}})^2 ,$$

which is linearly related to $\cos\theta^*_{\bar{\Lambda}}$. The solid curve represents the boundaries t_{min} and t_{max} of the kinematically allowed region for different values of the total centre-of-mass energy, $\sqrt{s}$. Like in other experiments, we observe strong negative polarization for reduced four-momentum transfer squared, $|t'| \geq 0.18$ $(GeV/c)^2$, where

$$-t' = -(t - t_{min}) = 2|p^*_{\bar{p}}||p^*_{\bar{\Lambda}}|(1 - \cos\theta^*_{\bar{\Lambda}}) .$$

The polarization distributions for different $\sqrt{s}$ values exhibit a zero-crossing point at the same value of $|t'|$ as indicated by the dashed line. For $|t'| \leq 0.18$ $(GeV/c)^2$ we observe positive polarization. Its strength appears to increase with decreasing $\sqrt{s}$. Note that the lowest $\bar{p}$ momentum shown corresponds to only a few MeV kinetic energy in the $\bar{\Lambda}\Lambda$ system.

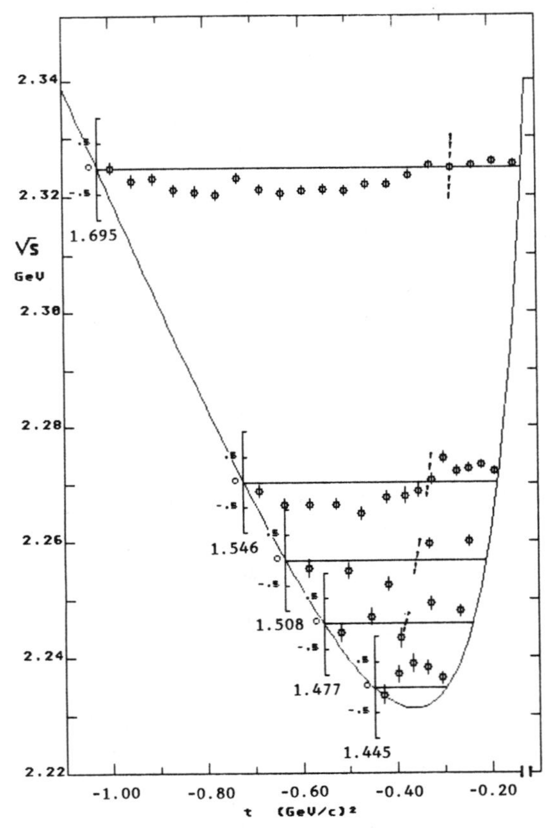

Fig. 3. Compilation of hyperon differential polarization data from $\bar{p}p \rightarrow \bar{\Lambda}\Lambda$.

The qualitative features of our polarization data are reproduced by various recent kaon-exchange and "quark – gluon" calculations. It has been argued [9] that the zero-crossing of the polarization is a consequence of strong P-wave dominance in $\bar{p}p \rightarrow \bar{\Lambda}\Lambda$, which would lead to a pattern with a characteristic $\sin2\theta^*_{\bar{\Lambda}}$ dependence. It seems, however, that the physical origin for the zero-crossing appearing at constant $|t'|$, and of the trend observed in the positive polarizations, has yet to be explored in more depth.

Because of the invariances discussed above, the polarizations of kinematically correlated Λ and $\bar{\Lambda}$ have been combined to a common distribution for each incident $\bar{p}$ momentum. However, the associated production of $\bar{\Lambda}\Lambda$ pairs and the simultaneous detection of their decays offer the possibility to perform a test of CP conservation in the non-leptonic hyperon decays [10]. The $\bar{p}p$ initial state, and thus also the $\bar{\Lambda}\Lambda$ final state, have a definite CP property, so that final state interactions cannot generate a misleading signal. There is no $\Lambda-\bar{\Lambda}$ mixing, and therefore any signal constitutes a measure of CP violation with the strangeness changing by one unit.

From the individual Λ and $\bar{\Lambda}$ distributions of αP as a function of the $\bar{\Lambda}$ centre-of-mass angle, one can extract the CP testing ratio $A = (\alpha + \bar{\alpha})/(\alpha - \bar{\alpha})$. The data obtained at 1.546 GeV/c incident $\bar{p}$ momentum yielded [11] the value $A = -0.07 \pm 0.09$. With the combined statistics of this measurement and another one performed at 1.695 GeV/c incident $\bar{p}$ momentum, corresponding to a total number of nearly 16000 events, we find

$$\langle A \rangle = \langle (\alpha + \bar{\alpha})/(\alpha - \bar{\alpha}) \rangle = -0.023 \pm 0.057 .$$

The error quoted here is only the statistical one.

With the experimental technique presently used in PS185, a sensitivity on A at the level of 10^{-3} could be reached. However, the values of A as predicted in the framework of the Standard Model are of the order of 10^{-4}. Alternate experimental approaches have to be considered in order to reach that level. The three-step reaction $\bar{p}p \to \bar{\Xi}^+\Xi^- \to \bar{\Lambda}\pi^+\Lambda\pi^- \to \bar{p}\pi^+\pi^+p\pi^-\pi^-$ is a unique and promising case [10], because it allows the polarization of the decay baryons – the Λ's and $\bar{\Lambda}$'s – to be determined. Such a measurement, which needs a high-intensity $\bar{p}$ beam of about 3.5 GeV/c momentum, could be performed at the FNAL $\bar{p}$ accumulator or at Super-LEAR proposed for CERN.

$\Lambda-\bar{\Lambda}$ SPIN CORRELATION

If one considers the decays $\Lambda \to p\pi^-$ and $\bar{\Lambda} \to \bar{p}\pi^+$ simultaneously, the double angular distribution of the decay baryons in the respective hyperon rest frames is given by

$$d^2N/[d(\cos\theta_{\bar{p}i})d(\cos\theta_{pj})] = (16\pi^2)^{-1}[1 + \alpha P_{\Lambda}\cos\theta_{py} + \bar{\alpha}P_{\bar{\Lambda}}\cos\theta_{\bar{p}y} + \alpha\bar{\alpha}\,\Sigma_{ij}(C_{ij}\cos\theta_{\bar{p}i}\cdot\cos\theta_{pj})] ,$$

where $\cos\theta_{\bar{p}i}$ ($\cos\theta_{pj}$) is the direction cosine of $\bar{p}$ (p) relative to the i (j) axis, with $i=\bar{x},\bar{y},\bar{z}$ ($j=x,y,z$). In particular, $\bar{y}$ (=y) denotes the direction normal to the $\bar{\Lambda}\Lambda$ production plane. The spin-correlation coefficients C_{ij} are normalized averages of products of three Λ and three $\bar{\Lambda}$ spin components,

$$C_{ij} = 9(\alpha\bar{\alpha})^{-1}\langle\cos\theta_{\bar{p}i}\cdot\cos\theta_{pj}\rangle .$$

The nine coefficients are not all independent. Parity conservation in $\bar{p}p \to \bar{\Lambda}\Lambda$ requires $C_{xy}=C_{yx}=C_{yz}=C_{zy}=0$, and, owing to charge-conjugation invariance, we have $C_{ij}=C_{ji}$. The only elements of the 3×3 matrix which can be non-zero are C_{xx}, C_{yy}, C_{zz} and C_{xz}. These are, however, dependent on the hyperon production angle.

The coefficients have been evaluated [7] adopting the "method of moments". Since this method requires an isotropic detector acceptance for the decay baryons, the data had to be corrected correspondingly. Figure 4 displays the $\Lambda-\bar{\Lambda}$ spin-correlation coefficients obtained, as a function of the $\bar{\Lambda}$ centre-of-mass angle, for 1.546 GeV/c incident $\bar{p}$ momentum. The conditions required from invariances mentioned above are fulfilled within experimental errors. The distributions shown correspond to a total number of 4063 events, whereas for $\sigma(C_{ij})\approx 0.1$ one would need about 5000 events per angular bin.

The spin correlations can be used to calculate the expectation value of the $\bar{\Lambda}\Lambda$ spin-zero projection operator,

$$F_S = 0.25(1 - \langle\vec{\sigma}_{\Lambda}\cdot\vec{\sigma}_{\bar{\Lambda}}\rangle) = 0.25(1 + C_{xx} - C_{yy} + C_{zz}) .$$

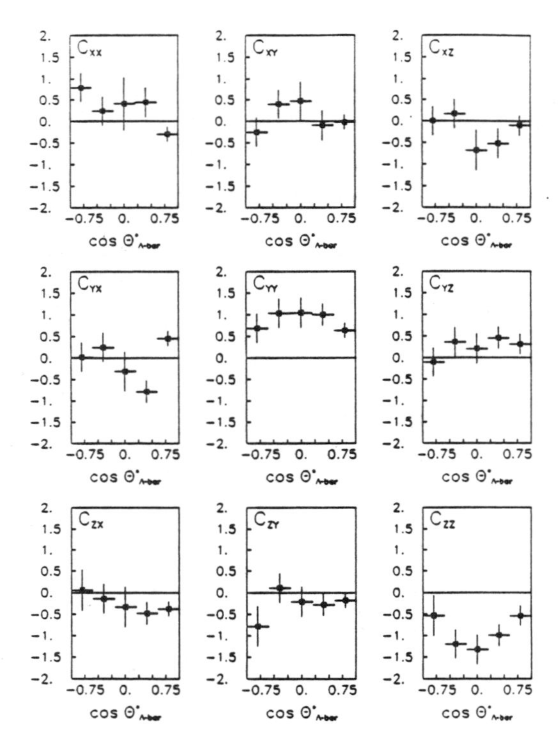

Fig. 4. Hyperon spin-correlation coefficients from $\bar{p}p \rightarrow \bar{\Lambda}\Lambda$ at 1.546 GeV/c incident $\bar{p}$ momentum.

One has $F_S = 0$ for a pure spin-one state, $F_S = 1$ for a pure spin-zero state, and $F_S = 0.25$ for uncorrelated Λ and $\bar{\Lambda}$ spins. From the data shown in figure 4 the average value $\langle F_S \rangle = -0.12 \pm 0.07$ has been calculated. This number is somewhat "unphysical". However, our simulations with uncorrelated $\bar{\Lambda}\Lambda$ pairs did not give the expected 0.25 for the singlet fraction, but yielded the value $\langle F_S \rangle_{MC} = 0.17 \pm 0.03$. We therefore conclude from our data that the $\bar{\Lambda}\Lambda$ pairs are preferably produced in a spin-one state. In the context of the static quark model picture, this should then also hold for the embedded $\bar{s}s$ quark pair.

References

1. D. Hertzog et al., these Proceedings.
2. See: L.G. Pondrom, Phys. Rep. **122** (1985) 57, and references therein.
3. B. Andersson et al., Phys. Lett. **85B** (1979) 417.
4. T.A. DeGrand and H.I. Miettinen, Phys. Rev. **D24** (1981) 2419.
5. B.E. Bonner et al., Phys. Rev. **D38** (1988) 729.
6. P.D. Barnes et al., Phys. Lett. **189B** (1987) 249, and K. Kilian, Nucl. Phys. **A479** (1988) 425c.
7. W. Dutty, Dissertation, University of Freiburg, 1988, unpublished.
8. See: Particle Data Group, Phys. Lett. **170B** (1986) 1, and references therein.
9. M. Kohno and W. Weise, Nucl. Phys. **A479** (1988) 433c, and I.S. Shapiro, these Proceedings.
10. N.H. Hamann, preprint FREI−MEP/88−02 (1988), and Proc. 3rd Conference on the Intersections between Particle and Nuclear Physics, Rockport (Maine), May 1988, to be published.
11. P.D. Barnes et al., Phys. Lett. **199B** (1987) 147.

ANTIPROTON-NUCLEUS INTERACTION

ELASTIC SCATTERING AND ATOMIC BOUND STATES

A.S. Jensen and the Intermediate Energy Theory Group
Institute of Physics, University of Aarhus
DK-8000 Aarhus C, Denmark

INTRODUCTION.

The antiproton-nucleus system is in general a very complicated many-body system. This is evident, since the statement already holds for the nucleus itself. Fortunately the extensive knowledge of the nucleus can be used to reduce the problem considerably. If the necessary assumptions and approximations are directed towards the specific applications of interest, it is often possible to use simple expressions arising from simple concepts. The validity range of the model should, of course, be established. Otherwise erroneous or dubious applications may unknowingly be carried out.

The goal is, of course, a microscopic description of the various phenomena. A very useful intermediate step, when data are available, is a systematic phenomenological analysis of the experimental quantities. We shall therefore first in sect.2 give an overall macroscopic description of the antiproton-nucleus elastic scattering data. Then, in sect.3, we shall develop a microscopic theory and apply it to first order. The second order corrections can be divided into two categories, medium effects treated in sect.4, and correlations discussed in sect.5. Finally, the often sufficiently accurate first order theory is in sect.6 applied to antiproton atomic-like bound states.

The formulation and the material is, apart from the experimental data, almost entirely collected from refs.1-5. Thus, if references seems necessary, but not given, they can presumably be found in these.

DATA AND PHENOMENOLOGICAL ANALYSES.

The data consist of the accurate LEAR measurements[6] of differential elastic scattering cross sections for antiprotons on ^{12}C, ^{16}O, ^{18}O, ^{40}Ca and ^{208}Pb. The projectile momenta are 600 MeV/c and, except for the oxygen isotopes, 300 MeV/c. An example is shown in fig.1, where the typical diffraction pattern is seen.

Antiproton-Nucleon and Antiproton-Nucleus Interactions
Edited by F. Bradamante *et al.*
Plenum Press, New York, 1990

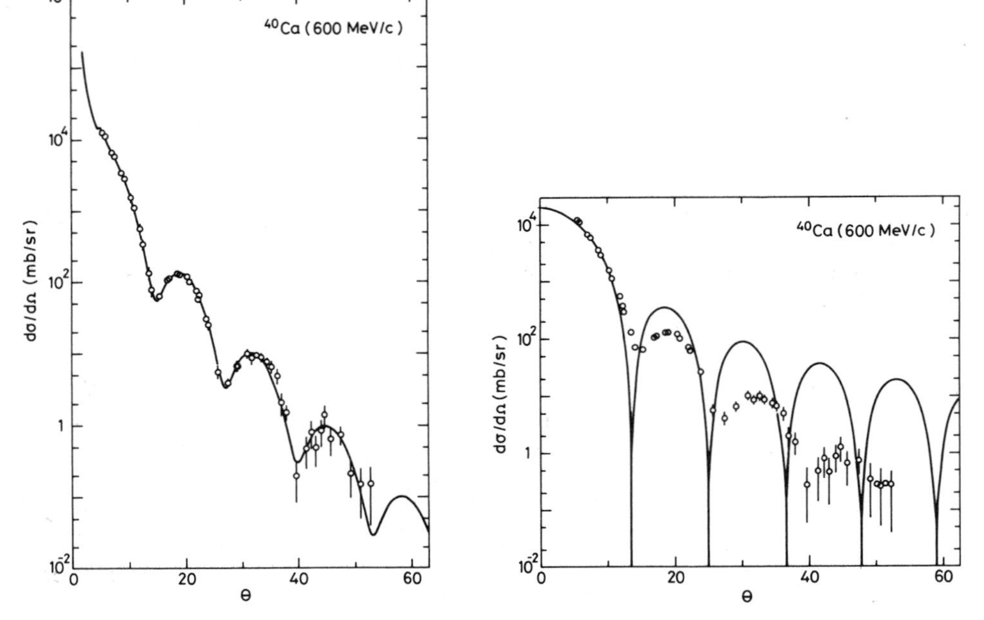

Fig. 1. Differential elastic cross section for 600 MeV/c antiprotons on ^{40}Ca. The solid curve is an optical model fit (left) and the black sphere result (right) is obtained for $R = 1.3$ fm $A^{1/3}$.

Mean Free Path and Annihilation Region

Let us imagine a cylinder situated in a medium of nucleons of density n. If it has an area σ_{tot} (the total antiproton-nucleon cross section) and a length λ (the mean free path of antiprotons in nucleons), the number of nucleons within the cylinder is on the average equal to 1, i.e.

$$\lambda \cdot \sigma_{tot} \cdot n = 1 \tag{2.1}$$

Using the parametrization[7]

$$\sigma_{tot} = [66 + 54000\ \text{MeV/c}/\hbar k_0]\text{mb} \tag{2.2}$$

for the antiproton laboratory momentum $\hbar k_0 = 400$ MeV/c and the nuclear matter density $n = 0.17\ \text{fm}^{-3}$, we find the very small value $\lambda = 0.3$ fm.

Let us now consider a beam of antiprotons moving perpendicular to the nuclear surface. The relative change dN/N of the number of antiprotons over a distance dr is then

$$dN/N = dr/\lambda(r) = \sigma_{tot} n(r) dr \tag{2.3}$$

where the mean free path λ (the decay constant) is a function of r through n(r) in eq.(2.1). Then

$$N(r) = N(\infty)\exp[-\int_r^{\infty}\sigma_{tot}n(r)dr]$$
$$\approx N(\infty)\exp[-n_0\sigma_{tot}e^{(c-r)/d}] \qquad (2.4)$$

where the tail of a Woods-Saxon distribution, with central density, radius and diffuseness parameters n_0, c and d, is used in the last approximation. Half of the antiprotons have disappeared from the incident beam at the distance

$$r-c \approx d\ \ln[dn_0\sigma_{tot}/\ln 2] \qquad (2.5)$$

which is about 0.5 fm. Thus the antiproton-nucleus reaction can be expected to take place in the tail of the nucleus.

Black Sphere Model

The differential elastic scattering cross section is extremely simple in the schematic, but fairly reasonable model where all antiprotons within the nuclear radius R are removed from the elastic channel but the others being left untouched:

$$\frac{d\sigma}{d\Omega} = \frac{1}{4k_0^{\ 2}}\left|\sum_{l=0}^{Rk_0}(2l+1)P_l\ (\cos\theta)\right|^2 \qquad (2.6)$$

The only parameter in this expression is k_0R, which also is fairly well known. An example of this black sphere scattering result is shown in fig.1. The main features are qualitatively, but not quantitatively reproduced.

Optical Model

An obvious idea for constructing an improved model is to introduce an antiproton-nucleus potential $V(\vec{r})$ and solve the Schrödinger equation with proper boundary conditions. The requirement of a Hermitean Hamiltonian immediately leads to probability conservation and zero absorption. Since this is in contradiction with experiments, we simply assume that V has an imaginary, W(r), as well as a real, U(r), part. This is called the optical model.

The interpretaion[8] of W can be understood by using r-independent U and W (the interior of a large nucleus). If the radial dependence of the wave function is $\psi(r,t=0) = \exp(i\vec{k}\cdot\vec{r})$, then the probability distribution is given by $|\psi(r,t)|^2 = \exp(2Wt/\hbar)$. Thus absorption occurs for $W < 0$ with a decay constant $\lambda = -2W/\hbar$.

For a given time dependence, $\exp(-i\varepsilon t/\hbar)$, of particles moving in the x-direction, the spatial wave function $\exp[(ik-\kappa)x]$ is a solution for specific values of k and κ (see ref.8). The probability distribution is now given by $|\psi(r,t)|^2 = \exp(-2\kappa x)$. Thus $\lambda = 1/2\kappa$ may be interpreted as the mean free path.

The short range nature of the strong interaction suggests a parametrization of U and W of similar shape to the nuclear density. Using Woods-Saxon forms we have depth-, radius- and diffuseness parameters for U and W, i.e. a total of 6 parameters. Relativistic corrections are necessary, but the Schrödinger equation is still used.

The data can then be extremely well reproduced as shown by the example in fig.1. In fact, many parameter sets are equally good, but about 1 fm outside the nuclear radius the resulting potentials are all equal. At this point, the imaginary part is roughly -14 MeV and the real part, which is somewhat less accurately determined, is about -4 MeV. The potentials may, apart from this region, differ significantly and still reproduce the cross sections. The real potential may even be repulsive in the center.

The general conclusions are that the cross sections are determined in the nuclear tail and the imaginary part is dominating. With radius and diffuseness parameters similar to those of the nuclear density, the depth parameters are in the intervals $W_0 = -(100 - 200)$ MeV and $V_0 = -(15 - 30)$ MeV.

MICROSCOPIC THEORY.

The system is described by the Hamiltonian

$$H = H_0 + V; \quad H_0 = H_A + K_0; \quad V = \sum_{i=1}^{A} v_{0i} \tag{3.1}$$

where H_A is the nuclear Hamiltonian, K_0 the kinetic energy operator of the antiproton and v_{0i} is the antiproton interaction with nucleon i. The differential cross section is (ignoring spin)

$$\frac{d\sigma}{d\Omega} = |f|^2; \tag{3.2}$$

$$f = -\frac{\mu}{2\pi\hbar^2} \langle \vec{k}' | T_E | \vec{k} \rangle \tag{3.3}$$

where μ is the reduced antiproton-nucleus mass, and the scattering matrix T_E (depending on the total energy E) is given by the operator equation

$$T_E = V + VG_E T_E \tag{3.4}$$

$$G_E = \lim_{\varepsilon \to 0} \frac{1}{E - H_0 + i\varepsilon} \tag{3.5}$$

KMT-formulation[9]

The nuclear wavefunction is antisymmetric with respect to permutations of nucleons. A general nuclear matrix element of the operator V between two nuclear states $|S_1\rangle$ and $|S_2\rangle$ is then

$$\langle S_1 | V | S_2 \rangle = \sum_{i=1}^{A} \langle S_1 | v_{0i} | S_2 \rangle = \sum_{i=1}^{A} \langle S_1 | v_{01} | S_2 \rangle = \langle S_1 | A v_{01} | S_2 \rangle \tag{3.6}$$

Thus, the operator V can everywhere be substituted by Av_{01}, provided only antisymmetric nuclear states are considered.

The "in medium" scattering operator τ_i is defined by

$$\tau_i = v_{0i} + v_{0i}G_E\tau_i \tag{3.7}$$

Using $i = 1$ in eq.(3.7) and the equivalent equation of eq.(3.4),

$$T_E = A(v_{01} + v_{01}G_ET_E) \tag{3.8}$$

we find by elimination of v_{01}

$$T' \equiv U^{(0)}(1 + G_ET') \tag{3.9}$$

$$T' \equiv \frac{(A-1)}{A} T_E \tag{3.10}$$

$$U^{(0)} \equiv (A-1)\tau_1 = \frac{A-1}{A} \sum_{i=1}^{A} \tau_i \tag{3.11}$$

Defining the projection operator P on the nuclear ground state $|0\rangle$

$$P \equiv |0\rangle\langle 0|;\ P^2 = P;\ Q \equiv 1-P \tag{3.12}$$

and inserting $P + Q = 1$ between G_E and T' in eq.(3.9), the equation can be rewritten as

$$T' = U\,(1 + G_EPT') \tag{3.13}$$

$$U \equiv U^{(0)} + U^{(0)}G_EQU \tag{3.14}$$

We then approximate U by $U^{(0)}$ implying that $U^{(0)}G_EQU$ is neglected.

Projecting eq.(3.13) on the nuclear ground state ($T_{00} \equiv PT'P$) and using eq.(3.12) we then get

$$T_{00} = U_{00}(1 + G_{00}T_{00}) \tag{3.15}$$

$$U_{00} \equiv PUP;\ G_{00} \equiv PG_EP \tag{3.16}$$

Eq. (3.15) is equivalent to solving the Schrödinger equation

$$[-\frac{\hbar^2\vec{\nabla}^2}{2\mu} + \langle 0|U_{00}|0\rangle]\Psi = E\Psi \tag{3.17}$$

where we have approximated the two-particle potential U_{00} by $U^{(0)}$ given in eq.(3.11).

The "free" scattering operator t_i is defined by

$$t_{0i} = v_{0i} + v_{0i}G_{E0}t_{0i} \tag{3.18}$$

$$G_{EO} = \lim_{\varepsilon\to 0} \frac{1}{E-K_0-K_i+i\varepsilon} \tag{3.19}$$

By elimination of v_{0i} from eqs.(3.7) and (3.18) we then obtain

$$\tau_i = t_{0i} + t_{0i}(G_E - G_{E0})\tau_i \tag{3.20}$$

The approximation $\tau_i = t_{0i}$ (neglecting $t_{0i}(G_E-G_{E0})\tau_i$) then defines the first order antiproton-nucleus optical potential $U_{00}^{(1)}$ (see eqs.(3.16), (3.14) and (3.11)) as

$$\langle 0|U_{00}|0\rangle = \frac{A-1}{A} \langle 0| \sum_{i=1}^{A} t_{0i}|0\rangle \equiv U_{00}^{(1)} \tag{3.21}$$

which then should be used in eq.(3.17).

Two-particle t-matrix

The antiproton is indicated by index 0 and the nucleons by i =1,2,....,A. The initial and final particle momenta are $(\vec{k}_0,\vec{k}_1)$ and $(\vec{k}'_0,\vec{k}'_1)$ and the related relative $(\vec{k})$, transfer $(\vec{q})$ and total momenta $(\vec{K})$ are given by

$$2\vec{k} \equiv \vec{k}_0 - \vec{k}_1; \quad 2\vec{k}' \equiv \vec{k}_0{}' - \vec{k}_1{}' \tag{3.22}$$

$$\vec{q} \equiv \vec{k} - \vec{k}' = \vec{k}_0 - \vec{k}_0{}'; \quad |\vec{k}| = |\vec{k}'| \tag{3.23}$$

$$\vec{K} \equiv \vec{k}_0 + \vec{k}_1 = \vec{k}_0{}' + \vec{k}_1{}' \equiv \vec{K}' \tag{3.24}$$

The plane wave matrix elements of the scattering operator are

$$\langle \vec{k}_0{}'\vec{k}_1{}'|t_{01}|\vec{k}_0\vec{k}_1\rangle = \langle \vec{k}'|t_{01}^{(0)}|\vec{k}\rangle\delta(\vec{K} - \vec{K}')/(2\pi)^3 \tag{3.25}$$

From general symmetry principles[10] we have the 5 possible terms

$$t(k^2,q^2) \equiv \langle \vec{k}'|t_{01}^{(0)}|\vec{k}\rangle = t_c + t_{ss}\vec{\sigma}_0\cdot\vec{\sigma}_1 + t_{so}\vec{n}\cdot(\vec{\sigma}_0 + \vec{\sigma}_1)$$

$$+ \; t_T(3\hat{q}\cdot\vec{\sigma}_0\hat{q}\cdot\vec{\sigma}_1 - \vec{\sigma}_0\cdot\vec{\sigma}_1) + t_{so2}(3\hat{n}\cdot\vec{\sigma}_0\hat{n}\cdot\vec{\sigma}_1 - \vec{\sigma}_0\cdot\vec{\sigma}_1) \tag{3.26}$$

$$\vec{n} = \vec{k} \times \vec{k}'; \quad \hat{n} \equiv \vec{n}/|\vec{n}|; \quad \hat{q} = \vec{q}/|\vec{q}| \tag{3.27}$$

For each isospin I we parametrize t_x (x = c,ss,so,T,R) by

$$t_x^I = \sum_i \frac{1}{m_i^2+q^2} \; \frac{a_x^{iI}+k^2 b_x^{iI}}{1+k^2/5\text{fm}^{-2}} \tag{3.28}$$

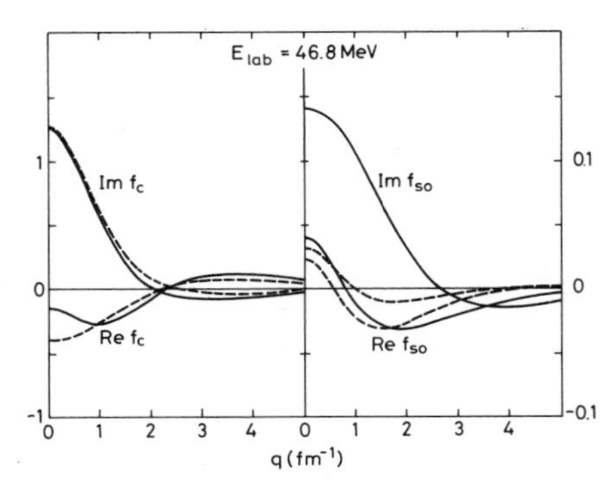

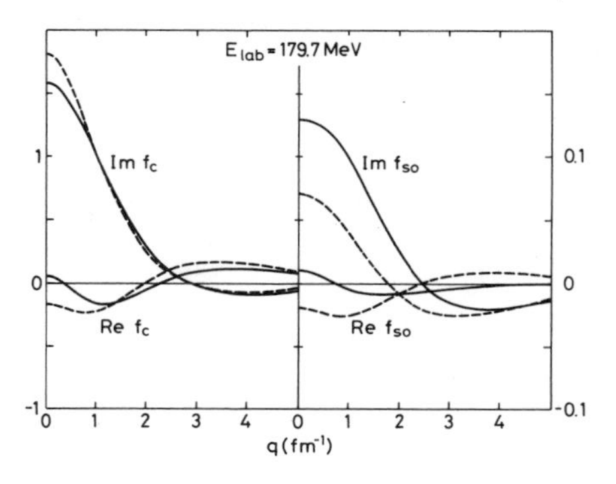

Fig. 2. Antiproton-nucleon scattering amplitudes for isospin 1 for two energies. The central and spin orbit amplitudes are in fm and fm^3, respectively. The full curve correspond to the Paris model and the dotted to the Dover-Richard interaction.

$$t^I_{SO2} = \frac{1}{2}(1 + \cos\theta)t^I_T + k^2\sin^2\theta t^I_R \qquad (3.29)$$

where θ is the scattering angle (between $\vec{k}$ and $\vec{k}'$) and eq. (3.29) ensures that $t_{SO2}(\theta=0) = t_T(\theta=0)$ as required by general symmetry principles[10].

From given isospin we can find t-matrix elements for neutrons, protons and charge exchange by

$$t^P_x = \frac{1}{2}(t^0_x + t^1_x); \qquad t^n_x = t^1_x; \qquad t^{CE}_x = \frac{1}{2}(t^0_x - t^1_x) \quad (3.30)$$

We choose about four values of m_i and the a's and b's are then obtained by fitting eqs.(3.28), in the energy interval of interest, to the appropriate interaction, which in our case is either the Dover-Richard or the Paris interaction (see fig. 2). They mostly differ in the spin dependent parts.

First Order Potential

The normalized Yukawa function ($\hbar$=c=1)

$$Y(r) = \frac{m^2}{4\pi r}\exp(-mr) \qquad (3.31)$$

has plane wave ($e^{i\vec{k}\cdot\vec{r}}/(2\pi)^{1.5}$)matrix elements

$$<\vec{k}_0'\vec{k}_1'|Y(|\vec{r}_0 - \vec{r}_1|)|\vec{k}_0\vec{k}_1> = \frac{m^2}{m^2+q^2}\,\frac{\delta(\vec{K}-\vec{K}')}{(2\pi)^3} \qquad (3.32)$$

This is easily seen after the change of variable:

$$\vec{r} \equiv \vec{r}_0 - \vec{r}_1; \qquad 2\vec{R} = \vec{r}_0 + \vec{r}_1 \qquad (3.33)$$

Thus for k^2 = constant, the coordinate-space central part of the t-matrix element is (see eqs.(3.26) and (3.28))

$$t_{01}^{(0)}(|\vec{r}_0 - \vec{r}_1|) = \Sigma d_i^c Y_i(|\vec{r}_0 - \vec{r}_1|) \tag{3.34}$$

$$d_i^c \equiv \frac{a_c^i + k^2 b_c^i}{1+k^2/5fm^{-2}} \cdot \frac{1}{m_i^2} \tag{3.35}$$

and the first order optical potential eq.(3.21), arising from the central part of t is now for neutrons and protons separately

$$U_{00}^{(1)}(\vec{r}_0) = \frac{A-1}{A} \sum_i d_i^c \int n(\vec{r}_1) Y_i(|\vec{r}_0 - \vec{r}_1|) d^3\vec{r}_1 \tag{3.36}$$

where $n(\vec{r}) = \Sigma|\varphi_k(\vec{r})|^2$ is the nuclear density.

Let us specialize to even-even spin zero nuclei. All t-matrix elements linear in nucleon spin $\vec{\sigma}_1$ vanish after averaging over all nucleons. Beside t_c, only the spin-orbit term therefore contributes. The plane-wave matrix element of $\vec{\nabla}_r Y(r) \times \vec{\nabla}_r \cdot \vec{\sigma}_0$ is

$$-\frac{\delta(\vec{K}-\vec{K}')}{(2\pi)^3} \frac{m^2}{m^2+q^2} \vec{k}' \times \vec{k} \cdot \vec{\sigma}_0 \tag{3.37}$$

and the coordinate-space spin-orbit part of t (see eq.(3.26)) is

$$t_{01}^{(0)}(\vec{r}) = \Sigma d_i^{so} \vec{\nabla}_r Y_i(r) \times \vec{\nabla}_r \cdot \vec{\sigma}_0 \tag{3.38}$$

The first order optical potential arising from t_{so} is then (see eq.(3.21))

$$\begin{aligned} U_{00}^{(1)}(\vec{r}_0) &= \frac{A-1}{A} \Sigma d_i^{so} \int d^3\vec{r}_1 Y_i(|\vec{r}_0 - \vec{r}_1|) \\ &\frac{1}{2} [\vec{\nabla}_{r_1} n(\vec{r}_1) \times \vec{\nabla}_{r_0} - \frac{1}{2}\vec{\nabla}_{r_1} \times \vec{j}(\vec{r}_1)] \cdot \vec{\sigma}_0 \end{aligned} \tag{3.39}$$

where the current density is

$$\vec{j}(\vec{r}_1) \equiv \sum_k [\varphi_k^*(\vec{r}_1) \vec{\nabla}_{r_1} \varphi_k(\vec{r}_1) - \varphi_k(\vec{r}_1) \vec{\nabla}_{r_1} \varphi_k^*(\vec{r}_1)] \tag{3.40}$$

and we used

$$\begin{aligned} &\vec{\nabla}_r \times \vec{\nabla}_r f(\vec{r}) = 0 \\ &\vec{\nabla}_r f(\vec{r}_0, \vec{r}_1) = \frac{1}{2}[\vec{\nabla}_{r_0} - \vec{\nabla}_{r_1}] f(\vec{r}_0, \vec{r}_1) \\ &\vec{\nabla}_{r_0} Y(|\vec{r}_0 - \vec{r}_1|) = - \vec{\nabla}_{r_1} Y(|\vec{r}_0 - \vec{r}_1|) \end{aligned} \tag{3.41}$$

Finally, partial integration, an assumption about spherical

symmetry ($r_0\vec\nabla_{r_0} = \vec r_0 \partial/\partial r_0$), and the substitution $\vec r_0 \times \vec\nabla_{r_0} \to i\vec l_0$ leads to

$$U^{(1)}_{00}(\vec r_0) = -\frac{i}{2}\frac{A-1}{A}\vec l_0\cdot\vec\sigma_0 \sum_i d_i^{so}\frac{1}{r_0}\frac{d}{dr_0}\int n(\vec r_1)Y(|\vec r_0-\vec r_1|)d^3\vec r_1 \tag{3.42}$$

for neutrons and protons separately. The last term in eq.(3.39) vanishes in our case.

Value of "k^2"

It is tempting to use an average value over all nucleons in the nucleus. For a spherically symmetric Fermi gas, we get (see eq.(3.22))

$$\langle k^2\rangle = \frac{1}{4}\langle(\vec k_0-\vec k_1)^2\rangle = \frac{1}{4}(k_0{}^2+\frac{3}{5}k_F^2) \tag{3.43}$$

where k_F is the Fermi momentum.

Let us assume $t_c = a + k^2 b$ and substitute $f_k(r_1)$ for k^2 in the resulting potential $V_k(r_0)$ from eq.(3.36)

$$V_k(r_0) = \int d^3\vec r_1\, n(\vec r_1)[a(|\vec r_0-\vec r_1|) + f_k(r_1)b(|\vec r_0-\vec r_1|)] \tag{3.44}$$

We can instead substitute

$$k^2 b \to -\frac{1}{2}[\vec\nabla_r^2 b(r) + b(r)\vec\nabla_r^2] \tag{3.45}$$

in the potential of eq.(3.21). A transformation of the resulting potential leads[3] to a Schrödinger equation with an energy dependent effective potential $V_E(r_0)$. Assuming $a(r) = a_0\delta(r)$, $b(r) = b_0\delta(r)$ and $V_k(r_0) = V_E(r_0)$ we find ($E = 1/2(\hbar^2 k_0{}^2/m)$)

$$\frac{\hbar^2}{2m}f_k(r_0) = \frac{1}{4(1+h)}[E-a_0 n + \frac{\hbar^2}{8m}(\tau-\vec\nabla^2 n)/n - b_0(\vec\nabla n)^2/(64n(1+h)] \tag{3.46}$$

where m is the nucleon mass and

$$\frac{\hbar^2}{2m}h = \frac{1}{4}b_0 n; \qquad \tau = \sum_k|\vec\nabla\varphi_k|^2 \tag{3.47}$$

In the nuclear center ($n = n_0$) for a Fermi gas,

$$\frac{\hbar^2}{2m}f_k(r<<R) = \frac{1}{4}(E - a_0 n_0 + \frac{3}{5}\frac{\hbar^2 k_F^2}{2m}) \tag{3.48}$$

and in the nuclear tail for a Woods-Saxon distribution for n

$$\frac{\hbar^2}{2m}f_k(r>>R) = \frac{1}{4}(E - \frac{\hbar^2}{2md^2}) \tag{3.49}$$

Thus, the antiproton potential ($a_0 n_0$) has to be subtracted from the energy (compare eqs.(3.43) and (3.48)). For large r we should apparently subtract $\hbar^2/(2md^2) \approx 68$ MeV. A better estimate may be found by using the tail behaviour of the single particle wavefunction[1]

$$\varphi_k(r_1) \sim \frac{1}{r_1} \exp[-r_1(2m|\varepsilon_k|/\hbar^2)^{1/2}] \quad (3.50)$$

The highest lying state ($\varepsilon_k \approx -8$ MeV) is then the main contributor in the nuclear tail and eq.(3.46) then gives

$$\frac{\hbar^2}{2m} f_k(r<<R) = \frac{1}{4}(E - 3|\varepsilon_{k=highest}|) \quad (3.51)$$

which results in a more modest reduction of about 24 MeV.

Nuclear Density

Electron scattering experiments give the charge density distribution. We then use the droplet[11] skin-thickness to find the neutron distribution. Since we treat the nucleons as point particles, we must correct for the finite size. Using the relation

$$n(\vec{r}) = \int n'(\vec{r}-\vec{r}')F(\vec{r}')d^3\vec{r}' \quad (3.52)$$

between density, point density (n') and nucleon form factor (F), we find

$$\langle r^2\rangle_{n'} = \langle r^2\rangle_n - \langle r^2\rangle_{nucleon} \quad (3.53)$$

which from the scaling assumption $n'(\vec{r}) = n(\vec{r}/s)/s^3$ gives

$$s = (1 - \langle r^2\rangle_{nucleon}/\langle r^2\rangle_n)^{1/2} \quad (3.54)$$

With $\langle r^2\rangle_{nucleon} = 0.68$ fm^2 we then have $n'(\vec{r})$ in the same form as $n(\vec{r})$. Here eq.(3.53) is essential and the detailed implementation is much less important.

Relativistic Corrections

The Schrödinger equation is the foundation, but two relativistic corrections are necessary. First in the transformation from antiproton-nucleon to antiproton-nucleus center of mass system. Secondly kinematical factors are needed on kinetic, potential and total energy in the Schrödinger equation.

MEDIUM EFFECTS

The second order expansion of τ in eq.(3.20) in terms of t is

$$\tau = t + t(G_E - G_{E0})t \quad (4.1)$$

where the diagonal plane-wave matrix elements of the propagators are

$$\langle \vec{k}_0\vec{k}_1|G_{E0}|\vec{k}_0\vec{k}_1\rangle = (E_0 - \frac{\hbar^2 k_0^{\ 2}}{2m} - \frac{\hbar^2 k_1^{\ 2}}{2m} + i\varepsilon)^{-1} \quad (4.2)$$

$$\langle \vec{k}_0 \vec{k}_1 | G_E | \vec{k}_0 \vec{k}_1 \rangle = \frac{\Theta(k_1 - k_F)}{E - \bar{\varepsilon}(\vec{k}_0) - \varepsilon(\vec{k}_1) + i\varepsilon} \tag{4.3}$$

The Pauli principle is included via the step function $\Theta(x)=1$ for $x>1$ and zero otherwise. The energies are given as

$$\bar{\varepsilon}(\vec{k}_0) = \frac{\hbar^2 k_0^{\ 2}}{2m} + V_{\bar{p}}(\vec{k}_0) \tag{4.4}$$

$$\varepsilon(\vec{k}_1) = \frac{\hbar^2 k_1^2}{2m} + V_N(\vec{k}_1) \tag{4.5}$$

$$V_{\bar{p}}(\vec{k}_0) = \int \langle \vec{k}_0 \vec{k}_1 | \tau | \vec{k}_0 \vec{k}_1 \rangle \, \Theta(k_F - k_1) d^3 \vec{k}_1 \tag{4.6}$$

We take V_N from realistic Skyrme interactions for bound nucleons and $V_N = 0$ for nucleons above the Fermi energy. Since the potentials vanish in the nuclear tail, the "in medium" propagator G_E approaches G_{E0} in the region where the antiproton scattering takes place. The first order ($\tau = t$) is our KMT result and the second order term in eq.(4.1) is therefore sufficient for our purpose. Consistently we can then use t in eq.(4.6) instead of τ which reduces the problem to perturbation theory instead of the difficult self-consistency problem.

The advantages are (i) no selfconsistency, (ii) the t-matrix is directly related to the two-particle data, (iii) all partial waves can be included, (iv) also quadratic spin-orbit can be included and (v) the method is simple and therefore transparent.

Method

The matrix elements $\Delta\tau$ of the correction to τ is first calculated in nuclear matter

$$\Delta\tau \equiv \langle \vec{k}' \vec{K} | t(G_E - G_{E0}) t | \vec{k} \vec{K} \rangle \tag{4.7}$$

Complete intermediate sets of $\vec{k}'', \vec{K}'' = \vec{K}$ are inserted between the operators and the t-matrix elements are calculated "half off-shell" for $k^2 = k'^2$ and

$$E_0 = \frac{\hbar^2 k_0^{\ 2}}{2m} + \frac{\hbar^2 k_1^{\ 2}}{2m}; \qquad E = E_0 + \bar{V}_N(k_1) \tag{4.8}$$

The intermediate states are for neutrons, $\bar{p}n$, and for protons, $\bar{p}p$ and $\bar{n}n$. Using the relation

$$\frac{\hbar^2}{2m}(k_0^{\ 2} + k_1^{\ 2}) = \frac{\hbar^2}{m}(k^2 + \tfrac{1}{4}K^2) \tag{4.9}$$

we then obtain

$$\Delta\tau = \int d^3\vec{k}''t(k^2,|\vec{k}''-\vec{k}'|^2)t(k^2,|\vec{k}-\vec{k}''|^2)$$

$$\left[\frac{Q(k'',K,k_F)}{\frac{\hbar^2}{m}(k^2-k''^2)+V_N(\vec{k}_1)-V_{\bar{p}}(\tilde{k}_0'')+i\varepsilon} - \frac{1}{\frac{\hbar^2}{m}(k^2-k''^2)+i\varepsilon}\right] \qquad (4.10)$$

where Q is the angle (between $\vec{k}''$ and $\vec{K}$) averaged value of Θ

$$Q(k'',K,k_F) = \langle\Theta(k_1''-k_F)\rangle \qquad (4.11)$$

and an analogous approximation is used for k_0'':

$$V_{\bar{p}}(\vec{k}''_0) \approx V_{\bar{p}}(\langle|\vec{k}'' + \tfrac{1}{2}\vec{K}|\rangle) \equiv V_{\bar{p}}(\tilde{k}_0'') \qquad (4.12)$$

Finally we average over all occupied nucleon momenta

$$\overline{\Delta\tau} \equiv \frac{3}{4\pi k_F^3}\int \Delta\tau\ \Theta(k_F-k_1)d^3\vec{k}_1 \qquad (4.13)$$

This function, which only depends on k_0, k_F and q, is now Fourier transformed with respect to $\vec{q}$ for fixed values of $k_F(R)=[3\pi^2 n(\vec{r}_0+\vec{r}_1)/2]^{1/3}$. The result, $\Delta\tau(r,k_F(R),k_0)$, is then integrated over the finite nucleus' density distribution to give the correction to the optical potential

$$\Delta V_{opt}(r_0,k_0) = \int\tilde{\Delta}\tau(|\vec{r}_0-\vec{r}_1|,k_F(R),k_0)n(\vec{r}_1)d^3\vec{r}_1 \qquad (4.14)$$

Numerical Results

The optical potentials for ^{208}Pb are shown in fig. 3 for the Paris interaction. The central potentials are flat in the center whereas the much smaller spin-orbit potentials peak at the surface. The second order correction to the central potential decreases from about 30% in the center towards zero with increasing r. The spin-orbit second order correction is also about 30% at its maximum. These corrections are small at 8 fm which is the crucial region for elastic scattering on ^{208}Pb.

The differential cross sections derived from these optical potentials are also shown in fig. 3. The experimental values are reproduced very well, but the medium corrections can hardly be seen in the calculated curves for $d\sigma/d\Omega$. The spin-dependent cross sections P (polarization) and Q (spin-rotation function) are also shown. They vanish for zero spin-orbit potential and fluctuate as function of angle.

In fig. 4, we show the cross sections for ^{12}C for two energies. The number of diffraction minima is seen to increase with energy. The experimental values are reproduced reasonably well. The medium corrections are still small, but they significantly improve the agreement at larger angles. The calculated polarization is too large for this interaction and unfortunately not measured across the first diffraction minimum where violent changes are predicted.

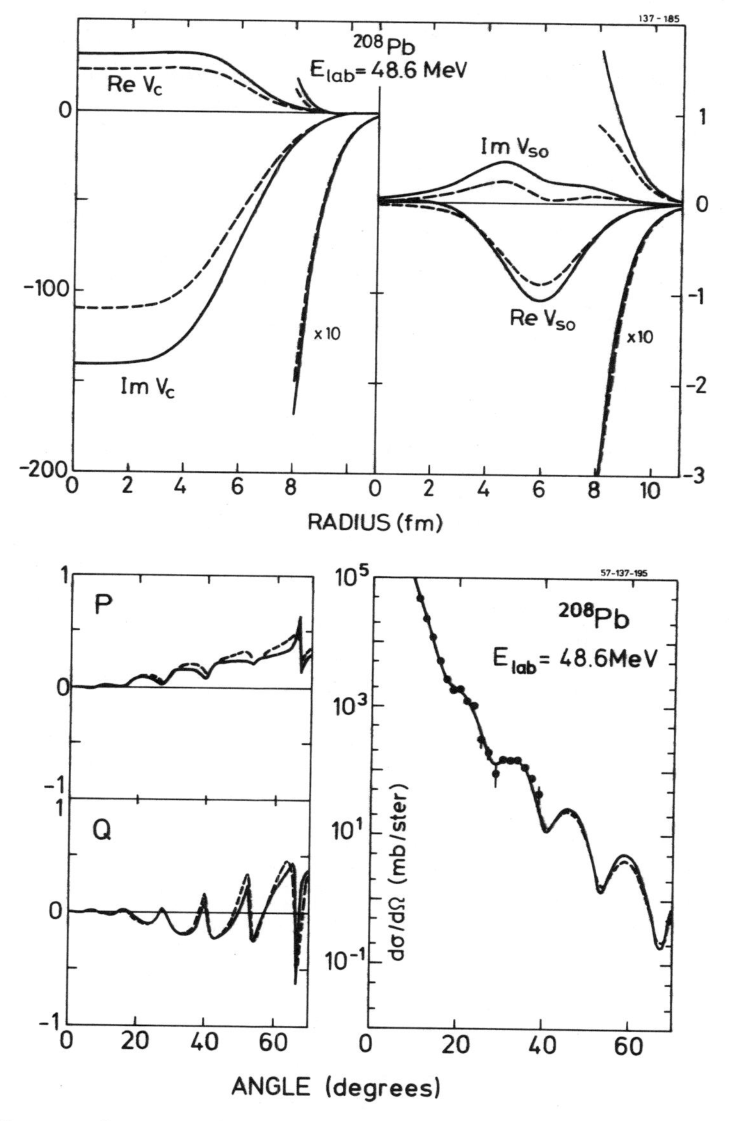

Fig. 3. The central and spin-orbit optical potentials in MeV for ^{208}Pb at 48.6 MeV. The curves show the real and imaginary parts with (full curve) and without (dashed curve) medium effects calculated with the Paris t-matrix elements. The tails are blown up by a factor of 10. The elastic differential cross section (dσ/dΩ), polarization (P) and spin rotation function (Q) with (full curve) and without (dashed curve) medium effects are also shown.

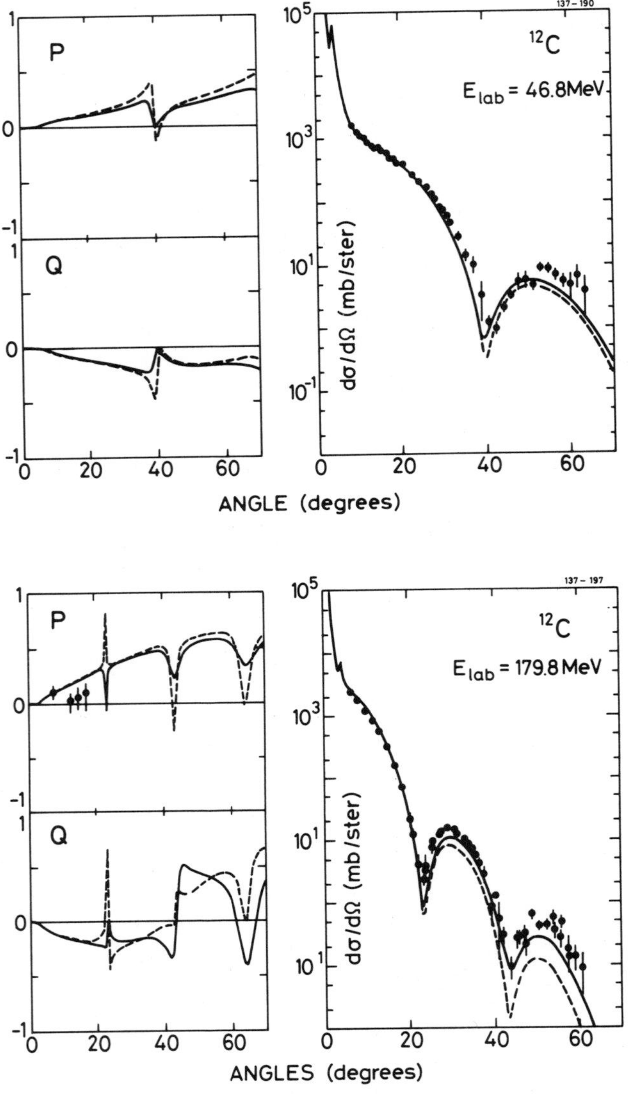

Fig. 4. Scattering cross sections for ^{12}C. (See fig. 3 for notation).

The remaining measured cross sections are all equally well reproduced by the theory. The medium effects always improve the agreement slightly, but significantly, at large angles (least for Pb at low energy). The polarization is more sensitive to the specific interaction (see fig. 2). Thus, more data for $d\sigma/d\Omega$ at even larger angles, and for the polarization, would be interesting for these reasons.

Analytical Approximations

The simplest possible assumptions are that the antiproton and the nucleons are at rest, $\vec{k}_a = \vec{k}_1 = 0$. Then the relative and total momenta are zero ($\vec{K}=\vec{k}=\vec{K}'=\vec{K}''=0$) and the Pauli operator simplifies from $\Theta(k''_1 - k_F)$ to $\Theta(k''-k_F)$. The scattering length approximation, $q^2=0$, in the t-matrix, i.e.

$$t(k^2,q^2) \approx t(0,0) = -\frac{\hbar^2 a}{2m\pi^2} \qquad (4.15)$$

reduces eq.(4.10) to

$$\Delta\tau = \frac{\hbar^2 a^2}{4m\pi^4}\int\left[\frac{1}{k''^2} - \frac{\Theta(k''-k_F)}{k''^2+\varkappa^2}\right]d^3\vec{k}'' = \frac{\hbar^2 a^2}{2m\pi^3}\left[2k + i\varkappa\ln\left(\frac{k_F-i\varkappa}{k_F+i\varkappa}\right)\right] \qquad (4.16)$$

$$\varkappa^2 \equiv \frac{m}{\hbar^2}(V_{\bar{p}} - V_N) \qquad (4.17)$$

To get an idea of the relative importance of Pauli- and binding energy effects, we (i) neglect Pauli effects by omitting Θ in eq.(4.16), i.e.

$$\Delta\tau_B = \frac{\hbar^2 a^2}{2m\pi^2}\varkappa; \qquad (\varkappa \equiv \varkappa_R + i\varkappa_I;\ \varkappa_R > 0) \qquad (4.18)$$

and (ii) neglect antiproton and nucleon binding energies ($\varkappa=0$), i.e.

$$\Delta\tau_P = \frac{\hbar^2 a^2}{m\pi^3}k_F \qquad (4.19)$$

The local density approximation leading from nuclear matter to finite nuclei consists of the substitutions

$$k_F \rightarrow k_F(r) = k_F(0)[n(r)/n(0)]^{1/3} \qquad (4.20)$$

$$\varkappa \rightarrow \varkappa(r) = \varkappa(0)[n(r)/n(0)]^{1/2} \qquad (4.21)$$

where we assumed that the potentials are proportional to $n(r)$. Since $\varkappa(r)/k_F(r)$ approaches zero for large r, the Pauli effect will dominate in the (perhaps extreme) nuclear tail and $\Delta\tau$ approaches $\Delta\tau_P$ in this region (see fig.5).

The approximantion in eq.(4.16) is an upper limit, because $t(k^2,q^2)$ is a decreasing function of q^2 and (at least for higher energies) also of k^2.

Improved Approximations

Let us consider the central part of t parametrized as

$$t_c(k^2,q^2) = \frac{m_c^2 t_c(k^2,0)}{m_c^2+q^2} \qquad (4.22)$$

The correction to the t-matrix from this term can be written

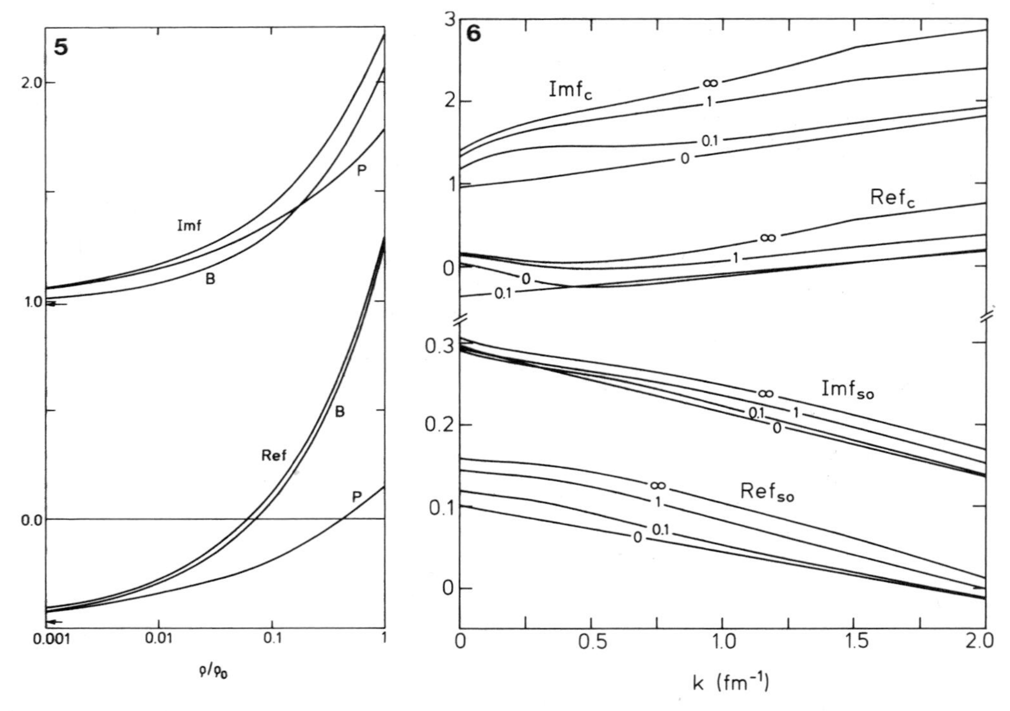

Fig. 5. The isospin averaged scattering amplitudes in fm as function of nuclear density ϱ in units of the saturation density $\varrho_0 = 0{,}17\ \text{fm}^{-3}$. The arrows indicate the values of the first order term which is added to the estimates of eqs.(4.17), (4.18) (labelled B) and (4.19) (labelled P). (left hand side)

Fig. 6. The total isospin averaged central scattering amplitudes (in fm), first order plus eq.(4.26), for various density ratios n/n_0 indicated on the curves. The strengths and masses are functions of k adjusted to reproduce the two-particle t-matrix. (right hand side)

$$\Delta\tau_c = \int_0^\infty k''^2 dk'' \int_{-1}^{1} d\mu \delta G \, Y_c^2 \tag{4.23}$$

where in forward direction $\vec{k}=\vec{k}'$

$$\delta G = \frac{2\pi m}{\hbar^2} \left[\frac{Q}{k^2-k''^2-\varkappa^2+i\varepsilon} - \frac{1}{k^2-k''^2+i\varepsilon}\right] \tag{4.24}$$

$$Y_c = \frac{m_c^2 t_c(k^2,0)}{m_c^2+k^2+k''^2-2kk''\mu} \tag{4.25}$$

By neglecting the Pauli effects (Q=1), we can find $\Delta\tau_c$ explicitly

$$\Delta\tau_c = \frac{m\pi^2}{\hbar^2} m_c t_c^2(k^2,0)\left\{\frac{1+2ik/m_c}{1+4k^2/m_c^2} - \frac{1+\kappa^2/m_c^2+2i\sqrt{k^2-\kappa^2}/m_c}{(1-\kappa^2/m_c^2)^2+4k^2/m_c^2}\right\} \tag{4.26}$$

which is shown in fig.6 as function of k for various nuclear densities.

CORRELATIONS

The second order term, $U^{(2)}$, neglected in the potential, see eq.(3.14), is

$$U^{(2)} = U^{(0)} G_E Q U^{(0)} \tag{5.1}$$

Projecting on the nuclear ground state and inserting complete sets between the operators we obtain[12] the plane-wave matrix element

$$U_{00}^{(2)}(\vec{k},\vec{q}) = \frac{A^2 m}{\hbar^2}\int \frac{t(k^2,|\vec{k}'-\vec{k}''|^2)d^3\vec{k}''}{k^2-k''^2+i\varepsilon} C(\vec{k}'-\vec{k}'',\vec{k}''-\vec{k})t(k^2,|\vec{k}''-\vec{k}|^2) \tag{5.2}$$

where we neglect 1/A terms and the binding potentials in the propagator, and where the correlation function[12] C is defined by

$$C(\vec{q}_1,\vec{q}_2) \equiv \varrho(\vec{q}_1,\vec{q}_2) - \varrho(\vec{q}_1)\,\varrho(\vec{q}_2) \tag{5.3}$$

$$\varrho(\vec{q}_1,\vec{q}_2) \equiv \langle 0|\exp[i(\vec{q}_1\cdot\vec{r}_1 + \vec{q}_2\cdot\vec{r}_2)]|0\rangle \tag{5.4}$$

$$\varrho(\vec{q}_1) \equiv \langle 0|\exp[i\vec{q}_1\cdot\vec{r}_1]|0\rangle \tag{5.5}$$

Changing integration variable in eq.(5.2) to $\vec{q}_1 \equiv \vec{k}''-\vec{k}$ ($2\vec{k}=\vec{k}_0-\vec{k}_1$), we get

$$U_{00}^{(2)} = -\frac{A^2 m}{\hbar^2}\int \frac{t(k^2,|\vec{q}-\vec{q}_1|^2)d^3\vec{q}_1}{q_1^{\,2}+\vec{q}_1\cdot(\vec{k}_0-\vec{k}_1)-i\varepsilon} C(\vec{q}-\vec{q}_1,\vec{q}_1)\ t(k^2,q^2) \tag{5.6}$$

Integrating (averaging) over $\vec{k}_1$ as usual, and using the low density case where $k_F \to 0$ in the local density approximation, is equivalent to omitting $\vec{k}_1$ in eq.(5.6). This is clear, since $k_1 < k_F$, but the proper scale for k_F depends on the specific case, and consequently it is harder to extract in general[1].

Short Range and Pauli Correlations

The correlation function $\tilde{C}$ in coordinate space for short range and Pauli correlations can be parametrized by [12]

$$\tilde{C}(\vec{x},\vec{y}) = \tilde{f}_c(|\vec{x}-\vec{y}|)\tilde{\varrho}(x)\tilde{\varrho}(y) \tag{5.7}$$

$$\tilde{f}_c(r) = (1-\frac{1}{3}\frac{r^2}{l_0{}^2})\exp[-\frac{1}{2}r^2/l_0{}^2] \tag{5.8}$$

$$f_c(q) = -\frac{1}{3}(2\pi)^{3/2}l_0{}^5q^2\exp[-\frac{1}{2}l_0{}^2q^2] \tag{5.9}$$

where "~" means Fourier transform of the corresponding momentum space function.

Assuming l_0 to be small enough to expand to lowest order, x=y, we have from eqs.(5.6)-(5.9)

$$U^{(2)}_{00}(\vec{q}) = -A^2\frac{m}{\hbar^2}[\int\varrho^2(x)\exp(i\vec{q}\cdot\vec{x})d^3\vec{x}]$$

$$\int\frac{t(k,^2|\vec{q}-\vec{q}|^2)f_c(\vec{q}_1)t(k,^2q_1{}^2)}{q_1{}^2+\vec{k}_0\cdot\vec{q}_1 - i\varepsilon}d^3\vec{q}_1 \tag{5.10}$$

By further parametrizing the t-matrix elements

$$t(k^2,q^2) = t_0\exp[-\frac{1}{2}\alpha q^2] \tag{5.11}$$

we obtain

$$U^{(2)}_{00}(q) = i\frac{mA^2}{\hbar^2}(2\pi)^3(\frac{l_0}{l})^5\frac{1}{k_0}g(k_0l)$$

$$t(k^2,\frac{1}{4}q^2)[\int\varrho^2(x)\exp(i\vec{q}\cdot\vec{x})d^3\vec{x}] \tag{5.12}$$

$$l^2 = 2\alpha + l_0^2 \tag{5.13}$$

$$g(y) \equiv \int_0^\infty dx(1-\frac{1}{3}x^2)(1-\exp(ixy))\exp(-\frac{1}{2}x^2) \tag{5.14}$$

and the corresponding second order correction to the optical potential

$$V^{(2)}_{opt}(r) = i\frac{m}{\hbar^2}\frac{l_0{}^5}{l^3}\frac{g(k_0l)}{k_0l}t_0\int d^3\vec{r}'\varrho^2(r')\tilde{t}(k^2,|\vec{r}-\vec{r}'|) \tag{5.15}$$

The result is shown in fig.7. Even in the center, where the assumptions might be violated, the correlation is still very small.

Center of Mass Correlation

The corresponding correlation function can be parametrized[12]

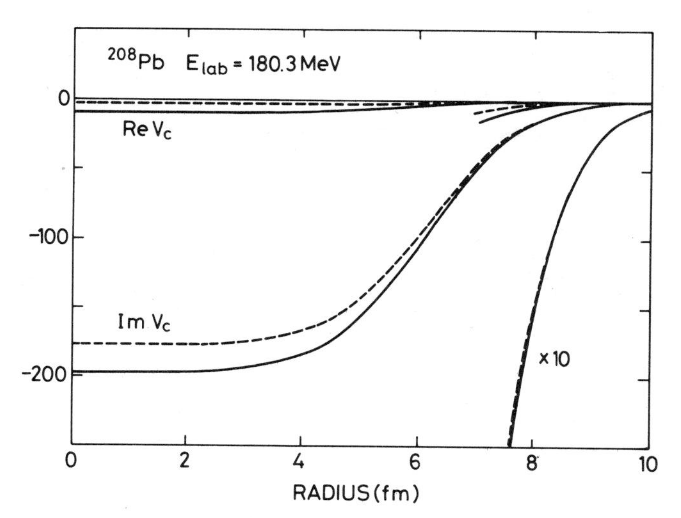

Fig. 7. The first order optical potential in MeV for ^{208}Pb at 180.3 MeV (dashed) is compared to the potentials where the correlation effects have been included (full). The tails are blown up by a factor of 10.

$$C(\tfrac{1}{2}\vec{Q}+\vec{q},\ \tfrac{1}{2}\vec{Q}-\vec{q}) = \frac{\beta}{A}(\tfrac{1}{4}Q^2-q^2)\exp[-\beta(\tfrac{1}{4}Q^2+q^2)] \tag{5.16}$$

Using the form $\varrho(\vec{q}) = \exp[-\frac{1}{2}\beta q^2]$ and eqs.(5.11) and (5.16) in (5.6) we get

$$U^{(2)}_{00}(q) = 3\frac{mA\beta}{\hbar^2}t^2(k^2,\tfrac{1}{4}q^2)\varrho^2(\tfrac{1}{2}\vec{q})(2\pi)^{1.5}g(k_0\lambda)/(ik_0\lambda^4) \tag{5.17}$$

with $\lambda = 2(\alpha+\beta)$. Comparing to eq.(5.12) we find

$$\frac{[U^{(2)}_{00}(q)]_{C.M}}{[U^{(2)}_{00}(q)]_{S.R}} = \frac{3\sqrt{8}}{A}\left(\frac{\sqrt{\beta}}{l_0}\right)^5\left(\frac{1}{\lambda}\right)^4\frac{g(k_0\lambda)}{g(k_0 l)} \tag{5.18}$$

which for the reasonable values $l_0 = 0.88$ fm, $\alpha = 0.8$ fm^2, $\beta = 0.8$ fm^2 $A^{2/3}$ turns out to be negligibly small in the surface. For very small A it may be significant, especially for larger nuclear densities.

ATOMIC-LIKE BOUND STATES

The total Hamiltonian for the nucleus and antiproton is

$$H_{tot} = H_A + K_0 + V_{coul}(r_0) + U^{(0)} \tag{6.1}$$

where $U^{(0)}$ is the two-particle ($\bar{p}$-nucleon) optical potential.

The total wave function is, except for the angular momentum coupling, assumed to be a product of nuclear (Φ) and antiproton ($R(r_0)\varphi(\Omega)$) wavefunctions:

$$\psi_{tot} = R_{\alpha lj,\beta JF}(r_0)[\varphi_{\alpha lj}(\Omega) \times \Phi_{\beta J}(\vec{r}_1 \ldots \vec{r}_A)]_{FF_z} \qquad (6.2)$$

The solution to the nuclear Schrödinger equation is known

$$H_A \Phi_{\beta JM} = E^A_{\beta J} \Phi_{\beta JM} \qquad (6.3)$$

and the total Schrödinger equation,

$$H_{tot}\psi_{tot} = E_{tot}\psi_{tot} \qquad (6.4)$$

then reduces to

$$[-\frac{\hbar^2}{2\mu}\frac{d^2}{dr^2} + \frac{\hbar^2 l(l+1)}{2\mu^2 r^2} + V_{coul}(r)$$

$$+ \langle \alpha lj;\beta J;F|U^{(0)}|\alpha lj;\beta J;F\rangle] = \varepsilon_{\alpha lj,\beta JF} R_{\alpha lj,\beta JF} \qquad (6.5)$$

where the matrix element of $U^{(0)}$ means expection values over all nuclear as well as antiproton angular coordinates. Eq.(6.5) is then the radial antiproton bound state equation. Spin-dependent solutions arise from the corresponding dependence of $U^{(0)}$.

Fine- and Hyperfine Structure; Light Nuclei

For even-even spin zero nuclei (J=0), ψ_{tot} is a true product wavefunction and the antiproton-nucleus radial optical potential (first order) is the same as described for scattering in the previous sections. Higher order corrections are negligible in the present case of a low density interaction region.

For odd nuclei we divide into an even-even closed shell core of spin zero plus one or several valence nucleons. The core part is treated as before. The valence part is complicated due to the angular momentum couplings. It has been calculated for a number of light nuclei[475].

The two-particle energy ($\langle k^2\rangle$ or f_k in eq.(3.46)) is chosen close to zero, since the interaction mainly occurs outside the nucleus. The, perhaps preferable, negative value is much harder to use, because the relevant t or K-matrix is conceptually and technically difficult to obtain.

Central, spin-spin, spin-orbit and tensor parts of the interaction are included. The latter only through its (dominant) diagonal part implying neglect of l-mixing.

The atomic state has an eigenvalue differing from that of the pure Coulomb interaction by a "shift" (real part) and a "half width" (imaginary part). Strong interaction fine structure arises from the spin-dependent forces when the nucleus is (or is treated as) a spin zero entity. Strong interaction hyperfine structure arises from spin-dependent forces coupling the nuclear and antiproton angular momenta.

The widths are large compared to the fine- and hyperfine splittings. In general, the average properties are therefore

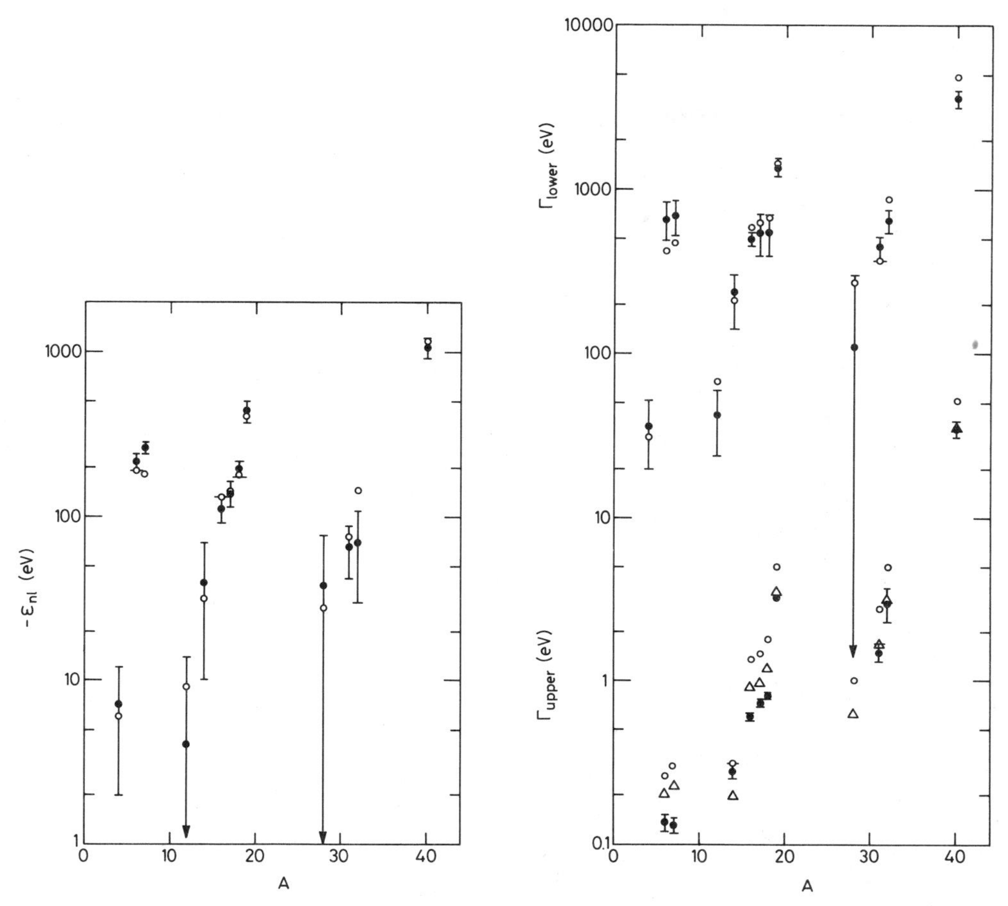

Fig. 8. Energy shifts and widths as function of nucleon number. The experimental black points are given with their uncertainties and the calculated values are the open circles. The widths are given for both sets of measured levels.

measured. In fig. 8, the observed shifts and widths are compared with calculations for nuclei below ^{40}Ca. The agreement is fairly good, especially for the lowest lying levels. the upper levels (n is one unit larger) have theoretical widths up to a factor of two too high. The shifts of these levels are not observed.

The fine- and hyperfine structure can be seen in fig. 9 for two typical cases. For ^{17}O the fine structure from the electromagnetic interaction is much larger than that of the strong force. For ^{7}Li the proportions are almost reversed. The hyperfine splittings are larger than (^{17}O) or about equal (^{7}Li) to the fine structure splitting. These spin-dependent features arise almost entirely from the spin-spin interaction (shifts) and from the spin-orbit potential (widths).

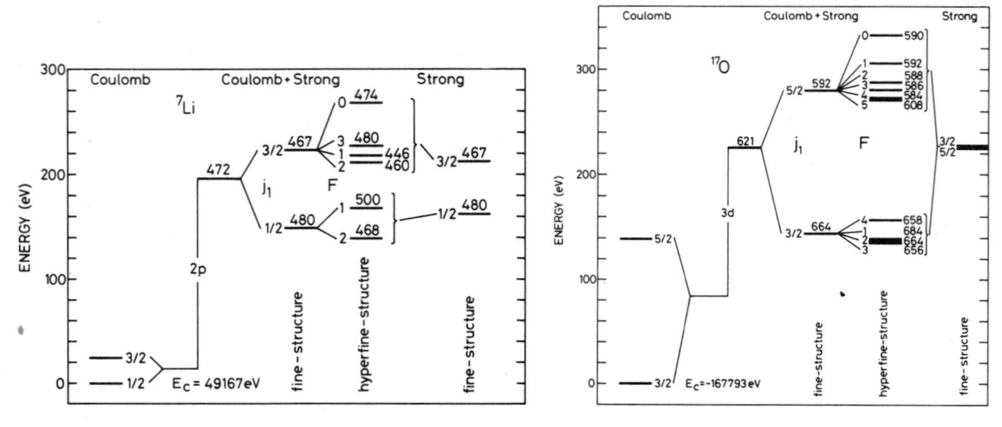

Fig. 9. The complete fine and hyperfine structure of the 3d and 2p levels for ^{17}O and ^{7}Li. The widths are given on top of or to the right of the levels which are labelled with their total angular momentum quantum number F.

Even-even Nuclei in General

The central and spin-orbit two-particle interactions contribute exclusively to the central and spin-orbit antiproton-nucleus optical potential. The Shcrödinger equation is solved and shifts and widths computed.

The results are shown in fig. 10, when the spin-orbit potential is ignored, for the average nuclear parameters of ref.11. Although the individual charge distributions from electron scattering are not used, the agreement with measurements is in general very good.

Including the spin-orbit potential we obtain the fine structure in fig. 11, where the differences between the values of the spin-orbit partners are shown. The lowest j-value (l-1/2) has the largest width. The shift differences have a peak structure whereas the width difference increases with proton number. The two measured[13] pairs of values for ^{138}Ba and ^{174}Yb are in eV $(\Delta\varepsilon, \Delta\Gamma)$ = (-99±143, 436±805), (58±26, 195±59) which fortuitously compares favourably with our computed average values of (21, 404) and (29, 253).

The imaginary potential arising from the Dover-Richard interaction can, in the contributing nuclear tail region, be parametrized by

$$W(r) = \frac{-W_c}{1+\exp[(r-R_c)/d_c]} - \langle \vec{l}\cdot\vec{s} \rangle \frac{W_{so}}{r}\frac{d}{dr}\frac{R_{so}^2}{1+\exp[(r-R_{so})/d_{so}]} \qquad (6.6)$$

with the parameters (W_c, R_c, d_c) = (73 MeV, 1.0 fm $A^{1/3}$, 1.0 fm) and (W_{so}, R_{so}, d_{so}) = (0.25 MeV, 1.1 fm $A^{1/3}$, d_c).

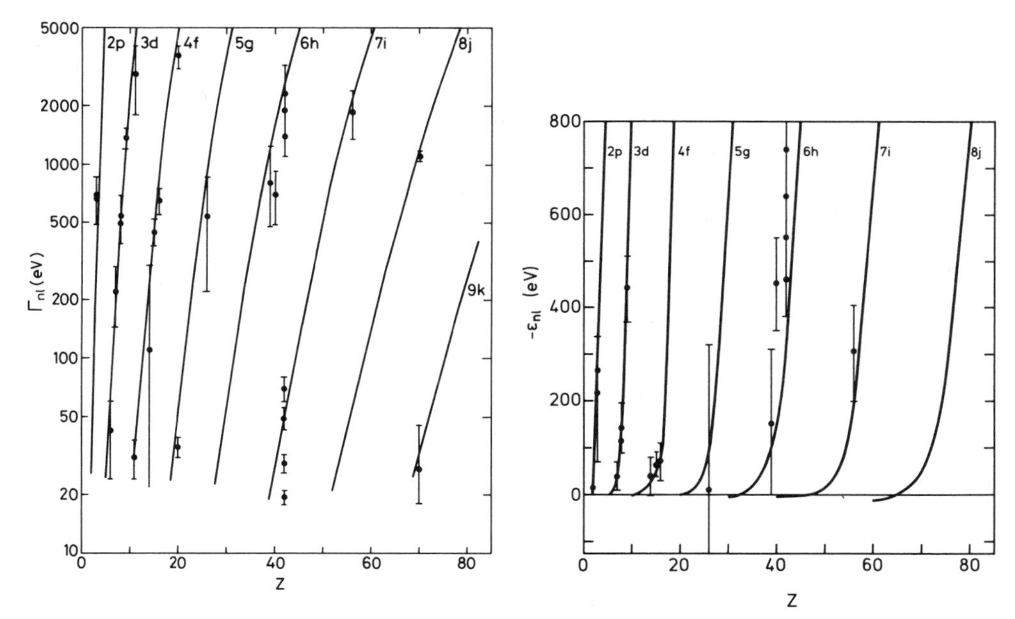

Fig.10. Energy shifts and widths as function of proton number. The experimental black points are given with their uncertainties and the calculated values are the continuous curves obtained with average nuclear parameters.

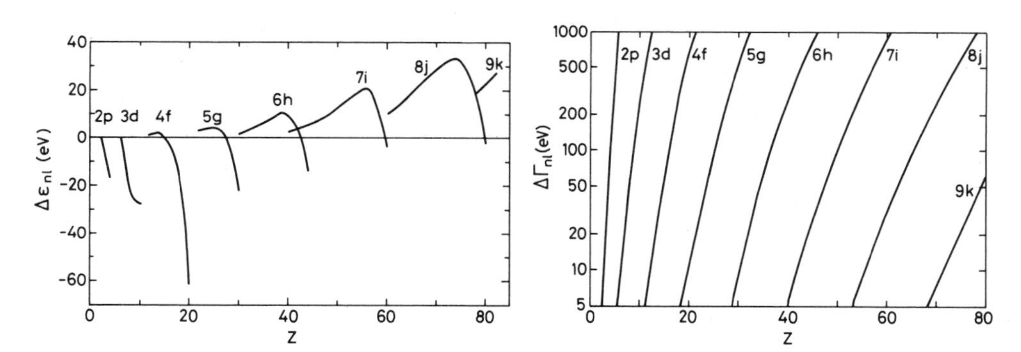

Fig.11. The difference in shifts and widths between the spin-orbit partners of the levels in fig.10.

The width is given by

$$\Gamma = -2 \int_0^\infty W(r) r^2 |R_n(r)|^2 dr \tag{6.7}$$

where R_n is the radial antiproton wavefunction of principal quantum number n. The integrand has a peak outside the nuclear surface, since W(r) decreases and the wavefunction increases exponentially in this region of space. In fig. 12 this peak structure is shown for various nuclei. Clearly only the nuclear tail region is important.

Using Coulomb wavefunctions reduced by a factor $\sqrt{x}$ in the contributing interval and expanding the central Woods-Saxon potential from eq.(6.6), as $\exp((R-r)/d)$, we obtain the approximation

$$\Gamma_c(n) = 2W_c x\ (1+ \frac{na}{2d_c})^{-2n-1} \exp(R_c/d_c) \qquad (6.8)$$

where $a = \hbar^2/(e^2 z\mu)$ is the Bohr-radius. With $x \approx 1/2$, Γ_c is a reasonably accurate approximation of the (main) contribution to the width from the central potential.

The fine structure width is also easily computed from eqs.(6.6) and (6.7). The difference in widths between spin orbit partners become

$$\Delta\Gamma(n) \equiv \Gamma_{so}(j = l+1/2) - \Gamma_{so}(j = l-1/2$$

$$= 2W_{so}\ x \frac{(1-1/2n)}{(1+ \frac{na}{2d_{so}})^{2n}} \frac{R_{so}^2}{nad_{so}} \exp(R_{so}/d_{so}) \qquad (6.9)$$

This is a decent approximation, but less accurate than eq.(6.8).

CONCLUSIONS

The antiproton-nucleus interaction has been investgated by elastic scattering and atomic bound state properties. Other processes like inelastic or charge exchange reactions have not been discussed in these lectures.

The essential interaction region for scattering is about 1fm outside the nuclear radius. The annihilation is very strong and the black sphere model contains the qualitative, but not quantitative features. In an optical model description, the imaginary part is dominating over the real part by almost an order of magnitude.

The microscopic KMT description is quantitatively good already to first order when a realistic antiproton-nucleon force is used. The second order effects vanish as the nuclear density decreases towards zero. Thus, they are small in the crucial tail region where the scattering takes place. They are consequently treated as perturbations. The medium effects are small, but detectable especially towards larger angles and for spin-dependent quantities like the polarization. The correlations are insignificant in the nuclear tail.

The atomic bound states are determined from the radial potential obtained by "folding" (first order) and by coupling the angular momenta. For even-even nuclei, where only central and spin-orbit interactions are relevant, we obtain reasonably good agreement with measurements. This also includes the recent fine structure results on Ba and Yb, although it may be somewhat fortuitous.

For light nuclei below ^{40}Ca, we calculated both fine and hyperfine structure for a number of nuclei. The behaviour depends on the individual nuclei in a very complicated manner. In general the hyperfine structure is larger (sometimes much larger) than the fine structure. They both arise mainly from the spin-spin and spin-orbit two-particle interaction. Due to the large widths, the splittings cannot be seen directly, but the interpretation should include these effects. It can be expected that strong interaction spin structure are largest for odd nuclei of high angular momentum.

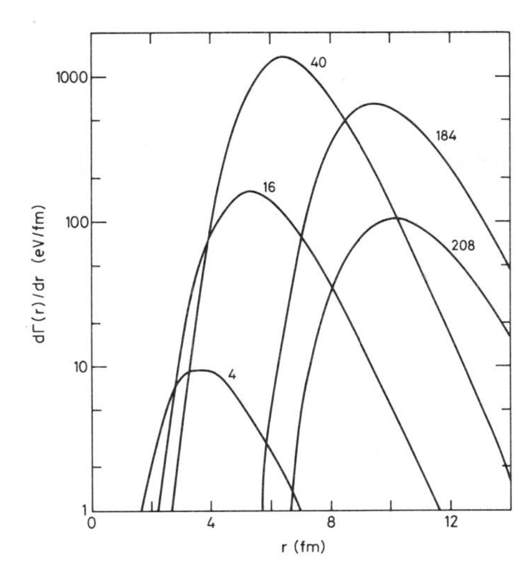

Fig.12. The integrand of eq.(6.7) as function of r for various nuclei with nucleon number on the curves. The area is thus the width.

REFERENCES

1) H. Heiselberg, Antiproton-Nucleus interaction, Ph.d.-thesis, Institute of Physics, University of Aarhus 1987, and H. Heiselberg, A.S. Jensen, A. Miranda and G.C. Oades to be published
2) H. Heiselberg et al. Nucl.Phys.A446(1985)637
3) H. Heiselberg et al. Nucl.Phys.A451(1985)562
4) O. Dumbrajs et al. Nucl.Phys.A457(1986)491
5) T.M. Jørgensen, Microscopic calculations of the fine- and hyperfine structure of antiproton atomic-like bound states, M.Sc.-thesis, Institute of Physics, University of Aarhus, 1988, and T.M. Jørgensen, A.S. Jensen, A. Miranda and G.C. Oades to be published.
6) D. Garreta et al., Phys.Lett.135B(1984)266, 139B(1984) 464, 149B(1984)64 and Nucl.Phys.A451(1986)541
7) C.I. Beard et al., Int.Phys.Conf. VII Eur.Symp. Antiproton interactions, Durham 1984
8) P.J. Siemens and A.S. Jensen, Elements of Nuclei, p.51, Addison-Wesley 1987.
9) A.K. Kerman, H. McManus and R.M. Thaler, Ann.Phys.8(1959) 551
10) N. Hoshizaki, Suppl.Prog.Theor.Phys.42(1968)107 and J. Bystricky et al. Jour.de Phys.39(1978)1
11) W.D. Myers and W.J. Swiatecki, Ann.Phys.55(1969) 395 and W.D. Myers, Nucl.Phys.A145(1969)387
12) A. Chameaux, V. Layly and R. Schaeffer, Ann.Phys.116 (1978)247
13) A. Kreissl et al. Z.f.Phys.A329(1988)235

DEEP ANTIPROTON ANNIHILATIONS ON NUCLEI

Johann Rafelski

Department of Physics
University of Arizona
Tucson, AZ 85721

INTRODUCTION

Hot nuclear matter can be created in $\overline{p}$ annihilations on nuclei: the antiproton penetrates deep inside the nucleus and the available energy is shared by a small number of nucleons participating in the reaction[1,2]. I consider here in detail recent strange particle production data obtained in 4 GeV/c $\overline{p}$Ta[3,4] annihilations and present an analysis of the experimental results in terms of a hot blob of quark gluon plasma. In particular, the observed spectra and total abundances of lambdas and kaons are consistent[5] with the hypothesis that (super cooled) quark matter phase has been formed at rather modest temperature T < 60 MeV.

When antiprotons with momenta of several GeV penetrate into nuclei, their range before annihilation can be several fermi. For example at momentum of 5 GeV/c we have an annihilation cross section σ of about 25 mb and hence the range $\lambda = (\sigma\nu)^{-1} = 2.8 fm$. It is natural to consider the kinetic and annihilation energy as being mostly shared by the nucleons in the matter tube forward of the antiproton. So formed nuclear fireballs will emerge from the target nucleus and disintegrate into individual hadrons, including the here interesting strange particles. For a number of reasons I have argued in the past that strange particles provide us with key information[6] about the possible formation of the quark gluon matter state of hot nuclear matter[7]. It should be noted that initial attempts to describe these data in terms of individual hadronic reactions have not been successful[8]. Also, it is clear that the particular processes of interest to us are strictly related to the presence of antiprotons and their annihilation, since in a parallel Dubna experiment[9] in which carbon nuclei of similar energy per nucleon were collided with heavy nuclei no strange particle anomalies were detected. However, in an experiment at CERN[10] a similar strange particle spectrum was obtained even at significantly lower $\overline{p}$ energy.

It is not impossible that one can, despite early disappointments[8] arrive at a qualitative explanation of the data employing hadronic degrees of freedom. However, in order to explain the absolute strange particle abundances and high $E_\perp$ particle spectra, it will be necessary to identify the proper degrees of freedom, which are as argued here, quarks and gluons. The key new property of the quark gluon plasma, (as quark matter is popularly referred to at finite temperature, QGP) is the loss of identity of individual nucleons, quarks, as well as gluons, the particles which serve to mediate their interaction are the only physical degrees of freedom present.

Antiproton-Nucleon and Antiproton-Nucleus Interactions
Edited by F. Bradamante *et al.*
Plenum Press, New York, 1990

ANNIHILATION DYNAMICS

We would like to find conditions under which the annihilation energy is shared in a rather small volume by several of the target nucleus' nucleons. Is the projectile antiproton momentum small, than due to the very high annihilation cross section the annihilation occurs directly on the nuclear surface. In such a reaction much of the annihilation energy escape backwards from the nucleus, and there is little orientation of the secondary particle flux. Thus we would like to increase the antiproton momentum to be so large, that the annihilation would occur well below surface (e.g. below first nuclear 'layer'), while the secondaries of the inelastic interaction are at the same time focused kinematically into a forward cone. But increasing the $\overline{p}$ energy will ultimately cause the antiprotons to penetrate deeper into the target nucleus, permitting initial annihilation products to escape in the forward direction. It is hard to predict which range of antiproton momentum will be best for the purpose of creating highly compressed and excited new forms of nuclear matter, except perhaps to observe that only when annihilation cross section drops below 30 mb does the antiproton penetrate deeper than one nucleon diameter into the nucleus; this occurs at a momentum of about 3-4 GeV/c. Thus the experiment performed by Miyano et al.[3,4] at 4 GeV seems to be well suited for the purpose of deposition of maximum excitation within the target nucleus.

Should due to the relatively high momentum of the projectile the reaction products stay mostly focused into a relatively narrow forward cone while travelling through the nucleus, there will be a substantial rescattering occurring on the nuclear matter in the forward path, leading to local distribution of annihilation energy among the available degrees of freedom and further transfer of the kinetic energy of the projectile into the cone of nuclear matter in the path of the reaction products. By such or an analog mechanism involving possibly also quarks, we would also distribute the annihilation energy and a large fraction of the kinetic energy of the projectile among the nuclear matter in the path of the impacting antiproton. Thus quarks form many target nucleons can participate in the annihilation reaction, even though the first reaction may occur at only one or perhaps two nucleons. One should be cautious about results about such reaction senarios derived from intra nuclear cascade (INC) calculations[11] , given the natural limitations of such an approach which ignores all subnuclear degrees of freedom. Another important difference of INC to the present approach is that INC calculations average over *all* annihilations, while annihilation events considered here are these which are self triggered by the presence of a strange particle in the final state.

We note that a tube of nucleons through the nuclear center contains $2A^{\frac{1}{3}}$ nucleons; in the case of annihilation on Ta we find[3] eleven nucleons in the forward tube. Considering further the opening of the intranuclear shower cone it is safe to presume that in central antiproton annihilations on heavy nuclei, triggered for central collisions, one should have frequently in excess of 11 nucleons in the effective target. Thus it is possible that a fireball of nuclear matter containing the annihilation and projectile kinetic energy will be formed, with baryon number being $O(10)$. Thus the issue is to select events with such fireballs. Normally, we could not identify this particular reaction channel among all other possible reactions in antiproton annihilations on nuclei. However, the formation of a thermalised fireball facilitates formation of strange particles[6]. Thus a suitable trigger is automatically set by looking at strange particle production[2]. When measuring the strange particle rapidity distribution one effectively determines the mass of the participating nuclear matter region from a simple kinematic considerations which essentially imply finding the frame of reference, if such exists, in which the spectrum of particles shows spherical symmetry. If a particle is emitted from a moving 'fireball' of hadronic matter, then its rapidity is simply the sum of the rapidity of the fireball moving along the projectile axis and the intrinsic rapidity of the particle. Hence we expect particle multiplicities to show a peaked distribution around the rapidity of the emitter, and furthermore the shape of the rapidity distribution to be as expected for an emitter of a certain 'temperature' which may be read off the transverse momentum distribution.

The rapidity of the fireball with respect to the laboratory is derived from the kinematics of the colliding objects and it depends in particular sensitively on the effective total mass of the target, i. e. the participating nuclear matter cone. In Fig. 1 the dependance of the rapidity of the fireball and the excitation energy per nucleon in the fireball on the number of participants in the target, A_t, is shown, computed for $P = 4$ GeV/c, i. e. $E_p = 4.11$ GeV. Hence the observation of a dn/dy distribution which is clearly peaked at a certain rapidity provides, as exploited by Miyano et al.[2] in itself a measure for the number of participating target nucleons. However, as observed above, we must also obtain a shape of the rapidity distribution which is consistent with the transverse momentum distribution.

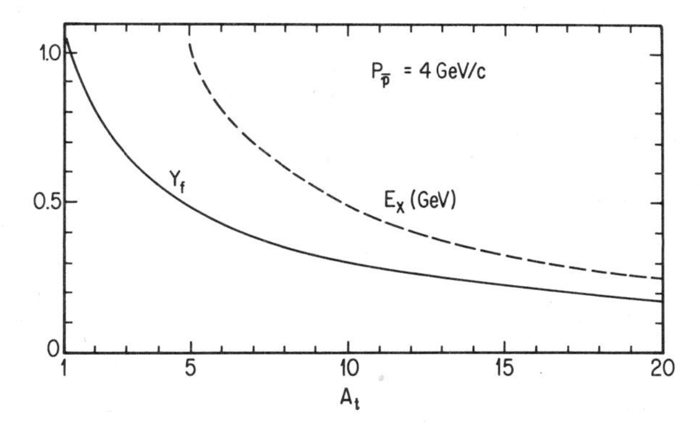

Fig. 1. Y_f, the rapidity of the central fireball as function of A_t, the number of target baryons. Dashed: E_x, the excitation energy in the fireball per baryon (A_t- 1) in GeV; all values computed for $P_{\bar{p}}$=4GeV/c.

The rapidity distribution of Λ particles is observed[2] to peak at $y = 0.25 (v/c = 0.245)$. Given the rapidity observed for the peak of the lambda distribution we obtain (along with Ref. 3) from Fig 1. that there were about 13 nucleons in the annihilation fireball. This implies further that the excitation energy per baryon number b (= 12) in the fireball is 0.38 GeV (that is, energy per baryon $E/b = 1.32$ GeV), as is also shown in Fig. 1. Further note that at the observed velocity of the fireball the radius of the Tantalum nucleus is transversed in time $t = R/v = 10^{-22}$ sec. This is a rather long live time for annihilation fireballs, but we must presume that the presence of other nonparticipating nuclear matter prevents the fireball from expanding, allowing it to leave the target nucleus. Indeed, this external constrain may permit the hot, excited portion of nuclear matter to seek a point of relative stability, which would be at the minimum of the energy per baryon curve for a given entropy per baryon. We will return further below to discuss this condition within the context of perturbative QCD-QGP equations of state.

STRANGE PARTICLE SPECTRA

Before proceeding with our theoretical analysis, it is worthwhile to record that it is not surprising that the data[3,4] shows a slight enrichment of the lambda abundance at the rapidity associated with the $\bar{p}$-annihilation on a single target nucleon. As there is a substantial cross section for annihilation on the large nuclear perimeter, we will also pick up in the data sample the more usual $\bar{p}$N-annihilation reactions into strangeness. In such reactions the expected Λ-rapidity is that for annihilation on one nucleon, and its rapidity distribution should therefore be centered around $y = 1.08$ for 4 GeV/c. Indeed, the data of Ref. (3,4) show a shoulder in the range $y = 1 - 1.1$, kinematically consistent with the usual annihilation reaction. The associated cross section is found consistent with the $A^{\frac{2}{3}}$ scaling, which is not the case for the anomalously large nuclear fireball component. Subtracting this 'direct' component from the spectrum[4] leaves more the 90% of the Λ-yield.

The overall yield of strange baryons (here Λ) originating mainly from the nuclear fireball region $y = 0.25$ is found not only to be unusually large, but also to be 2.4 times greater than that of kaons (here K_s). In usual $\bar{p}p$ annihilations the opposite is true with the K_s yield being 3.6 times greater. Thus in usual nucleon-antinucleon annihilations occurring primarily on nuclear corona we would expect to create a much greater kaon background (factor eight greater) than is seen for Λ's. This is fully born out by the experimental data reproduced in Fig. 2, which even suggests a greater yield of kaons than lambdas at $y = 1.1$. Remarkably, this ratio completely reverses when considering the entire rapidity range, proving a great difference in the underlying reaction mechanism. This phenomenon is easily understood. Assuming equilibration of particle abundance in the sense of relative chemical equilibrium[6], we expect that while practically all s-quarks are bound in Λ's, the equally abundant $\bar{s}$-quarks are found in Kaons. Half of these are initially found in $\bar{s}d = K^0$ kaons, and therefore only one quarter in K_s component. Hence naturally the Λ abundance from the fireball is expected to be up to 4 times stronger than the K_s signal! As we see, we do not yet need the presence of QGP, all that is required is strong rescattering of s-quarks into baryons -hence we *need* the presence of a baryon rich fireball.

Another interesting gross feature of the data is the widening of the rapidity K_s distribution towards smaller rapidity, a fact which is easily explained if a thermal source is assumed for the emitted particles. Namely, the width of the rapidity distribution depends on the mass of the emitted particles given their common temperature T. The normalized rapidity distribution for a Boltzmann emitter is:

$$N^{-1}\frac{dN}{dy} = \int \frac{d^2p_\perp dp_\|}{dy}|_{p_\perp} e^{-E/T} / \int d^3p e^{-E/T}, \tag{1}$$

where the transverse and longitudinal momenta in Eq. 1 are defined as usual. From this expression we find:

$$N^{-1}\frac{dN}{dy} = \frac{e^{\frac{m}{T}\cosh y}}{2K_2(\frac{m}{T})}(1 + \frac{2}{(\frac{m}{T}\cosh y)} + \frac{2}{(\frac{m}{T}\cosh y)^2}). \tag{2}$$

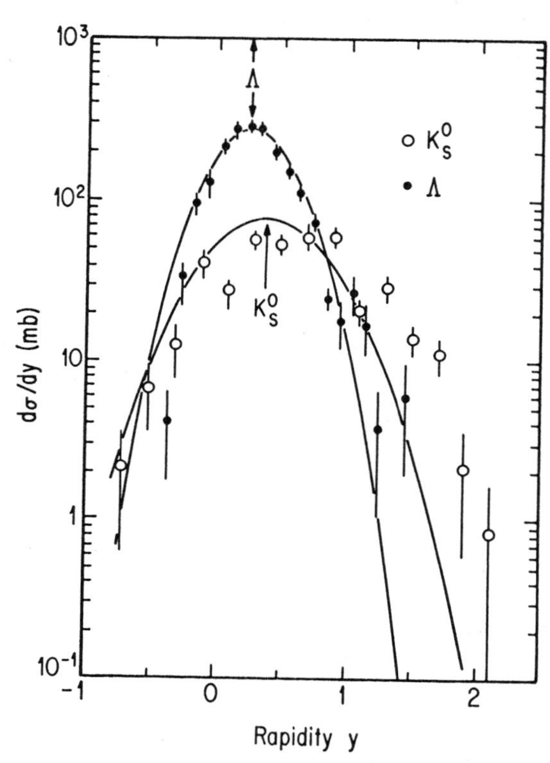

Fig. 2. Data of Miyano et al, Ref. 2: strange particle cross sections as function of rapidity are compared with thermal spectra. Λ's are emitted at $y_f = 0.25$, K_s at 0.4; kaons are renormalized to 75% of the observed yield.

It is self evident how one can improve the above expression to allow for presence of a second unresolved component, e.g. the Σ^0 which decays rapidly into Λ. In Fig. 2 the solid line for the Λ spectrum shows such a distribution computed for $T = 97$ MeV and $y_{CM} = 0.25$, these parameters as given in Ref. (3,4). The thermal yield was normalized to the total yield found by Miyano et al, 193 mb. I have here neglected the contribution to the Λ yield of the direct reactions, since the direct component contributes at the level of 5% only. However, due to the presence of an impressive background the direct component must be accounted for when considering kaons. Hence the thermal cross section must be some fraction of the total yield. In an eye ball estimate combined with the above expectation of 8 times greater K background I assign 75% of the yield to the thermal component, leaving the balance for the direct component in the $y = 1.1$ region. The important point is that now the same temperature parameter as determined for Λ's can be used in the Kaon rapidity spectrum. But in order to account for the absence of a strange baryon trigger when switching from Λ to K we have allowed the CM frame of Kaons to be at $y = 0.4$. This accounts for the fact that an 'average' fireball from which kaons originate is more baryon poor and hence using kaons as trigger we weight the sample of fireballs towards smaller baryon number (given the value of mean Kaon rapidity we obtain from Fig. 1 for an average Kaon associated event $< b >= 7$ and $E/b = 1.6$ GeV). The agreement of data with the rapidity distribution so obtained, c.f. Fig. 2 confirms the internal consistency of the hypothesis of a central fireball disintegrating into various particles, (data sample being triggered by strangeness). Further, the enhancement of the lambda abundance as compared to kaons can be easily understood if the particles are emitted from a baryon rich equilibrated environment, in which we can expect a major shift of strangeness into the baryonic channel[6]. It should be also mentioned that even though $T = 97$ MeV fits the rapidity Λ- spectra, it is just a 'mean' temperature. The transverse momentum distributions show two components, one at 60 MeV and one at 130 MeV, with relative strength 2.7, and I find for such a two component system nearly the same rapidity distribution as for the single component $T = 97$ MeV.

QUARK MATTER CONDITIONS

We can further estimate the statistical properties of the hadronic phase at the time individual particles decouple from each other. We will concentrate here on events triggered on by the presence of Λ's, since this data sample has much better statistics and is free of any ad hoc assumptions in its interpretation given above. Since we know now the energy distribution in CM frame containing presumably 12 baryons, we can easily compute the energy contained in unobserved pions by utilizing the energy conservation. When balancing the energy, we must remember that each Λ is accompanied by a $(\bar{s}q)$-kaon. We find that the energy per baryon is given in terms of baryon and pion yield as:

$$\frac{E}{b} = m_N + \frac{(m_\Lambda - m_N + m_K)}{b} + \frac{(b+2)}{b} 1.5T + \frac{n_\pi}{b} E_\pi. \tag{3}$$

where the second term is the energy of the strangeness, the third the thermal kinetic energy of baryons and two strange particles. Of the total internal excitation of the fireball consisting of $12 \times 380 MeV = 4.6$ GeV we find in baryons and strange particles 3.25 GeV at $T = 130$ MeV (which will turn out to be the temperature of the hadronic gas), with the balance of ~ 1.35 GeV being presumably in the pionic degrees of freedom. Indeed, in the most recent version of the data[4], there is mention of, on average, 1.11 negative tracks in strangeness triggered annihilation events. On the hypothesis that these are mostly π^-, I estimate considering the isospin asymmetry of the fireball that there are 0.71 π^+ and 0.92 π^0, hence in all 2.8 π, which just accounts for the missing energy. We also can estimate the entropy content of the fireball[6]; in the Boltzmann approximation:

$$S = [\frac{(M_N - \mu_b)}{T} + 2.5]b + n_\pi(\frac{m_\pi}{T} + 2.5). \quad (4)$$

With $\pi/b = 0.233$ we find about 5 units of entropy per baryon (at $\mu_b \sim M_N - T$, 3.5 units in baryon degrees of freedom, 1.5 units in pionic excitations). Any interpretation of the fireball in terms of some fundamental intrinsic structure should have smaller entropy per baryon in order to allow for some subsequent entropy production in the fireball evolution. (Note: for T> m_π the last parenthesis in eq.(4) should be replaced by 4.)

We now determine the properties of quark matter which would arise in the annihilation reaction under discussion. The perturbative QCD equations of state for the energy per baryon have been given repeatedly[7] and a systematic analysis of the properties of these equations has recently been carried out[12]. We seek the temperature of quark matter at which as function of baryon density a minimum appears in the energy per baryon expected for the fireball, 1.32 GeV. Taking the value of the QCD coupling constant to be $\alpha_s = 0.6$, and the B(ag) constant at $B = (171MeV)^4$, that is 0.112 GeV/fm^3, this turns out to be the case for $T = 53$ MeV. This minimum is located at a value of baryon density 2.75 times the normal nuclear density (and quark chemical potential of 395 MeV). The entropy per baryon is 4 units per baryon. For strange quark mass of 170 MeV the equilibrium ratio of strangeness to baryon number is 0.0044. For $T = 60$ MeV the minimum of E/b is already at 1.35 Gev. At this point the entropy per baryon has increased to 4.14 units, while the strangeness abundance nearly doubled to 0.008 per baryon. Variation of α_s in the range 0.5 - 0.7 and of B within a factor two leave all these qualitative results unchanged, in particular so when an increased value of α_s is associated with a reduced value of B, as required for internal consistency with other QCD and hadronic spectra studies.

WHY QUARK MATTER

It is thus possible to postulate the formation of (perhaps super cooled) quark matter at $T = 60 \pm 10$ MeV, $\mu_q = 400 \pm 50$ MeV, $S/b = 4.5 \pm 0.5$ in the $\bar{p}$-Ta annihilations, indeed perhaps to suggest that in most annihilations this is the reaction mechanism. In support of this hypothesis is:

1. The observation that alone the annihilation mechanism generates about 25 units of entropy, half of what we require in the associated quark fireball.

2. The observed large Λ - production cross section, 12 times greater than expected.

3. The total strangeness cross section σ_s being twice the $A^{\frac{2}{3}}$ scaled $p\bar{p}$ result. I estimate σ_s as the sum of the inclusive lambda cross section with twice the inclusive K_s cross section, less four times semi inclusive ($K_s\Lambda X$)-cross section, all as given in Ref. (4). For $\bar{p}$-Ta reactions I find 257 $\pm$ 20 mb, while for $\bar{p}$-p reactions 4.19 $\pm$ 0.15 mb results. Scaling the latter result with $A^{\frac{2}{3}}$ I find 136 $\pm$ 10 mb as the expected value.

4. The equilibrated quark matter abundance of strangeness, $<\bar{s}>=<s>= (0.004 - 0.02)b$ (in dependance on the Temperature of quark matter), where $b \sim 2A^{\frac{1}{3}}$ is the baryon content of the annihilation 'tube' leads to strangeness production cross section: $A(1.2fm)^2(0.04 - .016) = 84 - 210$ mb. INC cascade calculations, Ref. (8,13) do not explain these high strangeness and Λ yields.

5. The yield of $\bar{\Lambda}$ is at 1-2% of Λ and is consistent with the quark matter reaction picture developed here. This yield is much larger than expected[6] from an equilibrated hadronic gas at the temperature of 130 MeV. One estimates the

$\overline{\Lambda}/\Lambda$ yield by computing the likelihood of finding $\overline{sq}$ clusters ompared to sq. The third particle can be considered tobe supplied to neutralize the color of the cluster at hadronization. Since $< s >=< \overline{s} >$, we have to estimate the ratio

$(\overline{q}, q)$. Since each gluon is latently a $\overline{q} - q$ pair we find

$$\overline{\Lambda}/\Lambda \sim \overline{sq}x/sqx = \frac{2/\pi^2(3+8g)T^3}{2/\pi^2(\mu_q^3 + \mu_q(\pi T)^2)} = 11(\frac{T}{\mu_q})^3 \frac{1}{1+(\frac{\pi T}{\mu_q})^2}$$

For $\mu_q \sim$395MeV, T=55MeV as determined earlier, we find $\overline{q}/q = 0.025$, to be compared to $\overline{\Lambda}/\Lambda \sim 0.02 \pm 0.01$.

6. This state of quark matter satisfies all the physical constraints - indeed the small temperature of 60 MeV assures that the entropy is at the level of 80% of the hadronic gas value found in the hottest reaction component. This permits reheating as quark matter undergoes the transition to the hadronic gas phase.

7. I can predict the expected relative Ξ to Λ abundance along the chain of arguments or which $\overline{\Lambda}/\Lambda$ abundance was based .

$$\Xi/\Lambda \sim \frac{ssx}{sqx} = s/q \sim 0.012 - 0.025 \; (T \sim 53 - 60MeV)$$

as noted above in point 4.

Moreover, in my opinion, it is the natural fate of the annihilation process in which quark antiquark annihilations occur, to proceed via a fireball of quarks and gluons. While the postulation of quark matter formation at such low temperatures appears bizarre, the antiproton mechanism provides such a natural entrance channel into this phase, that it is well possible that this quark matter state is created in its super cooled form, i.e. it exists in the range of thermodynamic parameters where we would normally find hadronic gas. However, as it is the state of lower entropy, it is rather stable with respect to the global transition to the hadronic gas phase.

An anomalous abundance of strangeness could arise in the hadronic gas fireball if it 'cooks' for an unusually long time. Against such an interpretation speaks the fact that there is no sign of the anomaly in the C-Ta collisions at similar energy[9]. Thus we have here an effect clearly associated with the annihilation process. The essential point in support of the quark matter hypothesis[6] is that strangeness is produced more efficiently and hence more abundantly in the quark gluon plasma. This is particularly true at low temperatures considered here. In the quark matter phase the strangeness production time constant is 3×10^{-22} sec at $T = 60$ MeV. This time is of similar magnitude as the estimate for the fireball to transverse the nucleus and hence there could be approximate chemical equilibrium of strange quark abundance in the quark matter fireball should such be formed. There is no way this high s-abundance can be created in direct hadronic reactions, unless one is willing to introduce new reactions e.g. a chain such as

$$\overline{p}p \rightarrow \begin{cases} \rho + X \\ \qquad \rho + N \rightarrow K + \Lambda \end{cases}$$

with huge new strangeness production cross sections. Still, I believe that such an approach will miss the $\overline{\Lambda}$ and (yet to be reported Ξ) yields.

I have determined the properties of the fireball created in $\overline{p}$-Ta annihilations as follows: $E/b = 1.32$ GeV, $S/b < 5$. I have shown that it is possible to interpret the experimental data on strangeness production in antiproton annihilation on heavy nuclei in terms of the formation of a (super cooled) state of quark gluon matter. The relative and absolute strangeness abundance observed is consistent with such a hypothesis. The fireballs formed in such annihilations have the physical properties expected for the quark matter state.

REFERENCES

1. J. Rafelski, Phys. Lett. **B91**; 281 (1980);
J. Rafelski, H.- Th. Elze and R. Hagedorn, 'Hot Hadronic Matter in Antiproton Annihilation on Nuclei', in: "Proceedings of 5th European Symposium on Nucleon -Antinucleon Interactions", Bressanone, Italy, CLEUP - Padua (1980).

2. J. Rafelski 'Quark-Gluon Plasma in Antiproton Annihilation on Nuclei', in: "Physics at LEAR with Low Energy Cooled Antiprotons", U. Gastaldi and R. Klapisch, eds. Plenum Pub. Corp. (1984).

3. K. Miyano, Y. Noguchi, M. Fukawa, E. Kohriki, F. Ochiai, T. Sato, A. Suzuki, K. Takahashi, Y. Yoshimura, N. Fujiwara, S. Noguchi, S. Yamashita and A. Ono, Phys. Rev. Lett. **53**; 1725 (1984).

4. K. Miyano, Y. Noguchi, Y. Yoshimura, M. Fukawa, F. Ochiai, T. Sato, R. Sugahara, A. Suzuki, K. Takahashi, N. Fujiwara, S. Noguchi, S. Yamashita, A. Ono, M. Chikawa, O. Kusumoto and T. Okusawa, "Neutral Strange Particle Productions and Inelastic Cross Section in p+Ta Reaction at 4 GeV/c", KEK Preprint 87-160 February 1988.

5. J. Rafelski, Phys. Lett **B207**; 371 (1988);
J. Rafelski, Addendum to Proceedings of the LEAR workshop Villars 1987, C. Amsler et al. eds., Harwood Acad. Publ., Chur (1988).

6. J. Rafelski, Phys. Rep. **88**; 331 (1982);
J. Rafelski and B. Müller, Phys. Rev. Lett. **48**; 1066 (1982) and **56**; 2334(E) (1986); J. Rafelski, Nucl. Phys. **A428**; 215 (1984);
N.K.Glendenning and J. Rafelski, Phys. Rev. **C31** 823 (1985);
P. Koch, B. Müller and J. Rafelski, Phys. Rep. **142** 167 (1986).

7. B. Müller, "Physics of Quark Gluon Plasma" Springer Lecture Notes in Physics **225** (1985);
J. Rafelski and M. Danos in: "Springer Lecture Notes in Physics" **231**; pp. 361-455 (1985).

8. C.M. Ko and R.Yuan, Phys. Lett. **B192**; 31 (1987).

9. V.D. Toneev, H. Schulz, K.K. Gudima and G. Röpke, Sov. J. Part. Nucl. **17**; 485 (1986); [Fiz. Elem. Chastits At. Yadra **17**; 1093 (1986)] and references therein. See in particular page 511.

10. F. Balestra, M. P. Busso, L. Fava, L. Ferrero, D. Panzieri, G. Piragimo, F. Tosello, G. Bendiscioli, A. Rotondi, P. Salvini, A. Zenoni, Yu. A. Batusov, I. V. Falomkin, G. B. Pontecorvo, M. G. Sapozhnikow, V. I. Tretyak, C. Guaraldo, A. Maggiora, E. Lodi Rizzini, A. Haatuft, A. Halsteinslid, K. Myklebost, J. M. Olsen, F. O. Breivik, T. Jacobson, S. O. Sorensen, CERN PS179;Phys. Lett. **194B**; 192 (1987).

11. J. Cugnon and J. van der Meulen; "Antiproton and Antilambda Annihilations on Several Nucleons", Liege preprint 1988.

12. J. Rafelski and A. Schnabel, Phys. Lett. **B207**; 6 (1988).

ANTIPROTONIC ATOMS WITH A DEFORMED NUCLEUS

G.Q. Liu, A.M. Green and S. Wycech*

Research Institute for Theoretical Physics,
University of Helsinki, Siltavuorenpenger 20 C
SF- 00170 Helsinki, Finland
* Institute for Nuclear Studies, Hoża 69, Warsaw, Poland

ABSTRACT

The effect of a deformed nucleus on antiprotonic atoms is studied. This effect accounts for the wrong sign of the measured [1] shifts of the last observable transition in the $\bar{p}-^{174}$Yb atom.

1. INTRODUCTION

In a recent LEAR experiment[1] by the PS176 collaboration, the strong-interaction spin-orbit effects in the $\bar{p}-^{174}$Yb atom were measured for the *last observable* transition. After extracting the electromagnetic effects due to finite size and vacuum polarization corrections, resultant *attractive* shifts of $\Delta E_L = 341 \pm 43$ eV and $\Delta E_U = 283 \pm 36$ eV were obtained for the *last observable* transitions

$$E_L = E(n_0=9, l_0=8, j_0=\frac{15}{2}) \longrightarrow E(8,7,\frac{13}{2})$$

and

$$E_U = E(9,8,\frac{17}{2}) \longrightarrow E(8,7,\frac{15}{2})$$

with the widths for the $8j$ levels being $\Gamma_L = 1021 \pm 41$ eV and $\Gamma_U = 1261 \pm 41$ eV respectively. However, theoretical calculations [2,3,4,5] with realistic $\bar{p}-$ nucleus potentials produced good agreement for the widths, but all four groups predicted the *opposite* sign for the shifts with respect to the experimental values. The calculated *repulsive* shifts are in the range $\Delta E_L = -(80 \rightarrow 130)$ eV and $\Delta E_U = -(60 \rightarrow 120)$ eV. The predicted splitting between the two transitions is in the range $\Delta\delta = \Delta E_l - \Delta E_U = -(16 \rightarrow 71)$ eV in contrast to the measured 58 ± 26 eV.

Antiproton-Nucleon and Antiproton-Nucleus Interactions
Edited by F. Bradamante *et al.*
Plenum Press, New York, 1990

This difference in sign between the experimental and theoretical values is very important because an attractive shift means the strong interaction increases the binding of the $\bar{p}$–atomic levels, whereas a repulsive shift means the opposite. Except in the $\bar{p}-^{174}$Yb case, every measurement –all on spherical nuclei– so far yields ***repulsive*** shifts for $\bar{p}$–nucleus systems and experiments agree reasonably well with theoretical calculations.

This unusual experimental result for the $\bar{p}-^{174}$Yb system was studied by the present authors[6,7] and it was found that the sign conflict can be resolved by taking into account the deformation of the ^{174}Yb nucleus.

2. THE DYNAMIC E2 EFFECT

The quadrupole moment of a deformed nucleus causes the so called dynamic E2 mixing, which couples the nuclear ground state (I=0) with its first excited states (I=2). This coupling obeys the following selection rules: for the nuclear states,

$$\Delta I = 0, 2 \qquad (no\ I = 0\ to\ I = 0\ coupling)$$

and for $\bar{p}$ atomic states,

$$\Delta l = 0, 2$$

$$\Delta j = 0, 1, 2$$

Therefore, each basic state $(I = 0; n, l, j)$ can couple to five $I = 2$ states. For example, the state $(I = 0; n = 8, l = 7, j = \frac{13}{2})$ can be mixed with $(l, j) = (5, \frac{9}{2}), (5, \frac{11}{2}), (7, \frac{13}{2}), (7, \frac{15}{2})$ and $(9, \frac{17}{2})$.

The corrections to energy levels due to dynamic E2 mixing were calculated in two different ways: perturbation theory including only discrete $\bar{p}$–atomic levels and a full calculation taking into account both discrete and continuum $\bar{p}$ states. The perturbation calculation is done by simply summing over the discrete energy (n) levels for all five mixed $(I = 2)$ states. This produces corrections of –596 and –598 eV to the shifts of lower and upper lines respectively. In the full calculation the Green's function method is used. It is found by comparing the perturbation and full calculations that the $\bar{p}$ continuum states contribute about 4% to the correction due to deformation, i.e. at the 25 eV level. Therefore, corrections to the shifts become –620 and –622 eV respectively. When these corrections are removed from the experimental numbers 341 and 283 eV, ***repulsive*** shifts of –279 and –339 are obtained. The results of the perturbation [6] and full[7] calculations are given in table 1 together with the experimental values of ref.[1] and theoretical predictions from refs.[2,3,4,5].

Table 1. Comparison of the shifts.

Shift (eV)	*Uncorrected Expt.*[1]	*Corrected Expt. – Pert.*	*Corrected Expt. – Full*	*Theory*[2,3,4,5]
ΔE_L	341	–255	–279	$-(80 - 130)$
ΔE_U	283	–315	–339	$-(60 - 120)$

3. THE LINE SPLITTING AND A RELATIVISTIC EFFECT

The splitting between the two transitions provides a measure of the strength of the spin–orbit interaction. It is noticed in the previous section that the splitting is not much affected by the dynamic E2 effects. There is a relativistic effect caused by the deformation that could contribute to the splitting. It is of the form–see ref.[6]

$$H_{Def}^{FS} = \frac{\bar{e}}{4M^2}(1 + 2\frac{\bar{\kappa}e}{\bar{e}})\mathbf{p}\cdot\sigma V_{Def}\mathbf{p}\cdot\sigma + \frac{\bar{\kappa}\bar{e}}{4M^2}(p^2 V_{Def} + V_{Def}p^2).$$

This effect makes a contribution of $\Delta\delta = 7.5\ eV$ to the splitting and in the right direction to reduce the experimental–theoretical discrepancy of 60 eV and $-(16 \rightarrow 71)\ eV$. But it is still too small to get theory into agreement with experiment.

4. CONCLUSION

This study shows that taking into account the nuclear deformation resolves the sign conflict for the shifts between the experiment on $\bar{p}-^{174}$Yb atoms and the theory. There are still remaining discrepancies for the splitting and the magnitude of the shifts. Several reasons may play a role in theoretical estimates of the strong interaction shifts. For example, a) the accuracy of the value Q_0 –the measured quadrupole moment for the $\bar{p}-^{174}$Yb nucleus extracted under conditions different to those used in this paper, or b) the existence of a "neutron skin" at large distances. But there is no doubt that the nuclear deformation does account for the opposite sign between experiment and theory. As is suggested in ref.[6] the discrepancy of the splitting may be due to the inaccuracy of short range $\bar{p} - N$ potentials and could lead to a new understanding of the $\bar{p} - N$ spin-orbit interaction.

REFERENCES

[1] A. Kreissl et al., Z. Phys. A329 (1988) 235
[2] D.A. Sparrow, Phys. Rev. C35 (1987) 1410
[3] J. Wicht and H. von Geramb, Verhandlungen Deutsche, Phys. Ges (VI)22 (1987)E-5.8
[4] H. Heiselberg, private communication to ref.[1]
[5] W.B.Kaufmann and H.Pilkuhn, Phys. Lett. 166B (1986) 279
[6] A.M. Green, G.Q.Liu and S. Wycech, Nucl. Phys. A483 (1988) 619
[7] S. Wycech, G.Q.Liu and A.M. Green in preparation

ANTIPROTON-NUCLEUS SCATTERING

V.A. Karmanov

Lebedev Physical Institute
Leninsky Prospect 53
Moscow 117924, USSR

ABSTRACT

Analysing the experimental data on $\bar{p}A$ interaction obtained at LEAR we show that the low energy antiproton scattering from nuclei is a new effective tool for solving a number of problems of nuclear physics.

The experimental data[1a)] (LEAR) on the elastic and inelastic cross sections of the antiproton scattering from nuclei ^{12}C, ^{40}Ca and ^{208}Pb at energies ≈ 50 and 180 MeV were analysed in Refs. 2a-f in the framework of the Glauber-Sitenko approach[3]. The part of the calculations carried out in Refs. 2a-f together with the experimental data[1a)] is shown in Figs. 1 and 2. This analysis demonstrates that, in contrast to the proton case, the Glauber-Sitenko approach to the antiproton-nucleus scattering remains valid down to very low energies ≈ 50 MeV. This conclusion was later confirmed by the analysis[2g] of the $\bar{p}$-scattering from the nuclei ^{20}Ne, ^{27}Al and ^{64}Cu at energies ≈ 100 ÷ 350 MeV (data KEK and BNL[1b,c)]). The applicability of this approach at so low energies is ensured by extreme narrowness of the forward cone in $\bar{p}N$ scattering and by large value of the total $\bar{p}N$ cross section. The slope parameter B of the $\bar{p}p$ amplitude at 50 MeV is of the nuclear order of magnitude ($B \approx 1.4$ fm^2), whereas pN scattering at the same energies is practically isotropic. The total $\bar{p}p$ cross section $\sigma^{\bar{p}p}_{tot}$ is approximately by 5 times larger than σ^{pp}_{tot}. The corrections to the Glauber-Sitenko approach and the reasons for their suppression were investigated in Ref. 4a, the eikonal approximation in $\bar{p}A$ scattering was studied in Refs. 4b-d. It was found[4a] that different corrections (e.g. non-adiabatic and off-shell effects) cancel each other considerably. Owing to applicability of this transparent theoretical analysis (Glauber-Sitenko approach) the antiproton-nucleus scattering is rather informative in solving the following problems.

Antiproton-Nucleon and Antiproton-Nucleus Interactions
Edited by F. Bradamante *et al.*
Plenum Press, New York, 1990

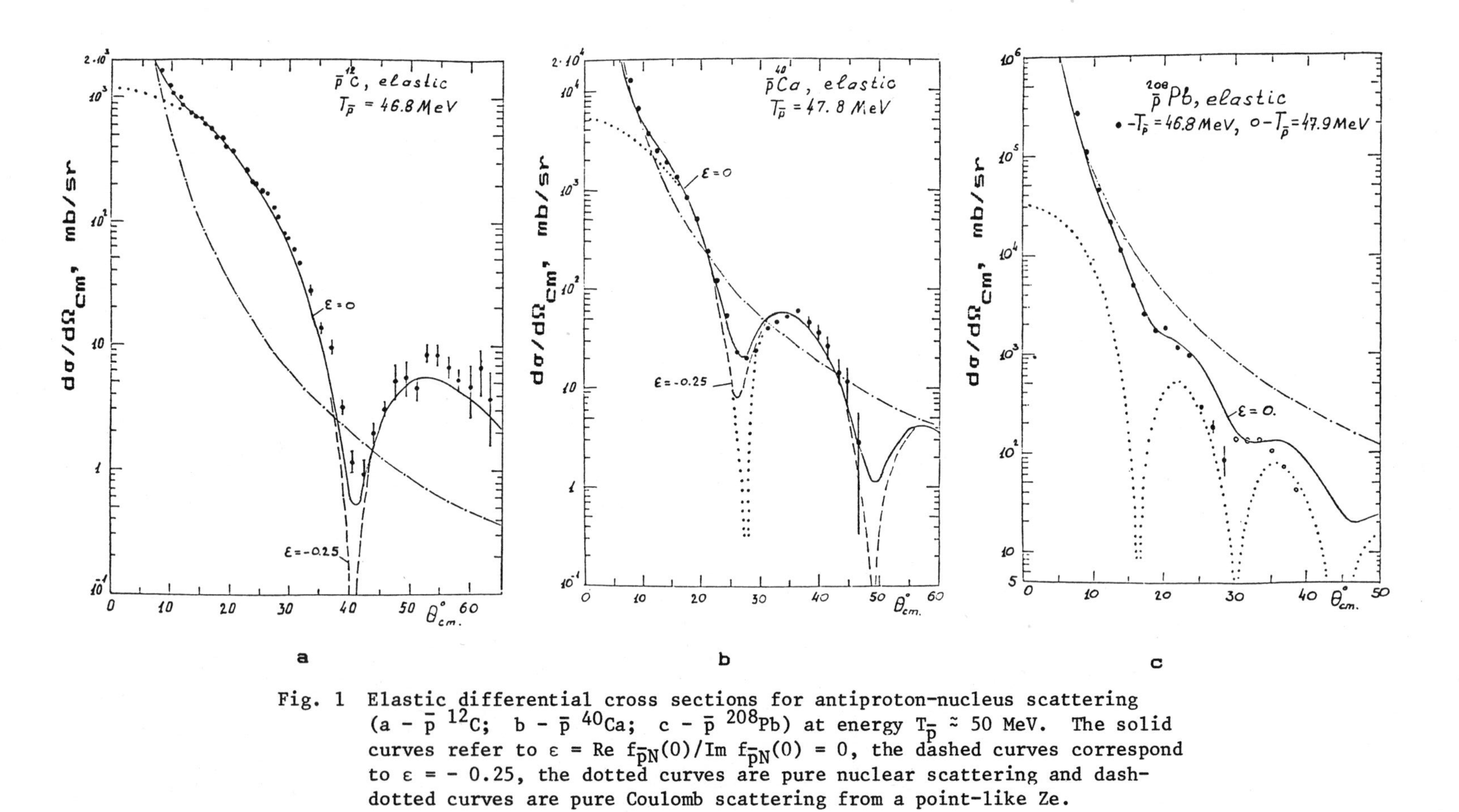

Fig. 1 Elastic differential cross sections for antiproton-nucleus scattering ($a - \bar{p}\,^{12}C$; $b - \bar{p}\,^{40}Ca$; $c - \bar{p}\,^{208}Pb$) at energy $T_{\bar{p}} \simeq 50$ MeV. The solid curves refer to $\varepsilon = \mathrm{Re}\, f_{\bar{p}N}(0)/\mathrm{Im}\, f_{\bar{p}N}(0) = 0$, the dashed curves correspond to $\varepsilon = -0.25$, the dotted curves are pure nuclear scattering and dash-dotted curves are pure Coulomb scattering from a point-like Ze.

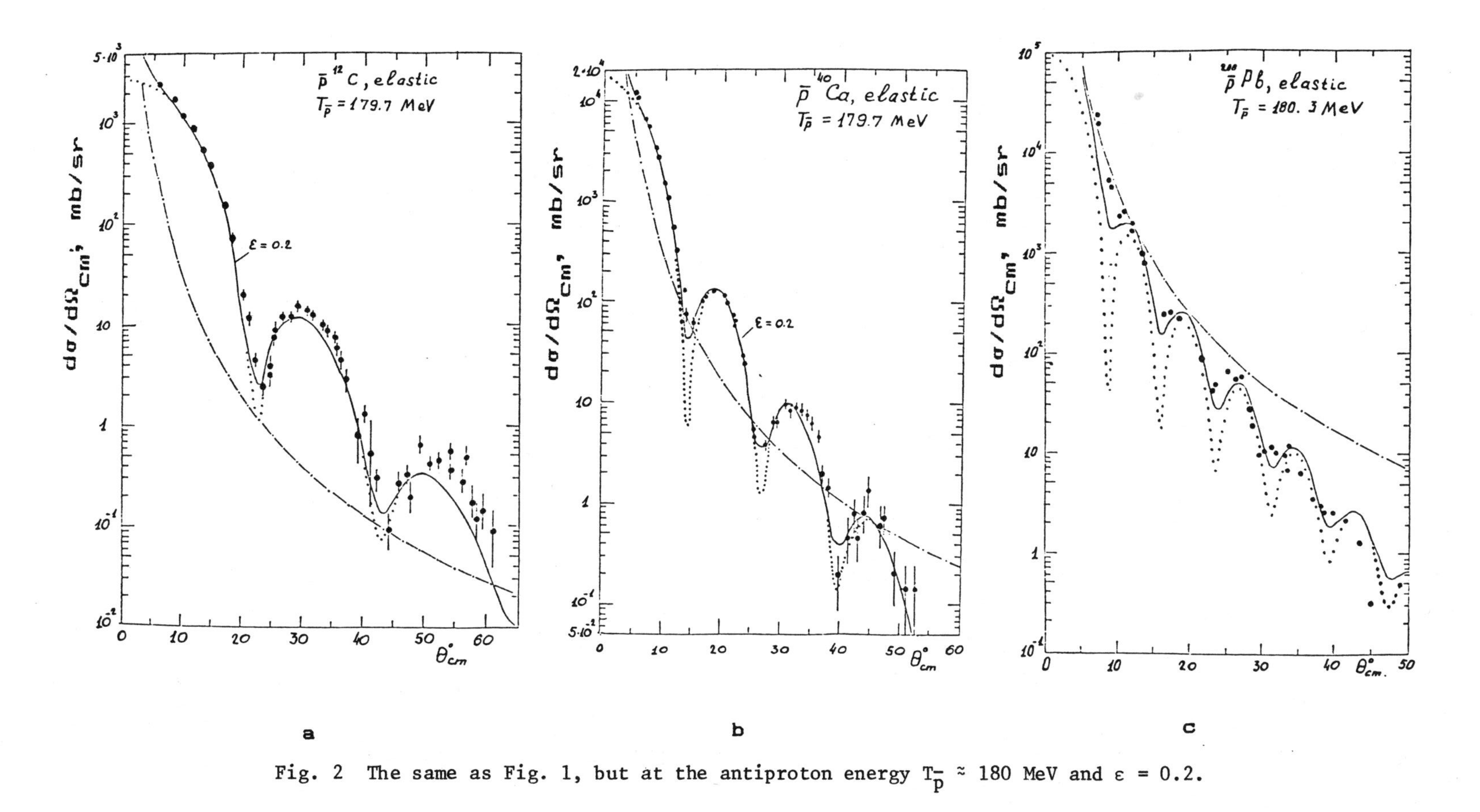

Fig. 2 The same as Fig. 1, but at the antiproton energy $T_{\bar{p}} \approx 180$ MeV and $\varepsilon = 0.2$.

1. INVESTIGATION OF THE NUCLEAR REACTION MECHANISMS

The validity of eikonal approximation for $\bar{p}A$ scattering at low energies gives information about accuracy of the adiabatic approximation, i.e. about corrections resulting from the motion of nucleons in a nucleus. In the case of the proton beam these two approximations can not be studied separately. It turned out that the adiabatic approximation has rather good accuracy in spite of the fact that the incident antinucleon momentum is close to momentum of the intranuclear nucleons.

In Ref. 2h the mechanisms of suppression (by one order of magnitude) of the cross section σ_b for the annihilationless nucleus break-up by the low energy antiprotons found for the $\bar{p}\,^4He$ interaction in Refs. 5 were investigated. It was shown that for the $\bar{p}D$ interaction this suppression is caused by narrowness of the forward cone in $\bar{p}N$ scattering (large value of B), whereas for more heavy nuclei this mechanism is quickly replaced by another one connected with large value of the $\bar{p}N$ annihilation cross section $\sigma_a(\bar{p}N)$. In the case of the $\bar{p}\,^4He$ interaction both mechanisms are equally important. This suppression of σ_b in comparison with σ_{el}, $\sigma_r = \sigma_{tot} - \sigma_{el}$ and σ_{tot} is illustrated by Fig. 3.

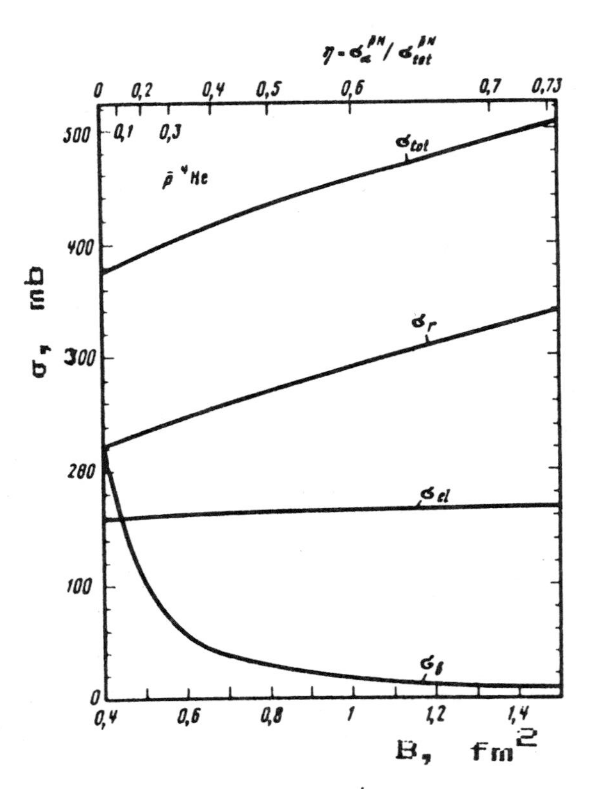

Fig. 3 The cross sections for $\bar{p}\,^4He$ interaction versus the slope parameter B of pN amplitude (down scale). The upper scale shows the relative values $\eta = \sigma_a^{\bar{p}N}/\sigma_{tot}^{\bar{p}N}$ (corresponding to variation of the $\bar{p}N$ annihilation cross section.

2. INVESTIGATION OF THE NUCLEAR STRUCTURE.

It was found[1a] that the spectrum of the nuclear level excitation by the low energy antiprotons has considerably smaller background from continuous spectrum than corresponding background for the case of incident protons. This fact also manifests itself in above-mentioned suppression of annihilationless break-up of nuclei by low energy antiprotons. Therefore the antiprotons can be used as a new convenient tool for investigation of the nuclear structure since they ensure more contrast picture of peaks corresponding to the nuclear level excitation. The inelastic (with excitation of 2^+(4.44 MeV) level) differential cross sections for $\bar{p}\,^{12}C$ scattering at $T_{\bar{p}} \approx 50$ and 180 MeV calculated in Ref. 2b agree with the experimental data[1a]. In Ref. 2 we predict also these cross sections for the case of the definite spin projections of excited nucleus $^{12}C^*(2^+)$ on the beam direction. They can be extracted from the angular distributions of γ-quanta emitted in the transition of the excited nucleus to the ground state.

3. FINDING THE $\bar{p}N$ AMPLITUDE PARAMETERS FROM NUCLEAR DATA

In Ref. 2e the real-to-imaginary ratio $\varepsilon \approx 0$ and $\varepsilon \approx 0.2$ for $\bar{p}p$ amplitude was extracted from $\bar{p}A$ data at energies 50 and 180 MeV respectively. These ratios coincide with direct measurements in $\bar{p}p$ scattering (see Fig. 4). The ratios $\varepsilon^{\bar{p}N}$ and $R = \sigma^{\bar{p}n}_{tot}/\sigma^{\bar{p}p}_{tot}$ were extracted from nuclear data, also in Ref. 6.

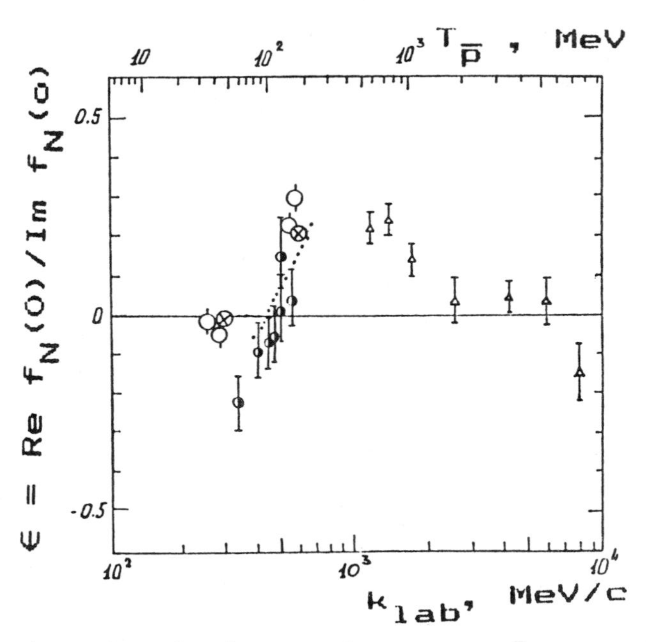

Fig. 4 Real-to-imaginary ratio ε for the $\bar{p}N$ scattering amplitude. ⊗-values obtained in Ref. 2 from $\bar{p}$-nuclear data.

CONCLUSION

All the facts discussed above prove that the low energy antiprotons are a new effective tool for investigations in nuclear physics.

REFERENCES

1a. D. Garreta et al.; Phys.Lett. 135B (1984) 256, 139B (1984) 464, 149B (1984) 65.
b. V. Ashford et al.; Phys.Rev. C30 (1984) 1080.
c. K. Nakamura et al.; Phys.Rev.Lett. 52 (1984) 731.
2. O.D. Dalkarov and V.A. Karmanov;
a. Pisma ZhETF, 39 (1984) 288 (JETP Lett. 39 (1984) 345).
b. ZhETF, 89 (1985) 1122 (JETP, 62 (1986) 645).
c. Elementarnye Chastitsy i Atomnoe Yadro (Particles and Nuclei), 18 (1987) 1399.
d. Phys.Lett. 147B (1984) 1.
e. Nucl.Phys. A445 (1985) 579.
f. Nucl.Phys. A478 (1988) 635c.
g. O.D. Dalkarov, V.A. Karmanov, A.V. Trukhov; Yad.Fiz. (Sov.J.Nucl.Phys.) 45 (1987) 688.
h. V.A. Karmanov, A.V. Trukhov; Pisma ZhETF (JETP Lett.) 47 (1988) 70, Preprint 67, Moscow, FIAN, 1988.
3a. R.G. Glauber; in "Lectures in Theor. Phys., vol. 1, pp.315-414, ed. by W.E. Brittin and L.G. Dunkam (Interscience, New York (1959)
b. A.G. Sitenko; Ukranian Phys. Journ. 4 (1959) 152.
4a. O.D. Dalkarov, V.M. Kolybasov, V.G. Ksenzov; Nucl.Phys. A397 (1983) 498.
b. V.P. Zavarzina, V.A. Sergeev; Czechoslovak Journ. of Phys. B36 (1986) 347.
c. Yad. Fiz. (Sov.J.Nucl.Phys.) 46 (1987) 486.
d. V.P. Zavarzina, A.V. Stepanov; Yad. Fiz. (Sov.J.Nucl.Phys.) 43 (1986) 854.
5a. Yu.A. Batusov et al.; JINR Rapid Communications No. 12-85, Dubna (1985) 6.
b. F. Balestra et al.; Phys.Lett. 194B (1987) 343.
6. G. Bendiscioli, A. Rotondi, P. Salvini and A. Zenoni; Nucl.Phys. A469 (1987) 669.

ANTIPROTONIC ATOMS

C J Batty

Rutherford Appleton Laboratory

Chilton, Didcot, OXON   OX11 OQX, UK

The study of antiprotonic-atoms has been the subject of much activity at LEAR with some 5 experiments being carried out in the period before the ACOL shut-down in 1986. Results from all of these experiments are now available. In the present lectures, the general topic of $\bar{p}$-atoms is considered with an emphasis on the results obtained at LEAR and their theoretical interpretation. The review falls naturally into two sections; $\bar{p}$-nucleus atoms and $\bar{p}$-nucleon atoms. First, however, a brief introduction is given to the general topic of exotic atoms. More detailed reviews are available elsewhere. (See ref. [1] and references therein).

1 EXOTIC ATOMS

An "exotic atom" is an atom with one of the electrons replaced by a negatively charged heavier particle such as a μ^-, π^-, K^-, $\bar{p}$ or Σ^-. In the formation process for the exotic atom, the incident negatively charged particle enters the target where it is slowed down to an energy $\sim$ 2 keV in a time of 10^{-11} to 10^{-9} sec. The particle now has a velocity similar to those of the electrons in the atom. It interacts with these electrons and is finally captured by the atom into a high Bohr orbit about the nucleus. This process typically takes 10^{-15} to 10^{-14} sec.

The captured particle now cascades down in a series of "hydrogen-like" atomic bound states. These transitions are initially accompanied by the emission of Auger electrons from the atom. In the later stages where, due to its much larger mass, the orbits of the captured particle are well inside those of the atomic electrons, the transitions are accompanied by the emission of X-rays. The time taken for this atomic cascade is typically 10^{-15} to 10^{-14} sec.

These atomic orbits can be classically described in terms of their principal quantum number n and orbital angular momentum ℓ where $0 \leq \ell < n$. In the Bohr model the binding energy E_B, and the radius r_n are then given by

$$E_B = -\mu c^2 \frac{(Z\alpha)^2}{2n^2} = -0.0266(\mu c^2)\frac{Z^2}{n^2} \text{ keV} \qquad (1)$$

$$r_n = \frac{\hbar^2}{\mu e^2}\frac{n^2}{Z} = \frac{27041}{\mu c^2}\frac{n^2}{Z} \text{ fm} \qquad (2)$$

Antiproton-Nucleon and Antiproton-Nucleus Interactions
Edited by F. Bradamante *et al.*
Plenum Press, New York, 1990

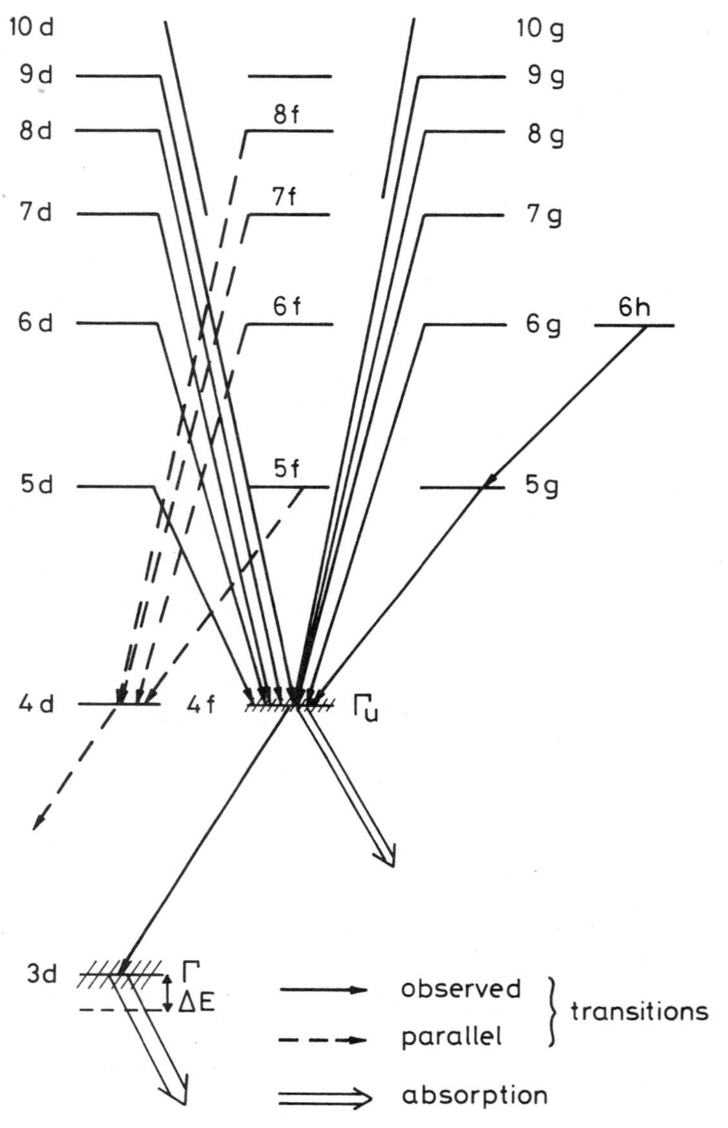

Fig. 1. Antiprotonic atom level scheme showing how the width (Γ_u) of the upper level can be determined.

where Z is the nuclear charge and μ is the reduced mass, in MeV/c^2, of the charged particle-nucleus system. For heavier antiprotonic atoms where $\mu \approx 938.3$ MeV/c^2 then $E_B \approx -25\ Z^2/n^2$ keV and $r_n \approx 28.8n^2/Z$ fm.

As the captured particle becomes more tightly bound the overlap of its wave function with the nucleus increases and for hadronic atoms, absorption of the hadron by the nucleus may take place. At some final n-value nuclear absorption will dominate, all hadrons reaching the level will be absorbed and the X-ray series will terminate. (For muonic atoms, the muon will reach its lowest orbit and then either decay or be captured by the nucleus through the weak interaction).

The strong interaction between the hadron and the nucleus causes a shifting of the energy of this last X-ray transition from its purely electromagnetic value whilst the nuclear absorption reduces the lifetime of the final atomic state and so the X-ray line is broadened. Precise measurements of the X-ray energy spectrum then allow the energy shift (ΔE) and width (Γ) associated with this final level to be determined. As the atomic number and size of the nucleus increase the absorption occurs from higher n-values. For example in $\bar{p}$-p atoms, X-ray transitions to n = 1 states have been observed whilst in $\bar{p}$-^{16}O and $\bar{p}$-^{174}Yb the transitions terminate at n = 3 and n = 8 respectively.

Although some hadrons will reach this final level, absorption also occurs from the next higher level although the effects on the level energy and width are generally too small to be directly measured. The width (Γ_u) of this upper level can frequently be deduced indirectly by measuring the decrease in intensity of the final X-ray transition (See Figure 1). The relative yield Y of this transition ($n + 1, \ell + 1 \rightarrow n, \ell$) is obtained by comparing the intensity of this transition with the sum of the intensities of the X-ray transitions feeding the upper ($n + 1, \ell + 1$) level.

$$Y = \frac{I\,(n+1,\ \ell+1 \rightarrow n,\ \ell)}{\sum_{i=n+2}^{\infty} I(i,\ell+2 \rightarrow n+1,\ \ell+1) + I(i,\ell \rightarrow n+1,\ \ell+1)} \tag{3}$$

$$= \frac{\Gamma_x}{\Gamma_x + \Gamma_u}$$

where Γ_x is width for X-ray transitions ($n + 1, \ell + 1 \rightarrow n, \ell$) which can be calculated. A number of corrections to this formula for parallel ($n + 1, \ell \rightarrow n, \ell - 1$) and for Auger transitions are required. Further details are given in ref. [2].

Measurement of the X-ray spectra thus allows, in favourable cases, three quantities, ΔE, Γ and Γ_u which are relevant to the strong interaction to be determined. By measuring X-rays from higher n-states, which are unaffected by the strong force and where the X-ray energies are determined only by the electromagnetic interaction, particle properties such as the mass and magnetic moment can be measured.

2 ANTIPROTONIC-NUCLEUS ATOMS

2.1 Experimental

The region of X-ray energies to be measured ranges from 2 to 20 keV for very light atoms such as $\bar{p}$-^{4}He through 10 to 90 keV for $\bar{p}$-O atoms to 300 to 450 keV for heavier atoms such as $\bar{p}$-Yb. For the lowest energies Si(Li) detectors are generally used whilst for the higher energy region experiments have used various types of Ge(Li) detectors. The LEAR

PS176/PS186 experiments could use up to 5 Si(Li) or Ge(Li) simultaneously and their experimental arrangement has been described in detail [3]. The PS174 experiment [4], which measured X-rays from $\bar{p}$-He atoms down to an energy of 1.8 keV used a Si(Li) detector fitted with a NaI(Tℓ) Compton shield. In both cases the dramatic improvement in the quality of antiproton beams now available with the LEAR facility has given very significant improvements in the results obtained.

2.2 Antiproton Mass

The antiproton mass has been determined from exotic atom data by measuring very precisely the energies of X-ray transitions in high n states (e.g., in the range n = 15 to n = 10 for $\bar{p}$-Pb) where the antiproton is well outside the nucleus so that strong interaction effects can be neglected. As we have already seen (eq. (1)), the measured transition energy is essentially proportional to the reduced mass of the antiproton-nucleus system, so that the antiproton mass can be determined by comparing measured X-ray energies with calculated values, corrected for a variety of QED effects. Values obtained in this way are given in Table 1.

Table 1. Measured antiproton mass

Mass (MeV/c^2)	Reference
938.179 ± 0.058	Hu et al [5]
938.229 ± 0.049	Roberson et al [6]
938.30 ± 0.13	Roberts [7]
938.216 ± 0.036	Mean

Taking the mass of the proton [8] 938.2796 ± 0.0027 MeV/c^2 gives almost a 2 s.d. difference for the quantity $(m(\bar{p}) - m(p))/m(p) = -(6.8 \pm 3.8)10^{-5}$. By CPT invariance this quantity should be zero. However much of the difference is due to the results from one experiment [5]. Dramatically improved measurements of the antiproton mass with accuracies of order 10^{-9} will eventually be available from the PS189 and PS196 experiments at LEAR.

2.3 Antiproton Magnetic Moment

The magnetic moment of the antiproton can be determined by a precision measurement of the fine structure splitting of an atomic level which is unaffected by the strong interaction. The situation is illustrated in figure 2. The fine structure splitting of a level with principal quantum number n and orbital angular momentum ℓ is given by

$$\Delta E_{FS} = (\mu_D + 2\mu_{anom}) \frac{(Z\alpha)^4}{2n^3} \frac{\mu c^2}{\ell(\ell+1)} \qquad (4)$$

where μ is the reduced mass of the antiproton-nucleus system, Z the nuclear charge, μ_D is the Dirac moment, equal to -1 for the antiproton and μ_{anom} is the anomalous (Pauli) moment in units of the nuclear magneton $\mu_{nm} = e\hbar/2m_p c$. The antiproton magnetic moment $\mu(\bar{p}) = \mu_D - \mu_{anom}$. As the splitting is proportional to Z^4, measurements with high Z targets are required. Although the transitions a and c of figure 2 dominate the observed X-ray spectrum (see figure 3), corrections are required for the transition labelled b and for the parallel (n = 11, ℓ = 9 → 10, 8)

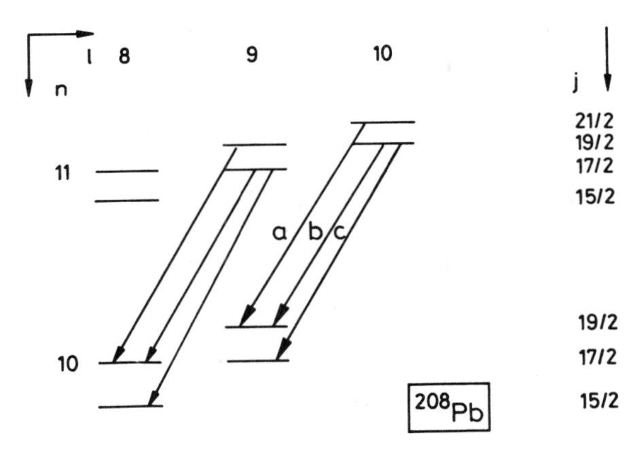

Fig. 2. Fine structure splitting of levels in $\bar{p}$-^{208}Pb due to the antiproton magnetic moment.

Table 2. Measured antiproton magnetic moment

Magnetic moment (μ_{nm})	Reference
- 2.791 ± 0.021	Hu et al [5]
- 2.817 ± 0.048	Roberts [7]
- 2.8005 ± 0.0090	Kreissl et al [9]
- 2.7995 ± 0.0082	Mean

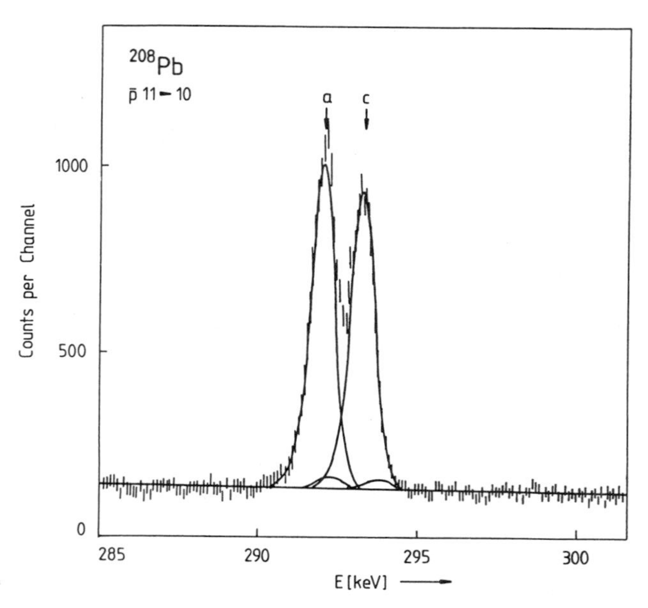

Fig. 3. Measured X-ray spectrum for the 11 → 10 transition in $\bar{p}$-^{208}Pb showing the fine structure splitting.

transitions. The latter are also shown in figure 3. Three experiments [5,7,9] to measure the magnetic moment have been made, with the latter experiment being carried out at LEAR. The results are listed in Table 2. The magnetic moment of the proton has been measured [8] to be $\mu(p) = (2.7928456 \pm 0.0000011)\mu_{nm}$ so that the relative difference in magnitude $(\mu(p) - |\mu(\bar{p})|)/\mu(p) = (-2.4 \pm 2.9)10^{-3}$. Equality of the absolute values of the magnetic moments of the proton and the antiproton is a test of CPT invariance.

2.4 Strong interaction effects

As we have already discussed (see section 1) strong interaction effects can be studied through measurements of the energy shift (ΔE) and broadening (Γ) of the final (n,ℓ) X-ray level together with, in some cases, an indirect measurement (eq. (3)) of the width (Γ_u) of the upper ($n + 1$, $\ell + 1$) level.

A summary of the available shift and width measurements for $Z > 2$ made in the pre-LEAR era has been given elsewhere [1,10]. 13 measurements each of energy shifts and of level widths together with 17 measurements of relative yields, which can be used to give values of Γ_u, had been made. These results covered a total of 8 levels and 18 different nuclei. However the quality of the data was poor with only 5 shift and 7 width measurements differing by more than 3 standard deviations from zero. (See figure 4). From the LEAR experiments, results of much higher quality are now available and measurements have been made for 2p and 3d levels in ^{3,4}He [4,11,12] and ^{6,7}Li [3], for 3d and 4f levels in ^{14}N ^{16,17,18}O, ^{19}F and ^{23}Na [3,13,14] for 6h and 7i levels in ^{92,94,95,98}Mo [15] for the 7i level in ^{138}Ba [16] and for the 8j level in ^{174}Yb [16]. These results, also plotted in figure 4, cover 6 levels and 9 nuclei with an emphasis on the

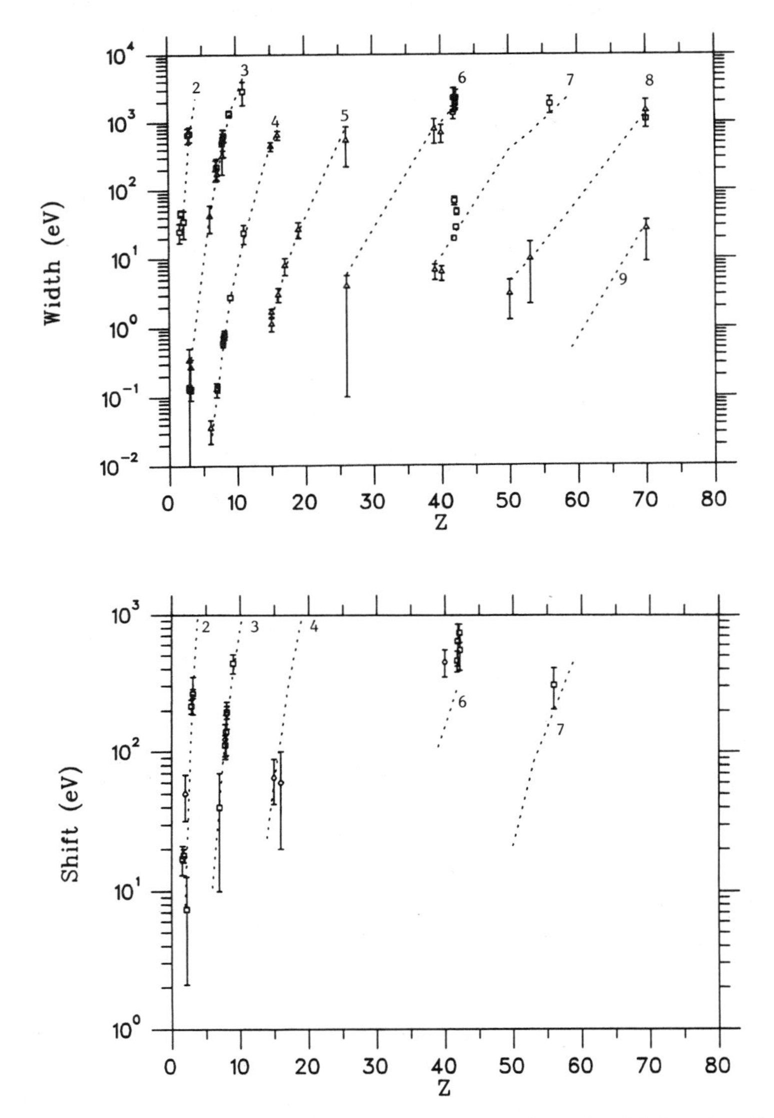

Fig. 4. Measured shifts (with reversed sign) and widths for $\bar{p}$-atoms. The dashed curve shows values calculated using the simple optical model (eq. (5)) with $\bar{a}$ = 1.53 + i2.50 fm.

study of isotope effects in light and medium weight nuclei. The isotope effects are particularly marked in the O isotopes where, for example, the shift increases by (74 ± 25)% from ^{16}O to ^{18}O and the width by (29 ± 12)%.

A variety of theoretical models have been used to calculate strong interaction effects in antiprotonic atoms. The early, pre-LEAR, results were analysed [10] using a simple optical model of the form

$$U = - \frac{2\pi}{\mu} \left(1 + \frac{\mu}{m}\right) \bar{a} \, \rho(r) \tag{5}$$

where μ is again the antiproton-nucleus reduced mass, m the mass of the nucleon, ρ(r) the nuclear density distribution and $\bar{a}$ a complex "effective" scattering length which determines the strength of the potential. A fit [10] to the early data with ρ(r) taken to be the same as the nuclear charge distribution then gave $\bar{a} = 1.53 \pm 0.27 + i(2.50 \pm 0.25)$ fm corresponding to a central potential at the origin $U(o) \approx -(150 + i\ 250)$ MeV. As expected the $\bar{p}$-nucleus interaction is found to be strongly absorptive.

It has been suggested [17] that antiprotonic atom data can be fitted with two families of potentials; one (S) with shallow imaginary and deep real parts and the other (D) with deep imaginary and shallow real parts. A detailed analysis [18] shows however that the S-potentials are spurious and when data for several nuclei are analysed simultaneously only one set of potentials are found corresponding to the set (D) given above.

More recent calculations [19] using a model of the form (eq. (5)) given above have shown that, providing realistic density distributions are used for ρ(r), the large isotope effects observed in ^{16}O, ^{18}O can be reproduced without the need to invoke [13] an enhanced $\bar{p}$-n interaction. However problems in fitting the 3d and 4f levels simultaneously could indicate some inherent limitation in the validity of the model.

An alternative description in terms of an optical potential in which the finite range of the antiproton-nucleon interaction is explicitly taken into account, is given by the folding model.

$$U(r) = \int V_{\bar{p}N}(|\underline{r} - \underline{r}'|) \, \rho(\underline{r}') \, d^3r' \tag{6}$$

The quantity $V_{\bar{p}N}$, which describes the complex antiproton-nucleon interaction potential, may be chosen to fit the free $\bar{p}$-N scattering lengths or taken to have some suitably chosen phenomenological form. Whilst the use of a form for $V_{\bar{p}N}$ which can be related to the free $\bar{p}$N interaction is an attractive feature of this model, which has had considerable success in reproducing the experimental data [20,21], its underlying physical derivation has been strongly criticised [22,23]. A problem with a phenomenological choice for $V_{\bar{p}N}$ is that for antiprotonic atoms measurements of only two (ΔE,Γ) or three (ΔE, Γ, Γ_u) quantities are available for each nucleus and this considerably restricts the information which can be obtained about the optical potential. Attempts to overcome this problem have been made [18] by making a combined analysis of antiprotonic atom and low-energy antiproton scattering data. A difficulty here is that it is then necessary to make some assumption about the energy dependence of $V_{\bar{p}N}$.

An alternative model, which has received a great deal of attention [23,24], is based on the use of a free space $\bar{p}N$ t-matrix or an effective g-matrix in the nuclear medium. Here the optical potential can be written in the general form

$$U(r) = \int \bar{T}_{\bar{p}N} (|\underline{r} - \underline{r}'|) \rho(\underline{r}') d^3r' \tag{7}$$

where $\bar{T}_{\bar{p}N} (|\underline{r} - \underline{r}'|)$ is the Fourier transform of the free two-body scattering matrix averaged over some energy interval for an appropriate combination of spin and isospin. A large number of calculations have been made with this type of model [e.g., 24,25] using a variety of forms for the two-body interaction. Generally the predicted shifts and widths are in agreement with the experimental values. However no one set of predictions gives a markedly superior fit to the measured values so that the relative quality of the theoretical calculations cannot be judged. This general topic of microscopic calculations for $\bar{p}$-atoms and $\bar{p}$-nucleus scattering is covered in much greater detail in the lectures by Jensen [26].

Finally we comment that in all these models some assumptions have to be made as to the form of $\rho(r)$. It is important to study the dependence of the strong interaction effects on the nuclear-shape parameters [27].

2.5 Strong interaction spin-orbit effects

As in the case of the nucleon-nucleon interaction, it is expected that the spin-orbit force will also play an important role in the $\bar{p}$-N interaction. Whilst low energy $\bar{p}$-nucleus scattering experiments have so far shown only rather small polarization effects, the spin-orbit term for the $\bar{p}$-nucleus interaction can also be directly investigated by measuring the strong-interaction shift and width in the separate atomic fine-structure levels. An experiment using the resolved fine-structure components of the "last observable" 9 → 8 transition in the measured X-ray spectrum of $\bar{p}$-^{174}Yb has been carried out [16] at LEAR.

The nucleus ^{174}Yb was chosen since it has zero spin, hence no hyperfine structure, and is an even-even nucleus so that there is no tensor force and the optical potential U(r) has central and spin-orbit terms only.

$$U(r) = U_C(r) + U_{LS}(r) \tag{8}$$

The principle of the measurements is illustrated in figure 5. In the presence of the Coulomb force only the n = 8, ℓ = 7 atomic levels have the electromagnetic fine-structure splitting due to the antiproton magnetic moment (See eq. (4)). The effect of the central strong interaction $U_C(r)$ is to broaden and shift the two components by equal amounts. However, due to their different total angular momentum j, the two fine structure states will be shifted and broadened by the spin-orbit potential $U_{LS}(r)$ by different amounts. By measuring the shifts ($\Delta E_{\ell+\frac{1}{2}}$, $\Delta E_{\ell-\frac{1}{2}}$) and widths ($\Gamma_{\ell+\frac{1}{2}}$, $\Gamma_{\ell-\frac{1}{2}}$) of the two fine structure states separately the contributions of $U_C(r)$ and of $U_{LS}(r)$ to U(r) (eq. (8)) can be separated. The results obtained for the 8j level of ^{174}Yb are given in Table 3.

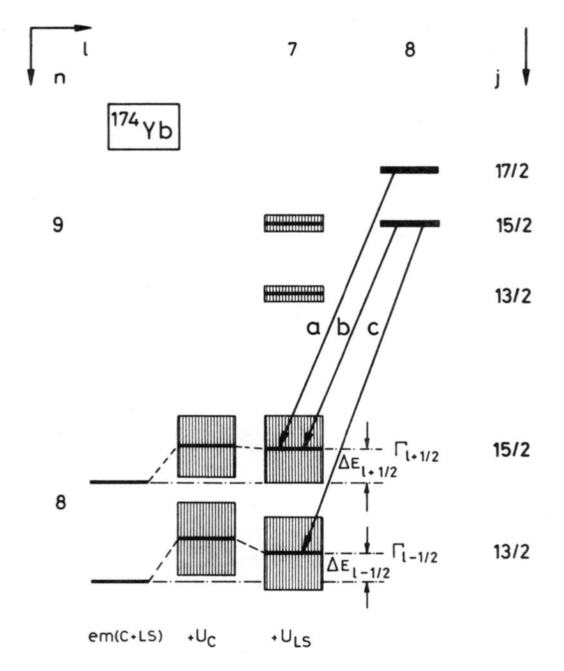

Fig. 5. Level scheme showing strong interaction effects on the fine structure splitting of the n = 8 level in $\bar{p}$-^{174}Yb.

Table 3. Strong-interaction effects in the separated 8j level of ^{174}Yb

$\Delta E_{\ell+1/2}$	$\Delta E_{\ell-1/2}$	δ_E	$\Gamma_{\ell+1/2}$	$\Gamma_{\ell-1/2}$	δ_Γ	Ref.
283±36	341±43	-58±26	1021±41	1216±41	-195±59	Expt [16]
- 60	-131	71	1072	1191	-119	[28]
-120	-160	40	780	862	- 82	[29]
- 63	- 79	16	996	1024	- 28	[26,30]

$\delta_E = \Delta E_{\ell+1/2} - \Delta E_{\ell-1/2}$ $\delta_\Gamma = \Delta\Gamma_{\ell+1/2} - \Delta\Gamma_{\ell-1/2}$

All values in eV.

Also shown in Table 3 are three sets of theoretical calculations [26,28,29,30] for the 8j levels in ^{174}Yb based on various models for the $\bar{p}$-N interaction. For the width values ($\Gamma_{\ell+½}$, $\Gamma_{\ell-½}$ and δ_Γ) the calculations of Sparrow [28] are in good agreement with the measurements. The other two sets [26,29,30] of calculated values are less satisfactory but in all cases the sign of δ_Γ is in agreement with that measured.

For the shift values the situation is confused. The measured mean energy shift $(\Delta E_{\ell+½} + \Delta E_{\ell-½})/2 = 312 \pm 39$ eV is positive (attractive) which means that the strong interaction increases the binding of the level. This result is unusual for $\bar{p}$ atoms where the energy shift ΔE is normally negative (i.e., repulsive). (See Figure 4). It has been suggested by Green et al [31] that part or all of this discrepancy could be due to dynamic E2 coupling of the atomic levels to the first 2+ nuclear state at 76.48 keV due to the large quadrupole moment of ^{174}Yb. This will give an additional attraction of $\approx$ 600 eV to each of the two transitions $\Delta E_{\ell+½}$ and $\Delta E_{\ell-½}$. When this attraction is removed the "corrected measured strong interaction" values become $\Delta E_{\ell+½} = -315 \pm 36$ eV and $\Delta E_{\ell-½} = -255 \pm 43$ eV respectively, in much better agreement with the theoretical estimates. The difference between the measured $\delta_E = -60 \pm 26$ eV and the calculated values however remains unexplained.

2.6 E2 nuclear-resonance effect

In the previous section we have briefly discussed the effect of the first 2+ nuclear level in ^{174}Yb on the $\bar{p}$-^{174}Yb 8j atomic level due to coupling via the large quadrupole moment. A related phenomena of considerable interest is the E2 nuclear resonance effect which occurs when an atomic deexcitation energy matches a nuclear excitation energy and the electric quadrupole coupling induces configuration mixing. The effect [32,33] has previously been observed in pionic [34] and kaonic [35] atoms but particularly large effects were predicted for $\bar{p}$-Mo atoms [33].

The situation for $\bar{p}$-^{94}Mo and $\bar{p}$-^{100}Mo is illustrated in Figure 6. In the case of ^{94}Mo the spacing between the antiprotonic $(n,\ell) = (7,6)$ and (5,4) levels (844.8 keV) is sufficiently close to the first $I^\pi = 2^+$ nuclear excitation energy (871.1 keV) to allow configuration mixing of the $(7,6{:}I^\pi = 0^+)$ and $(5,4{:}I^\pi = 2^+)$ atomic-nuclear states. The admixture of the lower (5,4) antiprotonic level causes an increase in the strong absorption width of the upper (7,6) level which reduces the $(n = 7) \rightarrow (n = 6)$ X-ray intensity compared with that, for example, in ^{92}Mo where the resonance does not occur as the first 2+ nuclear state is at 1.51 MeV. This is precisely what is seen in Figure 7, where antiprotonic X-ray spectra of five different Mo isotopes measured at LEAR by PS186 [15] are shown. Normalizing to the intensity of the 11 → 10 X-ray transition, the 7 → 6 transition has a measured intensity of 11.4 ± 1.5% in ^{94}Mo compared with a value of 33.9 ± 1.3% in ^{92}Mo.

The effect is seen even more dramatically in ^{100}Mo. As shown in Figure 6 the spacing between the (8,7) and (6,5) atomic levels (534.3 keV) is so close to the first 2^+ nuclear excitation energy (535.5 keV) that there is a very strong coupling between the $(8,7;\ I^\pi = 0^+)$ and $(6,5;\ I^\pi = 2^+)$ atomic-nuclear states. As a result the 8 → 7 X-ray transition in $\bar{p}$-^{100}Mo is strongly attenuated (see Figure 7) with an intensity of 5.4 ± 0.7% compared with 121.8 ± 2.2% in ^{98}Mo where there is no resonance.

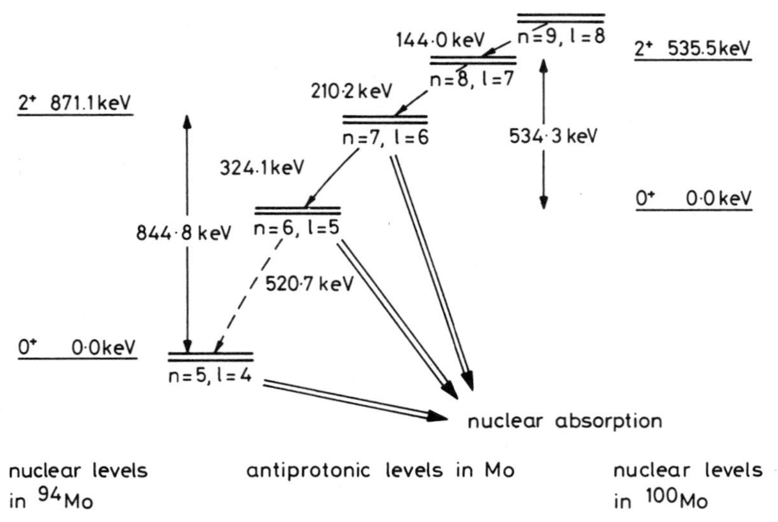

Fig. 6. Nuclear and $\bar{p}$- atomic levels in ^{94}Mo and ^{100}Mo for the E2 resonance effect.

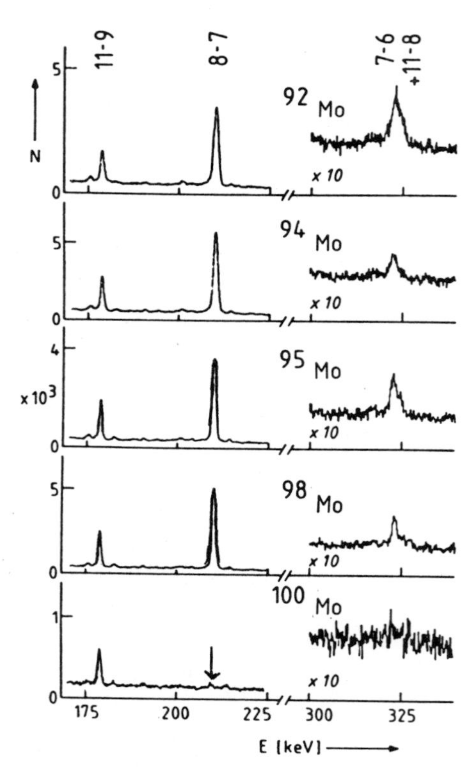

Fig. 7. X-ray spectra for $\bar{p}$-Mo atoms showing the E2 resonance effect for the 8 → 7 and 7 → 6 transitions.

As well as being an interesting example of coupling between nuclear and atomic effects, the results also give further quantitative information about the strong interaction. In the case of $\bar{p}$-Mo the (5,4) level is normally "hidden" by the very strong absorption from the (6,5) level which prevents direct X-ray transitions to the (5,4) level from being seen. However for those isotopes where the resonance effect occurs the width of the (5,4) level can be deduced from its admixture into the (7,6) level. Values for the n = 5 level obtained in this way are in good agreement with those predicted by simple optical model calculations. (See section 2.4).

3 ANTIPROTONIC-HYDROGEN ATOMS

In this section antiprotonic-hydrogen atoms will be discussed. This is a subject which has received considerable attention and three experiments (PS171, PS174 and PS175) were carried out at LEAR in the pre-ACOL era. Much of the theoretical and experimental work has focused on the $\bar{p}$-p system; the topic of $\bar{p}$-d atoms will, however, also be included in this section.

3.1 The atomic cascade

The general features of the atomic cascade in exotic atoms have already been briefly discussed in section 1. Here we concentrate on those aspects which are either particularly important or are unique to the antiprotonic hydrogen system.

The antiprotonic hydrogen atoms are formed when antiprotons stop in H_2 gas. The antiprotons in slowing down ionize the target gas molecules and are then captured to form $\bar{p}$-p atoms in a state of high n (principal quantum number) and high ℓ (angular momentum) [36]. Typically n ≈ 30. The atomic electron is then ejected by the Auger process. This highly excited $\bar{p}$-p atom then deexcites by Coulomb deexcitation (leading to rotational and vibrational excitation of H_2 molecules), by the external Auger effect, involving the ionization of neighbouring H_2 molecules, or by the emission of X-rays. The $\bar{p}$-p atom is relatively compact so that in collisions with neighbouring H_2 molecules it experiences intense and fluctuating electric fields. The Stark effect then gives mixing of the angular momentum states which allows the antiprotons to transfer to s and p states where they rapidly annihilate before reaching the low n states [37,38]. As a result the X-ray yields of transitions between low n states are reduced. Since the Stark effect depends on the collision rate with neighbouring H_2 molecules, the yield of X-rays is expected to vary with target gas density.

The effect is illustrated in Figure 8 where the yields of L X-rays (nd → 2p) predicted by a cascade calculation [39] are plotted as a function of pressure. [In the atomic spectroscopy notation, which is usually used for $\bar{p}$-p atoms, the subscript α, β, γ etc., indicates Δn = 1, 2, 3 etc., so that L_α X-rays are (3d → 2p), L_β (4d → 2p) etc]. The general decrease in the yield of X-rays due to the Stark effect is shown by the total yield of L X-rays which drops by almost two orders of magnitude as the gas density increases from 0.01 to $40\rho_{STP}$. Due to the relatively large P-state annihilation width for $\bar{p}$-p atoms the yield of K X-rays (np → 1s) is much lower. For example the 2p annihilation width is predicted to be in the range 20 - 40 meV compared with a radiative width for the 2p → 1s transition of 0.38 meV. From eq. (3) the relative yield for the K_α (2_p → 1s) X-rays compared to the total yield of L X-rays is therefore expected to be ∿ 1% and the yield of K_α X-rays as a function of gas density follows the shape of the L_{total} curve but a factor 100 lower. However the total yield of K X-rays over the same pressure range only drops by a factor ∿ 2 due to a slow increase in the summed yield of K X-rays with Δn≥3 (denoted by $K_{\geq\gamma}$).

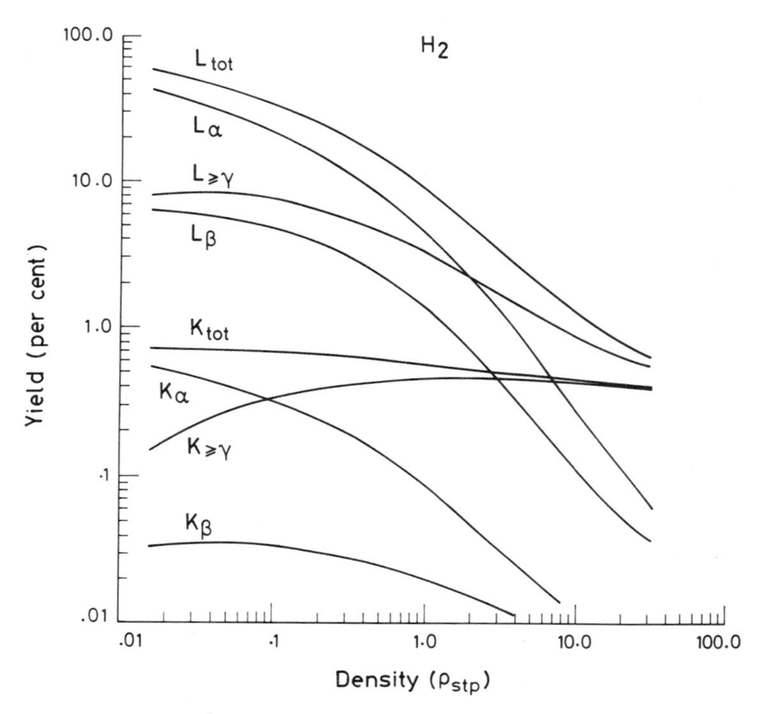

Fig. 8. Cascade calculation predictions for $\bar{p}$-p atomic K and L X-ray yields as a function of H_2 gas density. (See ref. [39]).

3.2 Experimental

One of the principal objectives of the experiments carried out at LEAR was to identify and to measure the shifts and widths of K X-rays from $\bar{p}$-p atoms. The antiproton-nucleon strong interaction broadens the 1s level and shifts the K X-ray energies with respect to their electromagnetic values. Theoretical models predict shifts (ΔE_{1s}) and widths (Γ_{1s}) for the 1s state in the range 0.5 - 1.0 keV. The energy shift is expected to be repulsive i.e., negative ΔE. However if baryonium bound states exist close to threshold, they could significantly alter these predictions [40]. The L X-ray measurements can also be used, together with the yield of K_α X-rays, to determine the strong interaction width (Γ_{2p}) of the 2p-state (See section 1).

The calculated electromagnetic (QED) energies for the K and L series X-rays from $\bar{p}$-p atoms are listed in Table 4. For the $\bar{p}$-d system the energies are increased by a factor $\sim 4/3$.

Table 4. Calculated electromagnetic energies for $\bar{p}$-p atoms

Transition	Energy (keV)	Transition	Energy (keV)
$\bar{p}$-p			
K_α (2p → 1s)	9.408	L_α (3d → 2p)	1.737
K_β (3p → 1s)	11.144	L_β (4d → 2p)	2.345
K_γ (4p → 1s)	11.752	L_γ (5d → 2p)	2.625
K_δ (5p → 1s)	12.032	L_δ (6d → 2p)	2.775
K_∞ (∞p → 1s)	12.534	L_∞ (∞d → 2p)	3.126

The L and K series X-ray energies to be measured lie in the range 1.7 to 3.2 keV and 9 to 13 keV. A number of different types of X-ray detectors are available for this energy range; e.g., proportional counters, Si(Li) and gas scintillation proportional detectors (GSPD). However it should be emphasised that the expected intensities of the K X-ray lines are low, typically less than 1%, so that reduction of background is of prime importance.

The predicted change in the relative yield of K X-ray transitions with density (see Figure 8) can be used to confirm the identification of weak K X-ray lines. At gas densities greater than ρ_{STP} the yield of the higher K-series ($K_{\geq\gamma}$) X-rays will dominate and only a single broadened X-ray peak corresponding to an unresolved complex of lines, each broadened by the strong interaction and with energies spaced over a range of 0.78 keV (see Table 4) is expected to be seen. At gas densities in the region of 0.1 ρ_{STP}, however, the yields of K_α and $K_{\geq\gamma}$ X-rays are comparable whilst the yield of K_β X-rays remains low. As the energy difference between the K_α and $K_{\geq\gamma}$ lines in the X-ray spectrum is approximately 2.7 keV (see Table 4), which is large compared to the strong interaction width of $\sim$ 0.5 to 1.0 keV, two distinct and well separated peaks should be seen with an X-ray detector of good energy resolution.

A further check on the origin of possible $\bar{p}$-p X-ray peaks can be made by filling the target with deuterium gas. As the lines expected from $\bar{p}$-d atoms are at significantly higher energies it is expected that the $\bar{p}$-d spectrum will show only a smooth background in the energy region of the $\bar{p}$-p X-rays.

A quantity which had a considerable influence on the design of the experiments carried out at LEAR is the fraction of antiprotons which stop in the H_2 gas target. Values of the stopping fraction for a 30 cm H_2 target, typical of the PS174 experiment and for the 1m target used by the PS171 (Asterix) experiment are given in Table 5.

Table 5. Stopping of antiprotons in H_2

Beam		Fraction (%) stopping in 30cm H_2				
Energy (MeV)	P (MeV/c)	Gas Density (ρ_{STP}) 0.25	0.5	1.0	2.0	10.0
46.8	300	2	5	9	17	72
21.1	200	24	46	78	100	100
5.3	100	80	100	100	100	100

1 m H_2 gas target at STP

Stops 20% at 300 MeV/c
Stops 86% at 200 MeV/c

With the initial LEAR beams at 300 MeV/c it was necessary either to use H_2 gas at a relatively high density (10 ρ_{STP}) in order to get a large number of antiprotons stopping in the target, as in PS174, or to use a long target (PS171) and to accept a reduced stopping fraction. Once 100 MeV/c beams become available it was possible to use the lower gas densities (0.25 ρ_{STP} in PS174) which are necessary if K_α X-rays are to be observed and the 2p-width measured. These comments do not apply to experiment PS175 which used a novel technique (cyclotron trap) in order to stop the antiprotons. The characteristics of the three experiments are summarised in Table 6.

Table 6. Characteristics of $\bar{p}$-p X-ray experiments

(a) PS171 (Asterix)

H_2 Gas 1 atm.
XDC (Granular proportional counter)
Resln. 2.4 keV at 8.5 keV
Acceptance 90% of 4π M-L-K X-ray coincidences
(Anti-)coincidence with charged particles

(b) PS174

H_2/D_2 gas 0.125 - 10 ρ_{STP}
Si(Li) detector 300 mm² Resln. 370 eV at 11 keV
GSPD detector 30 cm² Resln. 850 eV at 11 keV

(c) PS175

Cyclotron trap
H_2/D_2 gas 0.016 to 0.3 ρ_{STP}
Si(Li) detector 300 mm² Resln. 460 eV at 5.9 keV
(630 eV at 11 keV)*

Detector resolution quoted is FWHM
*Estimated value

The three experiments are seen to be complementary. Although the PS171 experiment had relatively poor energy resolution, its ability to detect M-L_α-K_α and L-K_α coincidences in anticoincidence with charged particles proved to be a powerful technique. The PS174 and PS175 experiments had detectors with good energy resolution and were able to make measurements over a range of target densities. The ability of PS175 to cover very low gas pressures and of PS174 to use Gas Scintillation Proportional Detectors with relatively large acceptance were important features of these experiments. Further experimental details together with a discussion of the results obtained are given in the following three sub-sections.

3.3 PS171 (ASTERIX) Experiment

In the ASTERIX experiment the antiprotons were stopped in a 90 cm long, 16 cm diameter, gaseous hydrogen target at STP. With a 105 MeV/c incident antiproton beam, essentially all the antiprotons stopped in the target gas (see Table 5). The target was surrounded by a cylindrical X-ray drift chamber [41] filled with 50% Argon + 50% Ethane gas and separated from the hydrogen target by a thin 6 μm aluminized Mylar window which was transparent to X-rays with energies greater than 1 keV. The target and central X-ray detector were then surrounded by seven cylindrical multiwire proportional chambers (MWPC's) and a solenoidal magnet providing a field of 0.8T. The momentum of charged particle tracks from $\bar{p}$-p annihilations in the target could be measured over a solid angle of about 65% of 4π.

The X-ray detector has been described in detail elsewhere [41]. Its particular features of interest for the detection of $\bar{p}$-p X-rays were its high geometrical acceptance (90% of 4π), low energy threshold with a detection efficiency greater than 5% between 1.0 and 16 keV and its ability to give three-dimensional localization of the X-ray conversion point so enabling multiple X-ray events to be resolved and also giving discrimination against charged particles. The detector resolution was 2.4 keV (fwhm) at 8.5 keV.

In the initial results published by this collaboration [42] single X-ray events in anticoincidence with charged particles were measured. This latter anticoincidence gives rejection of events where low energy Bremsstrahlung is emitted during the sudden acceleration of charged pions from the $\bar{p}$-p annihilation [43]. The observed spectrum contained a large broad bump in the energy region from 5 to 12 keV which was analysed in terms of ^{54}Mn, $\bar{p}$-Aℓ(11 → 10) and $\bar{p}$-Aℓ(10 → 9) contamination lines together with K_α and K_β lines of roughly equal intensity from $\bar{p}$-p atoms. Their best fit to the spectrum gave only very weak intensity for the higher K lines in disagreement with cascade calculations (see Figure 8) and later experiments [39]. The best fit measured strong interaction shift for the 1s state in $\bar{p}$-p atoms was $\Delta E = -0.5 \pm 0.3$ keV. Only an upper limit to the strong interaction width of $\Gamma_{1s} \leq 1$ keV was obtained.

In a later measurement [44] the very powerful X-ray coincidence technique was used and K_α X-rays from $\bar{p}$p atoms were identified by demanding coincidences with L X-rays populating the 2p-level. Again the background due to inner Bremsstrahlung was suppressed by selecting events annihilating into neutral states only, i.e., no charged particles. In the observed X-ray spectrum, see Figure 9, the K_α line is clearly seen at an energy of $E(K_\alpha) = 8.67 \pm 0.15$ keV. The measured strong interaction shift and width are $\Delta E_{1s} = -0.70 \pm 0.15$ keV and $\Gamma_{1s} = 1.60 \pm 0.40$ keV. These results are compared with those from other experiments in section 3.6.

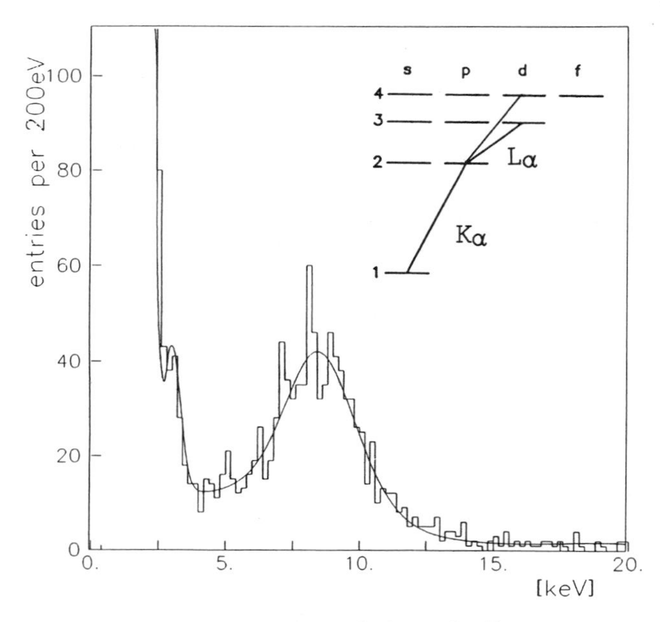

Fig. 9. X-ray spectrum for K_α transitions in $\bar{p}$-p atoms measured [44] by the PS171 collaboration.

3.4 PS174 Experiment

The PS174 experiment used a conventional gas target in which the pressure could be reduced (target density $\rho \leq \rho_{STP}$) or the gas cooled (target gas density $\rho > \rho_{STP}$). In the initial phases of the experiment a Si(Li) detector together with a 6μ Mylar window was used [39,45] to measure the energy spectrum of X-rays down to 1.5 keV. The observed spectra showed a relatively large continuum background, much of which was due to the Compton scattering of 50 - 500 keV γ-rays in the Si(Li) crystal. The γ-rays from the decay of π^0 from $\bar{p}$ annihilations shower in the surrounding material. To reduce this background the mass of the target assembly was kept low and a NaI(Tℓ) annulus around the Si(Li) crystal used as a Compton shield. Suppression factors of 3 - 5, depending on the X-ray energy were obtained.

The X-ray spectra measured [39] at H_2 gas densities of 0.25, 0.92, 2.0 and 10.0 ρ_{STP} are shown in Figure 10. The higher energy peak seen at all 4 gas densities was attributed to the complex of $K_{\geq\gamma}$ transitions (see section 3.1). Cascade calculations (see Figure 8) predict that the K_α line should only be seen at low gas densities. The lower energy peak, which was only clearly observed at 0.25 and 0.92ρ_{STP} and which had an energy separation consistent with that expected for the difference between K_α and $K_{\geq\gamma}$ transitions was therefore identified as the K_α line. The origin of these lines as being due to X-rays from $\bar{p}$-p atoms was confirmed by filling

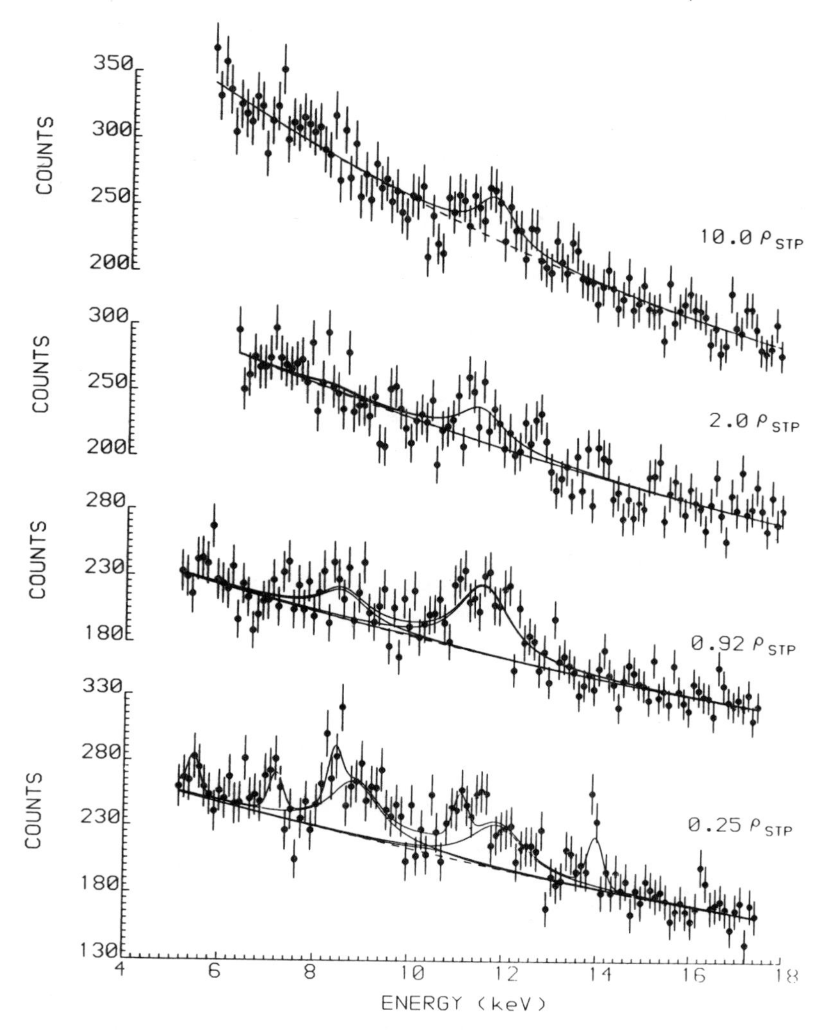

Fig. 10. X-ray spectra measured [39] for H_2 gas at densities of 0.25 and 0.92, 2.0 and 10.0ρ_{STP} by the PS174 collaboration using a Si(Li) detector.

the target with D_2 when the lines disappeared and only a smooth background remained. The additional structure seen in the spectrum at 0.25 ρ_{STP} is due to contaminant $\bar{p}$-N and $\bar{p}$-O X-rays.

Values for the strong interaction shift and width for the 1s state were obtained and are given in Table 7. Yields of the K and L X-ray lines were measured and are in good agreement with cascade calculations [39]. From the measured total yield of L X-rays and the yield of K_α X-rays a value for the 2p width of Γ_{2p} = 45 ± 10 meV was obtained [39].

In the later stages of this experiment two Gas Scintillation Proportional Detectors [46] were added to the experimental arrangement. These detectors have a number of significant advantages. Whilst their energy resolution (∿ 850 eV fwhm at 11 keV) is poorer than that of a Si(Li) detector (∿ 370 eV at 11 keV), their acceptance can readily be made about

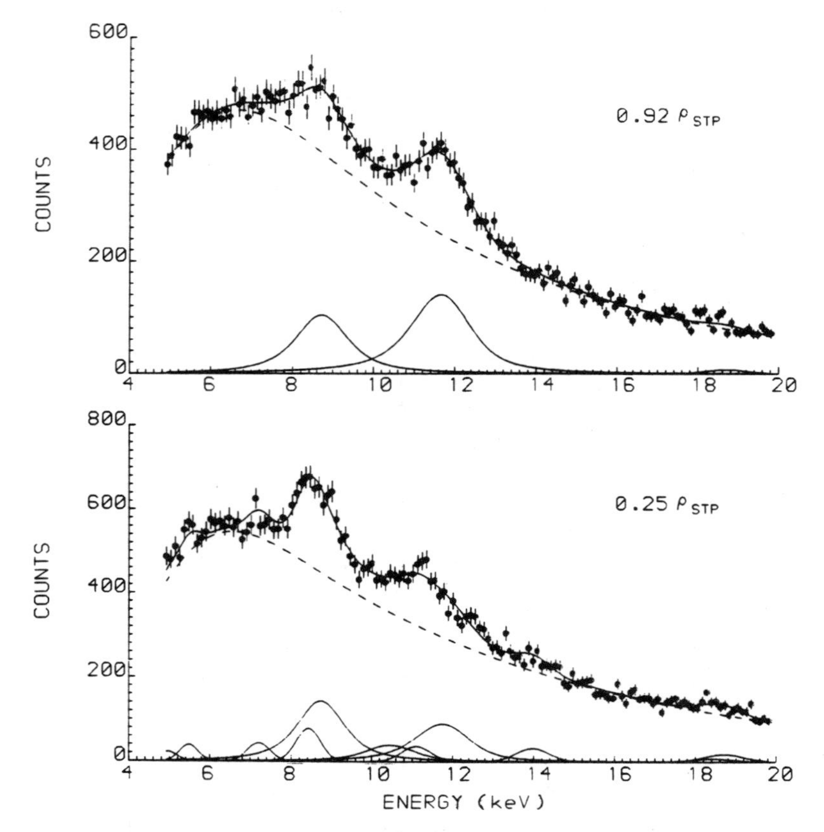

Fig. 11. X-ray spectra measured [47] for H_2 gas at densities of 0.25 and $0.92\rho_{STP}$ by the PS174 collaboration using a GSPD detector.

ten times as large. Moreover the smaller amount of material in the GSPD, 2cm Xe gas compared to 5mm Si in a solid state detector, results in a lower continuum background from Compton scattering of low energy γ-rays.

Spectra obtained with these detectors [47] for H_2 gas densities of 0.25 and 0.92 ρ_{STP} are shown in Figure 11 where peaks due to the K_α and $K_{\geq\gamma}$ X-ray transitions are clearly seen. (The spectrum at 0.25 ρ_{STP} again contains contaminant $\bar{p}$-N and $\bar{p}$-O lines). Measured shift and width values for the 1s level are presented in Table 7.

Measurements with the GSPD's and the target filled with D_2 gas [47] showed only a smooth spectrum with no evidence for $\bar{p}$-d K X-rays. Wycech et al [48] have predicted spin-averaged strong interaction shifts and widths ΔE_{1s} = - 2.14 keV, Γ_{1s} = 1.26 keV for the 1s state in $\bar{p}$-d atoms. From the measured $\bar{p}$-d spectra upper limits at 95% CL of < 0.08% and < 0.05% per stopped antiproton at 0.25 and 0.92 ρ_{STP} were obtained for a 2 keV wide line over an energy range corresponding to 1s state shifts of - 2.6 to 1.4 keV for the complex of $K_{\geq\gamma}$ lines. Cascade calculations for $\bar{p}$-d atoms [39,47] predict corresponding yields of 0.032% and 0.043% per stopped antiproton.

Table 7. **PS174 Results**
Energy shift and width of 1s state

Target density (ρ_{STP})	Energy shift ΔE_{1s} (keV)	Width Γ_{1s} (keV)
Si(Li) ref. [39]		
10.0	-0.65 ± 0.14	0.77 ± 0.46
2.0	-1.00 ± 0.25	1.10 ± 0.80
0.92	-0.86 ± 0.09	1.07 ± 0.34
0.25	-0.54 ± 0.12	1.19 ± 0.41
0.25	-0.76 ± 0.07	0.56 ± 0.34
Average	-0.75 ± 0.06	0.90 ± 0.18
GSPD ref. [47]		
0.92	-0.46 ± 0.11	0.94 ± 0.30
0.92	-0.72 ± 0.06	1.06 ± 0.18
0.25	-0.80 ± 0.04	1.20 ± 0.15
0.25	-0.69 ± 0.05	1.15 ± 0.18
Average	-0.73 ± 0.05	1.13 ± 0.09

3.5 PS175 (Cyclotron Trap) Experiment

Experiment PS175 uses a new [49,50] experimental technique, the cyclotron trap, in order to slow down antiprotons in a low density H_2 target. Antiprotons are radially injected into a chamber, filled with H_2 gas at low pressure and placed between the poles of a superconducting magnet which gives a weak focussing field as typically used in cyclotrons. The energy loss due to the target gas together with the focussing properties of the cyclotron field then causes the antiprotons to spiral towards the centre of the chamber where they stop and are captured to form $\bar{p}$-p atoms. At a target pressure of 60 mbar H_2, approximately 40% of the injected antiprotons stop in the centre. X-rays from the stopped antiprotons are then detected by a Si(Li) detector positioned in a warm bore hole along the magnet axis. The use of this technique allows a very low density target to be used where the Stark effect has little effect and where the K_α line should be clearly observed. (See Figure 8).

Due to a number of difficulties with the setting up and operation of the apparatus only results of limited statistical significance are available for the K X-rays. Figure 12 shows the spectrum [51] measured with H_2 at 30 mbar pressure. Also shown is the fit to the data showing the presence of fluorescent Cu X-rays and $\bar{p}$-O(8 - 7) and $\bar{p}$-O(7 - 6) contaminant lines as well as the broadened K_α and K_∞ lines. Fixing the separation between the two K lines to the QED value and assuming the widths of the two lines are the same then gave $\Delta E_{1s} = -0.66 \pm 0.13$ keV and $\Gamma_{1s} = 1.13 \pm 0.23$ keV. From the measured L X-ray and K_α yields a value for the width of the 2p-state $\Gamma_{2p} = 37 \pm 13$ meV was obtained. This experiment will continue after the ACOL shutdown when it is hoped to measure separately the shift and widths of the singlet and triplet S-states.

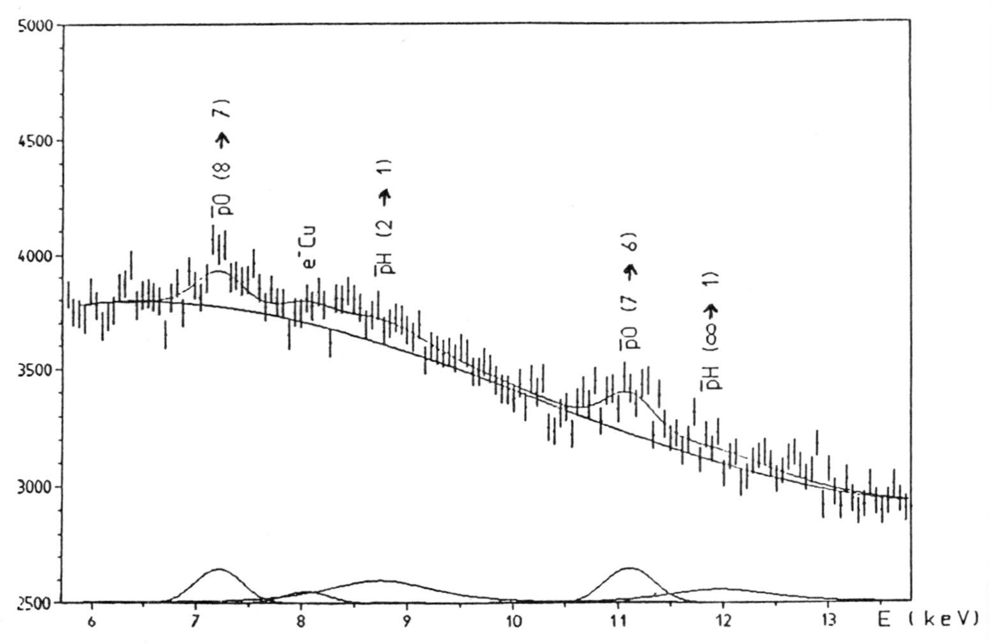

Fig. 12. X-ray spectrum measured [51] for H_2 gas at 30 mbar pressure by the PS175 collaboration using the "cyclotron trap".

3.6 Strong interaction effects in $\bar{p}$-p atoms

The strong interaction shift and width values for the 1s-state and the width values for the 2p-state obtained from the three LEAR experiments described above are summarised in Table 8 and in Figure 13 where they are compared with the predictions of a number of potential models [52-58]. Also shown are the predictions of the model of Shapiro et al [59,60] which includes P-states of a quasi-nuclear nature close to threshold in the nucleon-antinucleon system. Since the $^1S_0$ and $^3S_1$ states are not resolved experimentally and as the $^1S_0$ - $^3S_1$ splitting is expected to be small the theoretical values quoted are averaged over spin. In the experiment by the PS171 (ASTERIX) group [42,44] X-rays from $\bar{p}$-p atoms with only neutral particles in the annihilation final state were measured. Annihilation into only neutral pions in coincidence with a K_α X-ray will pick out the singlet S-state ($J^{PC} = 0^{-+}$) because of C-parity conservation and so a large contribution from the singlet state to the observed K_α line is expected. However, neutral annihilations involving uncharged kaons or radiative decays may proceed from the triplet state; their relative contribution to the K_α line is unknown but is expected to be small. The PS171 results may not therefore be directly comparable with the PS174 and PS175 results. However the results from the three experiments are in good agreement and we have also quoted mean values in Table 8. There is rather general overall agreement between the predicted theoretical values and those measured experimentally.

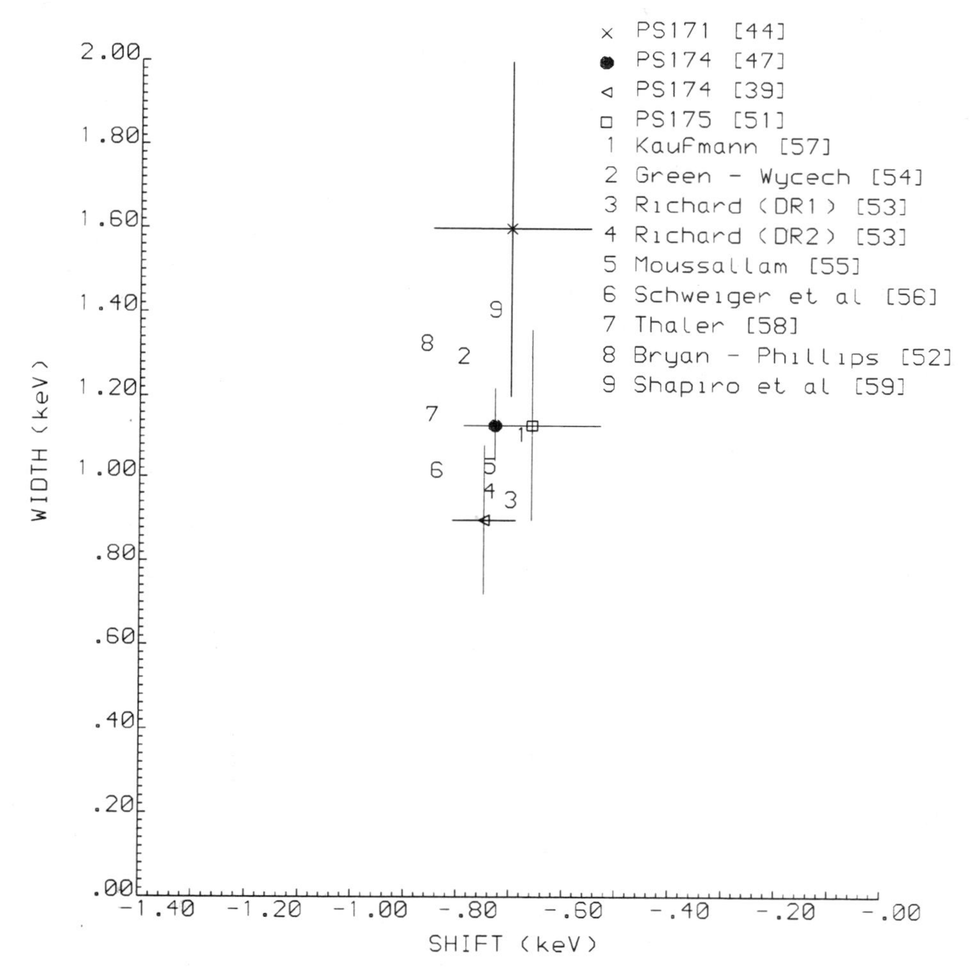

Fig. 13. Summary of measured strong interaction shift and width values for the 1s-state in $\bar{p}$-p atoms. Also shown are the predictions of a number of potential models.

Table 8. Strong interaction effects in $\bar{p}$-p atoms

	Energy shift ΔE_{1s} (keV)	Width Γ_{1s} (keV)	Width Γ_{2p} (meV)
EXPERIMENT			
PS171 [42]	-0.50 ± 0.30	< 1.0	
PS171 [44]	-0.70 ± 0.15	1.60 ± 0.40	
PS174 [39]	-0.75 ± 0.06	0.90 ± 0.18	45 ± 10
PS174 [47]	-0.73 ± 0.05	1.13 ± 0.09	
PS175 [51]	-0.66 ± 0.13	1.13 ± 0.23	37 ± 13
Mean value	-0.73 ± 0.04	1.11 ± 0.07	42 ± 8
THEORY			
Bryan & Phillips [52]	-0.87	1.31	
Richard & Sainio [53]	-0.71	0.93	27
	-0.75	0.95	24
Green & Wycech [54]	-0.80	1.28	
Moussallam [55]	-0.75	1.01	27
Schweiger et al [56]	-0.85	1.0	
Kaufmann [57]	-0.69	1.09	
Thaler [58]	-0.86	1.14	
Shapiro [59,60]	-0.74	1.39	39

The strong interaction shift and width for the 1s-state in $\bar{p}$-p atoms are related to the S-wave scattering length. The simplest relationship [61] is

$$\Delta E + i\frac{\Gamma}{2} = \frac{2(\mu c^2)^2\,\alpha^3}{\hbar c}\,a_c = 866.8\,a_c\ \text{eV fm}^{-1} \tag{9}$$

where μ is the reduced mass of the antiproton, α the fine structure constant and a_c the (complex) S-wave scattering length. The above formula is correct to first order in a_c/B where B is the 1s-state Bohr radius $B = \hbar c/\mu c^2\alpha = 57.6$ fm. Including second order terms [61,62] gives

$$\Delta E + i\frac{\Gamma}{2} = \frac{2(\mu c^2)^2\,\alpha^3}{\hbar c}\,a_c\,[1 - 3.154\,a_c/B] \tag{10}$$

for values of the effective range $r_0 \ll B$. Substituting in the above two formulae the mean measured values $\Delta E = -0.73 \pm 0.04$ keV and $\Gamma = 1.11 \pm 0.07$ keV (see Table 8) gives
$a_c = -0.842 \pm 0.046 + i(0.640 \pm 0.040)$fm and
$a_c = -0.824 \pm 0.042 + i(0.587 \pm 0.037)$fm respectively.

It should be noted that a_c is the scattering length defined in the presence of the Coulomb field ie., the so-called Coulomb-modified scattering length. The relationship between a_c and a_s, the purely hadronic S-wave scattering length, has been discussed by a number of authors [e.g., 58] who have attempted to connect these two quantities in a model independent way. An alternative approach is to fit the measured S-wave shift and width values using a simple optical model potential

$$U(r) = (V + iW)e^{-\left(\frac{r}{a_g}\right)^2} \tag{11}$$

by varying V and W for a given value of the Gaussian shape parameter a_g. Switching off the Coulomb potential and using the same values of V, W and a_g, the value of the S-wave scattering length a_s can be calculated [63]. Values of a_s obtained in this way for the mean shift and width values quoted in Table 8 are given in Table 9 for a number of values of a_g.

Table 9. Values of S-wave scattering length

a_g(fm)	Re a_s (fm)	Im a_s (fm)	ρ
1.0	-0.907 ± 0.062	0.858 ± 0.047	-1.06 ± 0.09
2.0	-0.939 ± 0.063	0.858 ± 0.056	-1.09 ± 0.10
3.0	-0.972 ± 0.060	0.869 ± 0.054	-1.12 ± 0.10

$$\rho = \frac{\text{Re } a_s}{\text{Im } a_s}$$

The values of a_s show only a small sensitivity to the value chosen for a_g and are in moderate agreement [47] with the predictions of potential models for the $\bar{p}$-p interaction.

A quantity which has aroused considerable interest recently is ρ, which is defined as the ratio of the real to imaginary parts of the forward amplitude f(0) for $\bar{p}$-p scattering. At zero energy this becomes

$$\rho(0) = \frac{\text{Re } f(0)}{\text{Im } f(0)} = \frac{\text{Re } a_s}{\text{Im } a_s} \tag{12}$$

The value of a_s determined above gives $\rho(0) = -1.09 \pm 0.10$. (The use of eq. (9) or (10) would give values $\rho(0) = -1.32 \pm 0.11$ or $\rho(0) = -1.40 \pm 0.11$ respectively). This value is compared with those determined from small angle Coulomb interference measurements [64-68] for $\bar{p}$-p scattering at momenta below 800 MeV/c in Figure 14. Of particular interest is the variation in ρ for momenta between 150 and 400 MeV/c which is not predicted by simple potential models. As an example the continuous curve in the figure shows values of ρ predicted [69] by the Paris potential [70].

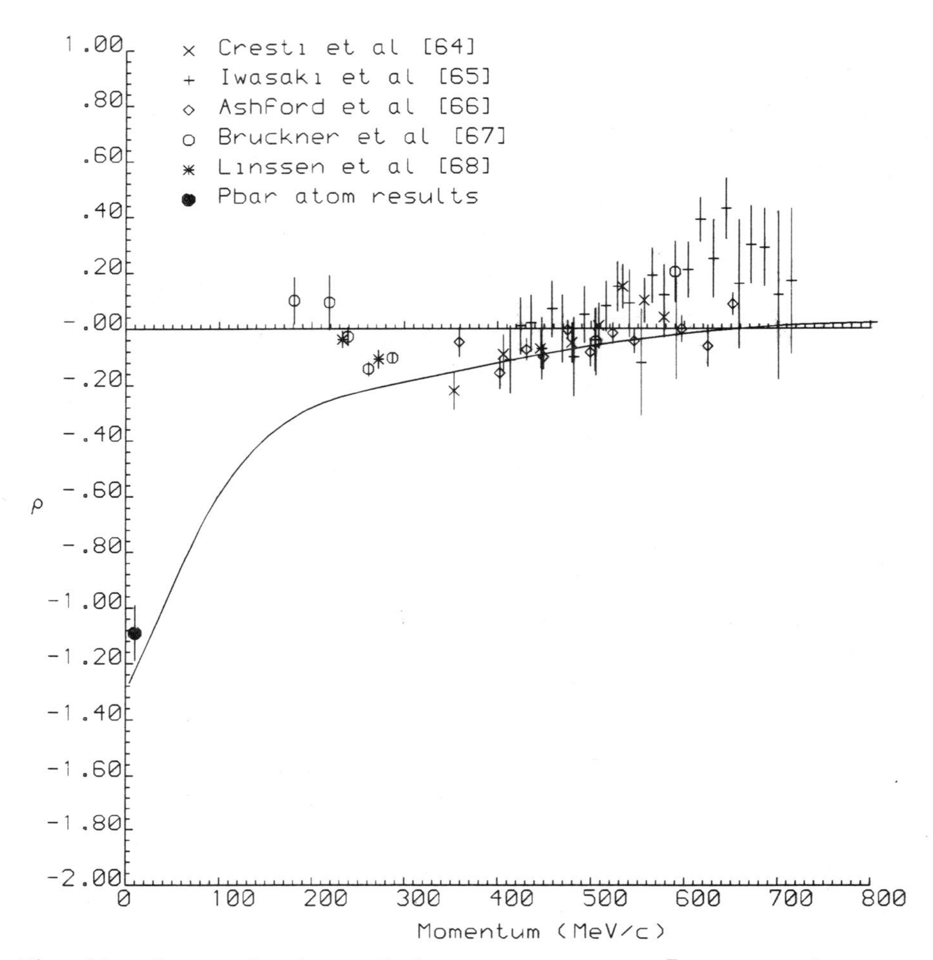

Fig. 14. Measured values of the ρ parameter for $\bar{p}$-p scattering over the momentum range 0 to 800 MeV/c. The continuous curve shows the predictions [69] of the Paris potential.

Effective range analyses [71,72] of low-energy antinucleon-nucleon scattering and atom data have suggested that the variation in ρ is due to S-P wave interference. Dispersion relation analyses [73] seem to require the introduction of a pole term in the region just below threshold. On the other hand the $\bar{p}$-p atom data and the variation in ρ has been used to strongly support the existence of P-wave quasi-nuclear states [59,60] in the nucleon-antinucleon system close to threshold. Measurements of strong interaction effects in $\bar{p}$-p atoms for singlet and triplet S-states separately and improved measurements of the 2p-state width would give important tests of the validty of these various suggestions.

References

1 C J Batty, Soviet Journal of Particles and Nuclei **13** (1982) 71.
2 G Backenstoss, J Egger et al., Nucl. Phys. **B73** (1974) 189.
3 H Poth, H Barth et al., Nucl. Phys. **A466** (1987) 667.
4 J D Davies, T P Gorringe et al., Phys. Lett. **145B** (1984) 319.
5 E Hu, Y Asano et al., Nucl. Phys. **A254** (1975) 403.
6 P Roberson, T King et al., Phys. Rev. **C16** (1977) 1945.
7 B L Roberts, Phys. Rev. **D17** (1978) 358.
8 Particle Data Group, Phys. Lett. **170B** (1986) 1.
9 A Kreissl, A D Hancock et al., Z. Phys. **C37** (1988) 557.
10 C J Batty, Nucl. Phys. **A372** (1981) 433.
11 C A Baker, C J Batty et al., (to be published).
12 R Bacher, P Blüm et al., Proceedings of Antiproton 86; 8th European Symposium on nucleon-antinucleon interactions (Thessaloniki), (World Scientific, 1987) p223.
13 Th. Köhler, P Blüm et al., Phys. Lett. **B176** (1986) 327.
14 D Rohmann, H Barth et al., Z. Phys. **A325** (1986) 261.
15 W Kanert, F J Hartmann et al., Phys. Rev. Lett. **56** (1986) 2368.
16 A Kreissl, A D Hancock et al., Z. Phys. **A329** (1988) 235.
17 C Y Wong, A K Kerman et al., Phys. Rev. **C29** (1984) 574.
18 C J Batty, E Friedman and J Lichtenstadt, Phys. Lett. **142B** (1984) 241 and Nucl. Phys. **A436** (1985) 621.
19 C J Batty, Phys. Lett. **B189** (1987) 393.
20 A Deloff and J Law, Phys. Rev. **C10** (1974) 2657.
21 J Thaler, J Phys. G: Nucl. Phys. **11** (1985) 689.
22 J R Rook, Nucl. Phys. **A326** (1979) 244.
23 A M Green and J A Niskanen, Prog. Nucl. Part. Phys. **18** (1987) 93.
24 O Dumbrajs, H Heiselberg et al., Nucl. Phys. **A457** (1987) 491.
25 A M Green and S Wycech, Nucl. Phys. **A377** (1982) 441.
A M Green, W Stepien-Rudzka and S Wycech, Nucl. Phys. **A399** (1983) 307.
26 A S Jensen, (These proceedings).
27 C J Batty, E Friedman, H J Gils and H Rebel, Advances in Nuclear Physics (to be published) KfK-Preprint 87-1.
28 D A Sparrow, Phys. Rev. **C35** (1987) 1410.
29 J Wicht, H von Geramb, (Private communication to authors of ref [16]).
30 H Heiselberg, (Private communication to authors of ref [16]).
31 A M Green, G Q Liu and S Wycech, Nucl. Phys. **A483** (1988) 619.
G Q Liu, (These proceedings).
32 M Leon, Phys. Lett. **50B** (1974) 425; Phys. Lett. **53B** (1974) 141.
33 M Leon, Nucl. Phys. **A296** (1978) 361.
34 J J Reidy, M Nicholas et al., Phys. Rev. **C32** (1985) 1646.
35 C J Batty, S F Biagi et al., Nucl. Phys. **A296** (1978) 361.
36 J S Cohen, Phys. Rev. **A36** (1987) 2024.
37 M Leon and H Bethe, Phys. Rev. **121** (1962) 636.
38 E Borie and M Leon, Phys. Rev. **A21** (1980) 1460.
39 C A Baker, C J Batty et al., Nucl. Phys. **A483** (1988) 631.
40 I S Shapiro, Phys. Rep. **35C** (1978) 129.
41 U Gastaldi, Nucl. Inst. Meth **157** (1978) 441.
U Gastaldi et al., Nucl. Inst. Meth **176** (1980) 99.
42 S Ahmad, C Amsler et al., Phys. Lett. **157B** (1985) 333.
43 R Rückl and C Zupancic, Phys. Lett. **150B** (1985) 225.
44 M Ziegler, S Ahmad et al., Phys. Lett. **B206** (1988) 151.
45 T P Gorringe, J D Davies et al., Phys. Lett. **162B** (1985) 71.
46 W J C Okx, C W E van Eijk et al., IEE Trans. Nucl. Sci. **N533** (1986) 391.
W J C Okx, C W E van Eijk et al., Nucl. Inst. Meth. **A252** (1986) 605.

47 C W E van Eijk, R W Hollander et al., Nucl. Phys. **A** (to be published).
48 S Wycech, A M Green and J A Niskanen, Phys. Lett. **152B** (1985) 308.
49 L M Simons, Ettore Majorana International Science Series: Fundamental Symmetries (Plenum Press, New York) **31** (1987) 89.
50 L M Simons, Physica Scripta **T22** (1988) 89.
51 R Bacher, P Blüm et al., Ettore Majorana International Science Series: Fundamental Symmetries (Plenum Press, New York) **31** (1987) 125.
52 R A Bryan and R J N Phillips, Nucl. Phys. **B5** (1968) 201 and **B7** (1968) 481.
53 J M Richard and M E Sainio, Phys. Lett. **110B** (1982) 349.
54 A M Green and S Wycech, Nucl. Phys. **A377** (1982) 441.
55 B Moussallam, Z. Phys. **A325** (1986) 1.
56 W Schweiger, J Haidenbauer and W Plessas, Phys. Rev. **C32** (1985) 1261.
57 W B Kaufmann, Phys. Rev. **C19** (1979) 440.
58 J Thaler, J. Phys. G: Nucl. Phys. **9** (1983) 1009.
59 O D Dalkarov, K V Protasov and I S Shapiro, Lebedev Physical Institute Preprint 37 (1988).
60 I S Shapiro, (These proceedings).
61 T L Trueman, Nucl. Phys. **26** (1961) 57.
62 W Stepien-Rudzka and S Wycech, Nukleonica **22** (1977) 929.
63 C J Batty, Nucl. Phys. **A411** (1983) 399.
64 M Cresti, L Peruzzo and G Sartori, Phys. Lett. **132B** (1983) 209.
65 H Iwasaki, H Aihara et al., Nucl. Phys. **A433** (1985) 580.
66 V Ashford, M E Sainio et al., Phys. Rev. Lett. **54** (1985) 518.
67 W Brückner, H Döbbeling et al., Phys. Lett. **158B** (1985) 180.
68 L Linssen, C I Beard et al., Nucl. Phys. **A469** (1987) 726.
69 M Lacombe, B Loiseau et al., Phys. Lett. **124B** (1983) 443. B Moussallam (Private communication).
70 J Côté, M Lacombe et al., Phys. Rev. Lett. **48** (1982) 1319.
71 I L Grach, B O Kerbikov and Yu A Simonov, Phys. Lett. **B208** (1988) 309 and ITEP preprint 210 (1987).
72 J Mahalanabis, H J Pirner and T-A Shibata, Nucl. Phys. **A** (to be published) and CERN Report TH4833/87.
73 P Kroll and W Schweiger. Proceedings 4th LEAR Workshop, Villars sur Ollon, Nuclear Science Research Series (Harwood Academic) **14** (1988).

ANALYSYS OF $\bar{p}$-NUCLEUS ELASTIC SCATTERING AND REACTION CROSS SECTIONS WITH A GLAUBER MODEL

Aldo Zenoni

Sezione INFN di Pavia
27100 PAVIA
ITALY

INTRODUCTION

In the recent years many new $\bar{p}$-proton and $\bar{p}$-nucleus data at low energy (E<200 MeV) have been obtained in some laboratories, in particular at the LEAR facility of CERN. Among them there are high precision elastic scattering[1] and reaction cross section[1,2,3] data on several nuclei.

A Glauber model of the $\bar{p}$-nucleus interaction can be utilized for the analysis of these data in order to obtain two kinds of results:
i) the determination of the $\bar{p}$-neutron scattering amplitude parameters;
ii) the degree of accuracy of the theory at low energy.
The investigation on these two points was performed by fitting $\bar{p}$-^{12}C, ^{40}Ca and $\bar{p}$-^{12}C, ^{16}O, ^{40}Ca elastic scattering data at 300 and 600 MeV/c respectively, considering the parameters of the $\bar{p}$-neutron scattering amplitude as free[4]. The same model was also used to reproduce $\bar{p}$-nucleus reaction cross section data.

THE GLAUBER THEORY

In the framework of the Glauber theory[5] the hadron-nucleus scattering amplitude can be expressed as function of the initial and final state wave functions of the nucleus, and of the projectile-free nucleon amplitudes which are contained, through the individual nucleon profile functions, in the overall nucleus profile function; the overall phase shift function is simply the sum of the phase shift functions for collisions with individual nucleons. The theory consists essentially of the eikonal approximation and of the adiabatic approximation, the so called "frozen" nucleus, in which rescattering and off-shell effects are neglected.

It has been successfully applied to high energy hadron nucleus collisions, where the basic conditions for the validity of the theory are clearly satisfied, and recently it has surprisingly revealed to be an effective means also for analizing low energy $\bar{p}$-nucleus scattering data. The validity of the eikonal approximation for low energy antiprotons can be ensured by the narrow forward cone in the $\bar{p}$-nucleon scattering, while the rescattering effects result to be completely cancelled by off-shell effects[6,7].

Antiproton-Nucleon and Antiproton-Nucleus Interactions
Edited by F. Bradamante *et al.*
Plenum Press, New York, 1990

Indeed elastic and anelastic scattering data of antiproton on several nuclei were well reproduced by Dalkarov and Karmanov[8] by means of a Glauber model. In these calculations the elementary amplitudes of the antiproton on the proton and the neutron were parametrized in the usual way as function of three parameters:

$$f_j(q) = (K/4\pi)\ \sigma_j\ (\ \alpha_j + i)\ e^{-\ \beta^2{}_j q^2/2}$$

where K and q are the laboratory incident and transferred momenta, j stands for neutron or proton (j=n,p), σ_j is the $\bar{p}$-nucleon total cross section, $\beta^2{}_j$ is the slope parameter and $\alpha_j = \mathrm{Re}f_j(0)/\mathrm{Im}f_j(0)$. The values of these parameters were taken from experiments, with some assumptions concerning the neutron parameters which are rather unknown. At 300 MeV/c the comparision between experimental and predicted cross sections, expecially at the diffractive minima, allowed to distinguish the best between two possible values of the real to imaginary ratio of the $\bar{p}$-nucleon scattering amplitude (α_p was assumed equal to α_n).

DETERMINATION OF $\bar{P}$-NEUTRON SCATTERING AMPLITUDE PARAMETERS

The efficacy of the Glauber approach in analizing low energy $\bar{p}$-nucleus data suggests a more ambitious program. Indeed the $\bar{p}$-neutron scattering amplitude parameters, which are almost completely unknown, can be completely determined by means of a fit on the differential elastic $\bar{p}$-nucleus cross sections. The $\bar{p}$-neutron parameters are left as free parameters of the fitting procedure. Information on the degree of accuracy of the theory at low energy can also be deduced by the comparision of the results of the fit, for different nuclei, at the same energy.

The model

The same Glauber model which describes successfully the $\bar{p}$-nucleus scattering data at 1 GeV was used[5]. The model practically consists of the eikonal approximation to the first order, assuming for the nucleus the independent particle model without antisimmetrization of the wave function and taking into account the Coulomb interaction in the average Coulomb field approximation.

The most relevant effects not accounted for in the model are: non eikonal effects, Pauli and spin correlations, effects due to the approximate treatment of the Coulomb interaction and of the center of mass constraint for heavy nuclei. These effects are expected to be small at low transferred momenta; their role in the data analysis can be put in evidence by the statistical compatibility among the best fit parameters determined separately for the different nuclei.

The input parameters

The input parameters of the model are relative to the $\bar{p}$-proton scattering amplitude and the nuclear shapes. The parameters of the $\bar{p}$-proton scattering amplitude come from $\bar{p}$-proton scattering experiments[9,10]. For the light nuclei ($A<=16$) harmonic oscillator single particle densities were used while for the heavy nuclei a Wood-Saxon type density was employed. The density of the point like nucleons was deduced from nuclear charge distribution data.

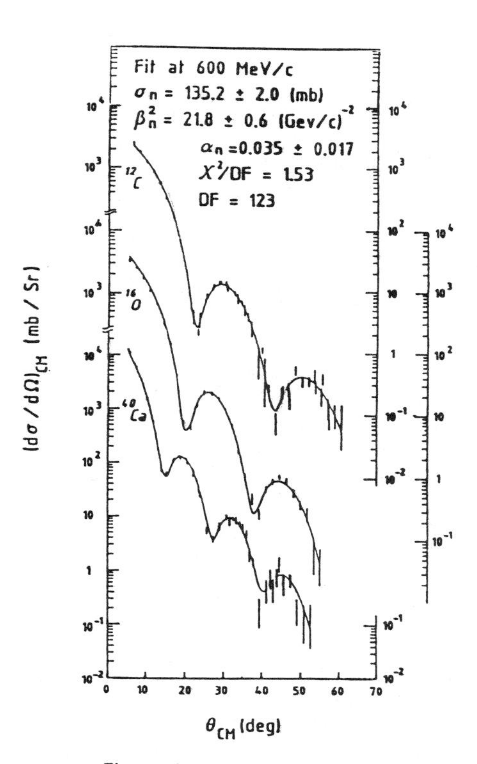

Fig. 1 Overall fit at 600 MeV/c

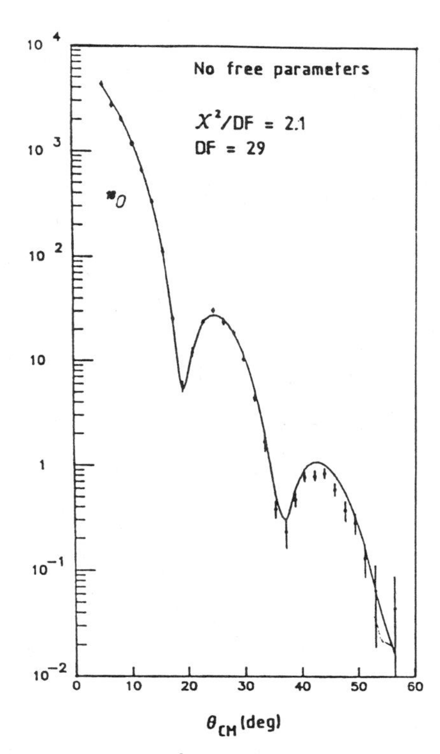

Fig. 3 $\bar{p}$-^{18}O differential elastic cross section at 600 MeV/c

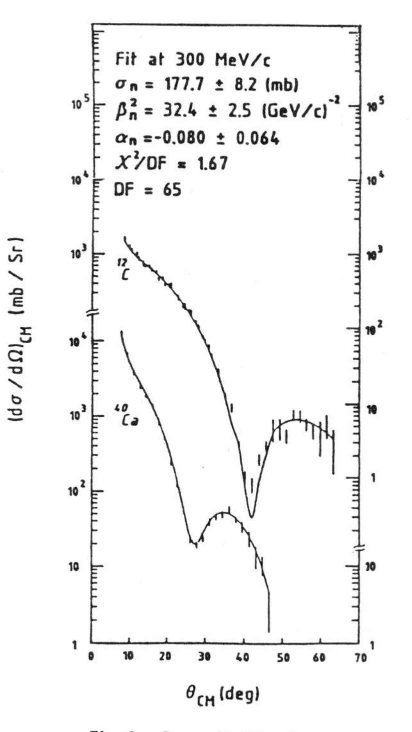

Fig. 2 Overall fit at 300 MeV/c

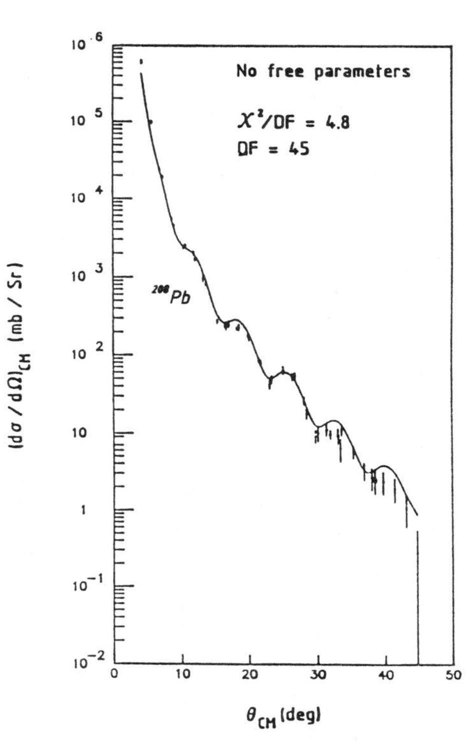

Fig. 4 $\bar{p}$-^{298}Pb differential elastic cross section at 600 MeV/c

Data analysis

Each nucleus, at the two different energies, was fitted separately in order to check the statistical compatibility among best fit parameters. At 600 MeV/c the fit on each nucleus is good and the best fit parameters are consistent. Hence the overall fit (Fig.1) gives a reliable estimate of the $\bar{p}$-neutron parameters at this energy. At 300 MeV/c the fit on each nucleus is good but only the values of the total cross section (σ_n) and the slope ($\beta^2{}_n$) are statistically consistent, whereas the values of the real to imaginary ratio (α_n) are incompatible. The inaccuracy of the model at this energy allows only the reliable determination of the two former parameters by means of the overall fit. In Fig.2 the result of the contemporary fit on the two nuclei is shown; the theoretical predictions are particularly inaccurate around the diffraction minima of the distributions.

In order to test independently the result at 600 MeV/c, where the model seems to work better, the differential elastic cross sections for ^{18}O and ^{208}Pb are reproduced using the best fit values for the $\bar{p}$-neutron parameters (Fig.3 and 4). The agreement with the data is remarkably good with the exception of the points at high transferred momenta. Preliminary data on $\bar{p}$-^4He elastic scattering are also fairly well reproduced[11].

Comparision with experimental data on the $\bar{p}$-neutron amplitude parameters

The $\bar{p}$-neutron amplitude parameters obtained by the fitting procedure at the two energies (apart from the α_n parameter at 300 MeV/c) can be compared with the available experimental data, which consist only of the recent data on $\bar{n}$-proton total cross section[12] σ_n, whereas no experimental data are available for the two other amplitude parameters $\beta^2{}_n$ and α_n. As shown in Fig.5 at 600 MeV/c, where the experimental points are not available, a good agreement is suggested, whereas at 300 MeV/c the value of σ_n seems to be slightly understimated.

A MORE CONSTRAINED FIT AT 300 MeV/c

The results exposed above show that the theory holds well at 600 MeV/c whereas its validity seems to be questionable at 300 MeV/c. At the lowest energy, in order to explore in more detail the consistency of the theory a second, more constrained fit, can be attempted taking into account two facts: i) at 300 MeV/c experimental data on the total $\bar{n}$-p ($\bar{p}$-n) cross section do exist, ii) at the same energy the experimental situation concerning the value of the real to imaginary ratio (α_p) for the $\bar{p}$-proton interaction is rather confused, and some authors[13] suggest that its value could be near to zero (in the previous calculation the value $\alpha_p = -0.132$, deduced from a fit on the existing experimental data, was used).

Hence a second fit at 300 MeV/c was performed fixing the value of the total $\bar{p}$-neutron cross section at the experimental value (214±13±12)mb, and the value of the real to imaginary ratio for the proton (α_p) at zero. The other two parameters of the $\bar{p}$-neutron amplitude were left free. As result of the fit the slope parameter seems to be very well determined at the value $\beta^2{}_n = 25.2\pm1.1\ (\mathrm{Gev/c})^{-2}$, since the values obtained for the two nuclei separately are very close, whereas the α_n parameter, which is extremely sensitive to the minima of the distributions, shows the same inconsistency as before. In conclusion the behaviour of the real to imaginary ratio at this energy seems to be independent on the proton input parameters. This result suggests that, at 300 MeV/c, the Glauber theory could have found a low energy limit of applicability.

It is important to remark that the values of the two parameters σ_n and β^2_n are strictly inversely correlated; so that increasing the value of σ_n a corresponding decreasing of the value of β^2_n is obtained. Consequently, the best fit value obtained for β^2_n depends on the value assumed for the total cross section (the experimental value is affected by a considerable statistic and sistematic error). In any case the consistency is not destroyed changing the value of σ_n.

ANALYSIS OF $\bar{P}$-NUCLEUS REACTION CROSS SECTIONS

A different indication seems, still, to come from the analysis of the $\bar{p}$-nucleus reaction cross sections with the Glauber theory. Indeed the same model, with the best fit parameters for the $\bar{p}$-neutron amplitude at 600 MeV/c and 300 MeV/c and a proper set of parameters at 200 MeV/c, was used to reproduce the reaction cross sections on different nuclei down to 200 MeV/c[3,14].

Fig.6 shows that the agreement between experimental data and theoretical predictions of the theory is remarkably good at the three energies. The Neon cross sections seem to make exception, but it is necessary to remark that the experimental values reported are lower limits, probably understimated of about the 5% [3]. Applying this correction to the experimental values a very good overall agreement is obtained.

CONCLUSIONS

At 600 MeV/c the validity of the Glauber approach for the $\bar{p}$-nucleus scattering has been shown and the $\bar{p}$-neutron scattering amplitude parameters have been determined. At 300 MeV/c the value of the slope parameter has been well determined fixing the total $\bar{p}$-neutron cross section at its experimental value, whereas the inaccuracy of the model at this lower energy prevents the possibility of determining the value of the real to imaginary ratio for the $\bar{p}$-neutron scattering amplitude. This inaccuracy seems to be independent on the values of the $\bar{p}$-proton amplitude parameters and, rather, to be the evidence of the low energy limit of applicability of the Glauber theory to the analysis of $\bar{p}$-nucleus data.

On the contrary the Glauber theory reproduces well the $\bar{p}$-nucleus reaction cross sections on several nuclei down to 200 MeV/c. This fact is not in contradiction with the previuos conclusions since the values of the reaction cross sections predicted by the theory depend on the gross features of the differential elastic cross sections and not on the details of the distributions, like the depth of the minima. The inaccuracy of the model affects, at low energy, the determination of the value of the α_n parameter, which is sensitive to the minima of the scattering distributions, but doesn't affect the determination of the total reaction cross section.

In any case the model proved to be, at both energies, an effective tool for parametrizing the diffractive behaviour of the $\bar{p}$-nucleus scattering distributions.

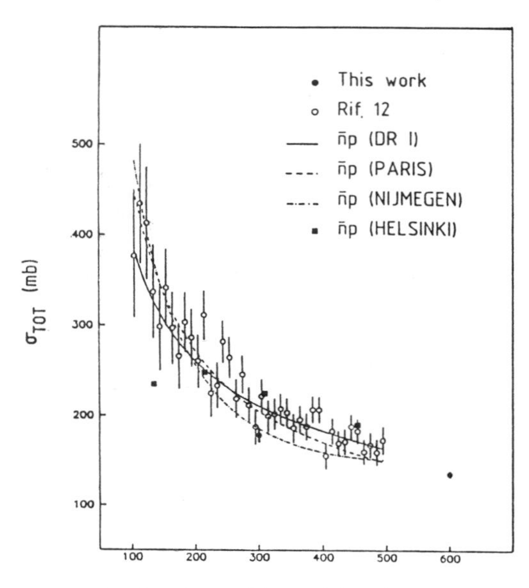

Fig. 5 $\bar{n} - p$ ($\bar{p} - n$) **total cross section**

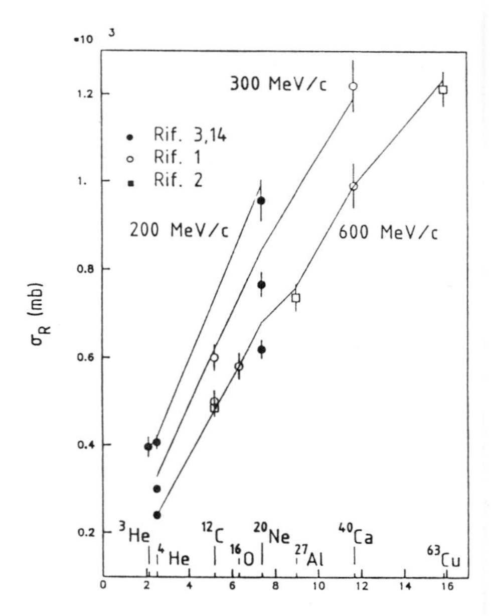

Fig. 6 Reaction cross section vs $A^{2/3}$

REFERENCES

1. D. Garreta, Proc. of Third LEAR Workshop, Tignes 1985, (Editions Frontieres) p.599
2. K. Nakamura et al., Phys.Lett. 52 (1984) 73
3. F. Balestra et al., Nucl.Phys. A452 (1986) 573
4. G. Bendiscioli et al., Nucl.Phys A469 (1987) 669
5. R. Glauber and G. Matthiae, Nucl.Phys. B21 (1970) 135
6. V.B. Mandelzweig and S.J. Wallace, Phys.Rev. C25 (1982) 61
7. V.B. Belyaev and S.A. Rakityanskii, Sov.J.Nucl.Phys. 42 (1985) 867
8. O.D. Dalkarov and V.A. Karmanov, Phys.Lett. 147B (1984) 1
   Nucl.Phys. A445 (1985) 579
9. L.A. Kondratyuk and M.G. Sapozhnikov, Yad.Fiz. 46 (1987) 89
10. H. Iwasaky et al, Nucl.Phys. A433 (1985) 580
11. Yu.A. Batusov et al., IX International Seminar on High Energy Physics Problems, Dubna 14-19 June 1988
12. T. Armstrong at al., Phys.Rev. D36 (1987) 659
13. O.D. Dalkarov and V.A. Karmanov, Proc. of IV LEAR Workshop, Villars-sur-Ollon 1987 (Harwood Academic Publishers) p.679
14. F. Balestra et al, CERN-EP/88-83

ANTIPROTON-HELIUM ANNIHILATION AROUND 1 MeV

Bergen, Brescia, Dubna, Frascati, Oslo, Pavia, and Torino Collaboration

Presented by E. Lodi Rissini
Dipartimento di Automazione Industriale
University of Brescia and INFN, Sezione di Pavia
Via Valotti 9, Brescia, Italy

ABSTRACT

The $\overline{p}$ - ^4He annihilation cross sections have been measured in the energy range from 0.85 to 1.55 MeV (40 ÷ 54 MeV/c) in the PS-179 experiment at the LEAR facility of the CERN using a streamer chamber in a magnetic field. The σ_{an} (0.85 ÷ 1.55 MeV) mean value has been found to be (1541 ± 262 ± 92) mbarn.

Our data have been found to be obtained at the lowest $\overline{p}$ energy achieved till now, by one order of magnitude lower than any previous value (about 20 MeV). The obtained cross section being about 4 times as large as that of 19.6 MeV (192.8 MeV/c) suggests that at these very low energies σ_{an} behaves like 1/v.

MAIN CHARACTERISTICS OF THE APPARATUS

We report the measurement of the annihilation cross sections of antiprotons on ^4He in the kinetic-energy range from 0.85 to 1.55 MeV (40÷54 MeV/c).

The present data integrate the results obtained by our collaboration in the PS-179 experiment at the LEAR facility of CERN at the three momenta 192.8, 306.2 and 607.7 MeV/c, and at rest [1-5].

Antiproton-Nucleon and Antiproton-Nucleus Interactions
Edited by F. Bradamante *et al.*
Plenum Press, New York, 1990

Annihilation events were detected with a self-shunted streamer chamber (90x70x18 cm^3) in a magnetic field (0.409 T in this measurement) filled with helium at atmospheric pressure; the chamber served both as a target and as a detector. The events were recorded on pairs of photographic films. The gas temperature (around 25°C) and pressure were continuously under control during the runs, and variations did not exceed some parts in a thousand. The experimental apparatus is described in detail in ref.[6], and the procedure for event reconstruction is discussed in detail in ref [7].

The high quality of the LEAR $\bar{p}$ beam (initial momentum 105 MeV/c, $\Delta p/p \cong 10^{-4}$) and the low density of the gas target permitted a first investigation of annihilations in flight at a very low energy (lower than 60 MeV/c) where hitherto there existed no experimental values, despite the issue being of interest in many speculations concerning the $N\bar{N}$ and nuclear physics [8-12]. It is noteworthy that annihilation is the only inelastic reaction occurring at such a low energy.

The events were detected in a central fiducial region of the sensitive volume, the detection efficiency being close to 100%. We recognized 97226 pictures, each with a sole $\bar{p}$-^{4}He annihilation event in which the antiproton was at rest, generally.

Antiprotons from LEAR lose energy in the beryllium and mylar windows, in the thin scintillation counters and in some centimeters of air before the entrance to the gas target. The kinetic energy of the antiprotons at the entrance to the chamber was spread around 2 MeV (61.3 MeV/c), see Fig.1.

The beginning of the fiducial volume (X_0) was set to be 22.5 cm from the entrance to the target for obtaining the data presented here. In Fig.2 the antiproton momentum distribution at this point is shown, the average value being 48.5 MeV/c.

The $\bar{p}$ - ^{4}He annihilations at rest occur within quite a large volume in the central region of the streamer chamber.

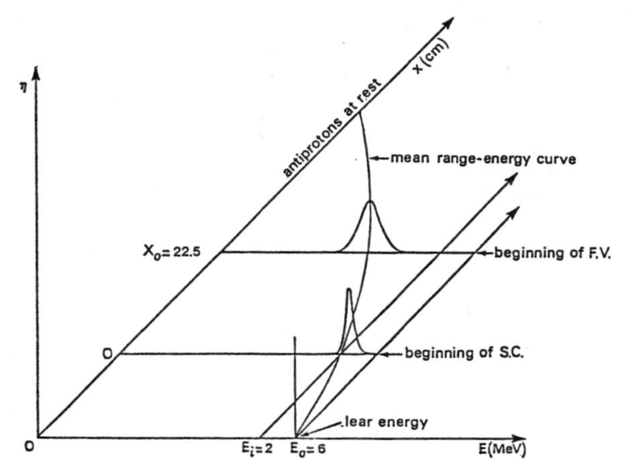

Fig.1 - Spread of kinetic energy of beam antiprotons with energy losses in the streamer chamber.

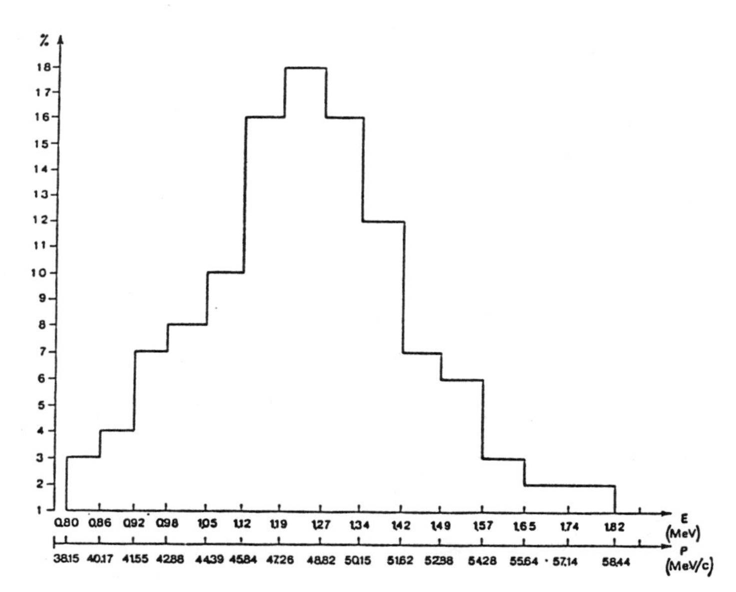

Fig.2 - Antiproton momentum (kinetic energy) distribution at the beginning of the fiducial volume at 22.5 cm from the entrance to the streamer chamber.

Moreover, inspection of Fig.1 reveals that at a given coordinate, $x > X_o$, annihilations both at rest and in flight can occur, the former being by far the most probable (by a factor of about 10^3).

EVALUATION OF THE CROSS SECTION AVERAGED OVER AN ENERGY INTERVAL

We checked carefully all antiprotons tracks by analyzing their curvature in magnetic field by comparison with a reference one. About 2.5 10^3 antiprotons were selected and carefully measured many times.

34 annihilations are obtained to be in flight with momenta between 40 and 54 MeV/c.

Care must be taken to avoid error in momentum fit due to elastic scattering, especially in the last centimeters of the tracks, wich results both in lowering or increasing the real momentum value, obtained by the evaluation of τ (curvature radius) and $d\tau/dx$ related to E and dE/dx.

The experimental error in momentum evaluation for the selected 34 in flight annihilation events is estimated, from repeated measurements, to be $\sigma_m = (1.49 \pm 0.91)$ MeV/c.

For the purpose of extracting data from our thick target, the approssimation used is:

$$\sigma\ (E_2 \div E_1) \cong \frac{N_{ev}\ (E_2 \div E_1)}{\rho \sum n_i(\bar{p})\ X_i}$$

where $n_i(\bar{p})$ is the number of antiprotons in the i channel and $N_{ev}\ (E_2 \div E_1)$ is the number of annihilation in flight with $\bar{p}$ kinetic energy between E_1 and E_2.

The exact expression for the tick target yield , Y, for a target of tickness X (in units of nuclei/cm^2) is given by :

$$Y(X) = \int_{X_0}^{X} \int_{E_F}^{E_0} \sigma(E)\ .\ \eta(E,x)\ dE\ dx$$

where the units for Y are reactions per incident beam particle, E is the energy in MeV, and X, x and and dx are measured in nuclei/cm^2. $\eta(E,x)$ is a normalized distribution function which gives the fraction of the beam particles of energy E (within dE) at a depht in the target of x, see fig.1. In our case, using the $\bar{p}$ momentum distribution in fig.2, we evaluated the contribution of the various antiprotons channels to the mean cross section value $\sigma(E_2 \div E_1)$ between two energies $E_1 > E_2$. To do this, we have for every antiproton in the i-channel with kinetic energy $E_{oi} > E_2$, the yield:

$$Y_i(X_i) = \int_{X_{oi}}^{X_i} \sigma(E_2 \div E_1)\, dx = \sigma\,(E_2 \div E_1)\, X_i$$

where $X_{oi} = X_o$ (the beginning of the fiducial volume) if $E_{oi} < E_1$, $X_{oi} > X_o$ if $E_{oi} > E_1$ and deduced from the condition $E_{oi}(X_{oi}) = E_1$, see Fig.3. The X_i is given by:

$$X_i = \int_{E_2}^{E_{oi} \leqslant E_1} \frac{1}{dE/dx}\, dE$$

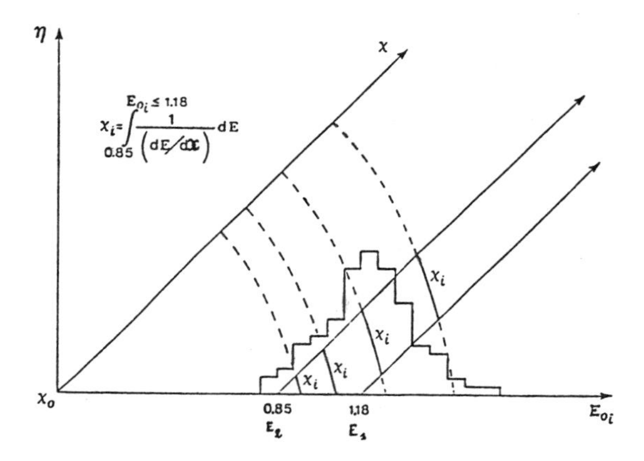

Fig.3 - Illustration of X_i, the determination of the target thickness passed by antiprotons in the i-th channel of the kinetic energy distribution.

$$\sigma\,(E_2 \div E_1) \cong \frac{N_{ev}\,(E_2 \div E_1)}{\rho \sum_i n_i(\bar{p})\, X_i}$$

In table 1 we present the cross sections with the statistical and systematic errors. To obtain the latter, we have also evaluated the cross sections assuming two other different fiducial volumes, starting at 20 and 25 cm from the entrance window of the streamer chamber. We have measured one

$\bar{p}$ track every 50 pictures to evaluate the beam momentum distribution function at various x coordinates in the streamer chamber (0÷20÷22.5÷25 cm from the entrance window). We also introduced modifications of the beam profile corresponding to the error in momentum.

TABLE 1. $\bar{p}$ - ^4He annihilation cross section (mbarn)

σ (0.85 - 1.18 MeV)	σ (1.18 - 1.55 MeV)	σ (0.85 - 1.55 MeV)
1431 ± 315 ± 86	1752 ± 491 ± 193	1541 ± 262 ± 93

In Fig.4 the cross section mean value σ (0.85 ÷ 1.55 MeV) is plotted together with other results on $\bar{p}p$, $\bar{p}d$, $\bar{p}$ ^3He and $\bar{p}$ ^4He annihilations in flight: one can see that the $\bar{p}$ 4He annihilation cross section at 200 MeV/c amounts to about 0.25 of the one obtained in this work, showing a dependence of 1/v.

We note that in the interval 200-600 MeV/c the annihilation cross section on ^4He is close to twice that of hydrogen.

A good interpolation to our present data, at 200 and 300 MeV/c is obtained with the relation:

$$\sigma_{AN} (\bar{p}\ ^4He) \cong 70 + \frac{70}{p_{\bar{p}}\ (GeV/c)} \text{ mbarn}$$

to be compared with the theoritical one by Dover and Richard in H at low energies:

$$\sigma_{AN} (\bar{p}p) \cong 38 + \frac{35}{p_{\bar{p}}\ (GeV/c)} \text{ mbarn}$$

and with that for experimental $\bar{n}p$ data by Amstrong et al.[13] in the range (100 ± 500 MeV/c):

$$\sigma_{AN}(\overline{n}p) \cong 41 + \frac{29}{p_{\overline{p}}\ (GeV/c)}\ \text{mbarn}$$

This last expression fits also the value for $\overline{n}p$ at 50 MeV/c obtained by the experiment E795[14] but not the value at 20 MeV/c.

A carefully analisys of the measurements at these very low energies is necessary.

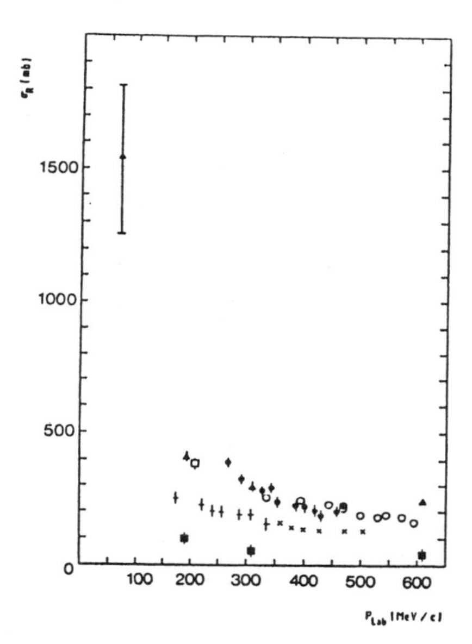

Fig.4 - Reaction cross section behavior for different nuclei: **o** $\bar{p}$ ^{2}H, R.Bizzarri et al ; ● $\bar{p}$ ^{2}H., T.e.Kalogeropoulos et al; x - $\bar{p}$ H, V.Chaloupka et al.; + - $\bar{p}$ H, W.Bruckner et al.; □ $\bar{p}$ ^{3}He, F.Balestra et al.; ▲ $\bar{p}$ ^{4}He, F.Balestra et al..

REFERENCES

1. F.Balestra et al. Phys.Lett. B149 (1984) 69.
2. F.Balestra et al. Phys.Lett. B165 (1985) 265.
3. F.Balestra et al. Nucl.Phys. A465 (1987) 714
4. F.Balestra et al. Phys.Lett. B194 (1987) 343
5. F.Balestra et al. Nucl.Phys. A474 (1987) 651
6. F.Balestra et al. Nucl.Instr. and Meth A234 (1985) 30
7. F.Balestra et al. Nucl.Instr. and Meth A257 (1987) 114
8. V.I. Nazaruk, Sov. J.Nucl.Phys. 46 (1987) 51

9. C.B. Dover, AIP Conf.Proc Nn 150 (1986) 272
10. A.M.Green and J.A.Niskanen, Progress in Particle and Nuclear Physics (Pergamon, New York) (1986)
11. C.Dover and J.M.Richard, Phys.Rev C21 (1980) 1466.
12. J.Mahalanabis et al. CERN-TH4833/87. In press in Nucl.Phys (1988)
13. T.Amstrong et al., Phys. Rev. D36 (1987) 659.
14. L.Pinsky, Physics at LEAR with low energy antiprotons. 4th LEAR workshop. Villars-sur-Ollon (1987), pp.255.

ISOSPIN EFFECTS IN ANTIPROTON-NUCLEUS ANNIHILATION

Giorgio Bendiscioli

Dipartimento di Fisica Nucleare e Teorica-Pavia (Italia)
and INFN - Sezione di Pavia

1. INTRODUCTION

The antiproton-neutron ($\bar{p}n$) system is in a pure isospin state with I=1 and the antiproton-proton ($\bar{p}p$) system is in a mixture of states with I=0 and I=1. If we indicate with $f_I(\theta)$ the spin averaged scattering amplitude for the state with isospin I, for the $\bar{p}n$ and the $\bar{p}p$ systems we have the elastic scattering and charge exchange amplitudes:

$$f^{el}_{\bar{p}n}(\theta) = f_1(\theta) \quad 1.1$$

$$f^{el}_{\bar{p}p}(\theta) = (1/2)\ (f_o(\theta) + f_1(\theta)) \quad 1.2$$

$$f^{ce}_{\bar{p}p}(\theta) = (1/2)\ (f_o(\theta) - f_1(\theta)) \quad 1.3$$

By indicating with σ, σ^{el}, σ^{ce} and σ^{a} total, total elastic, total charge exchange and annihilation cross sections, respectively, the annihilation cross sections for the $\bar{p}n$ and $\bar{p}p$ systems are given by:

$$\sigma^a(\bar{p}n) = \sigma_1 - \sigma^{el}_{1} = \sigma^a_1 \quad 1.4$$

$$\sigma^a(\bar{p}p) = \sigma(\bar{p}p) - (\sigma^{el}(\bar{p}p)+\sigma^{ce}(\bar{p}p)) = \frac{1}{2}(\sigma^a_{o}+\sigma^a_{1}) \quad 1.5$$

Hence:

$$R^a = \frac{\sigma^a(\bar{p}n)}{\sigma^a(\bar{p}p)} = \frac{2\sigma^a_1}{\sigma^a_o+\sigma^a_1} = \frac{2\ \sigma^a_1/\sigma^a_o}{1+\sigma^a_1/\sigma^a_o} \quad 1.6$$

and

$$\sigma^a_1/\sigma^a_o = \frac{R^a}{2-R^a} \quad 1.7$$

Eq.(1.7) allows the relative intensities of the I=1 and I=0 interactions to be evaluated, when the value of R^a is known experimentally.

Antiproton-Nucleon and Antiproton-Nucleus Interactions
Edited by F. Bradamante *et al.*
Plenum Press, New York, 1990

We note that if the $\bar{N}N$ interaction is independent of the isospin, i.e. $\sigma^a_1/\sigma^a_o=1$, then $R^a=1$ (and viceversa) (see Fig.1). Consequently, if R^a is found experimentally to be other than 1, it indicates that the $\bar{N}N$ interaction depends on the isospin.

The previous considerations concern annihilation on free nucleons, but in the following we shall consider also annihilation on nucleons bound in mass A nuclei. For this purpose, we define the ratio

$$R^a_b = \frac{\sigma^a_b(\bar{p}n)}{\sigma^a_b(\bar{p}p)} \qquad 1.8$$

where b means bound and $\sigma^a_b(\bar{p}n)$ and $\sigma^a_b(\bar{p}p)$ are cross sections for annihilation on single nucleons bound in the nucleus.

This paper reports a review below 600 MeV/c of the data on R^a and R^a_b obtained from $\bar{p}p$ and $\bar{n}p$ annihilation experiments, from Glauber theory analyses of $\bar{p}$-nucleus data and from $\bar{p}$H-2, $\bar{p}$He-3 and $\bar{p}$He-4 annihilation experiments.

2. DATA ON R^a

(a) There are more sets of data on the $\bar{p}p$ annihilation in the interval 300-600 MeV/c [1-7] and only one below 300 MeV/c [8]. Some data are reported in Fig.2. We stress that all the data displayed concern total $\bar{p}p$ annihilation cross sections. Because most of the original papers report cross sections for $\bar{p}p$ annihilation into charged pions or for $\bar{p}p$ inelastic interaction (i.e., annihilation plus $\bar{p}p\rightarrow\bar{n}n$ charge exchange), we have corrected the data by adding the cross section for annihilation into neutral mesons [3,4,5,6] or by subtracting the cross section for the charge exchange interaction [2,8].The cross sections for the reaction $\bar{p}p\rightarrow$ neutrals (annihilation into neutrals plus charge exchange) are reported in Refs. 1,2 and 9 and those for charge exchange in Refs. 10-11. The difference between these two quantities gives the cross section for the annihilation with production of neutral mesons. This cross section decreases from about 7 mb down to about 4 mb between 300 and 600 MeV/c. The points of Ref. 8 are corrected considering the charge exchange data of Ref. 12, which differ somewhat from those of Ref. 10 below 250 MeV/c.

The lines in Fig. 2 were obtained by best fit formulas reported in

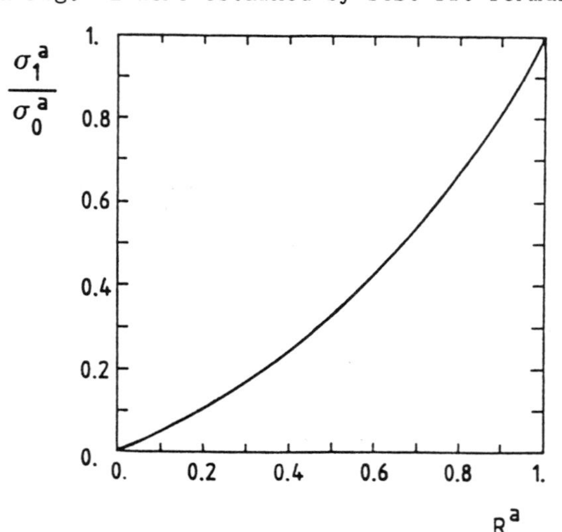

Fig. 1 - σ^a_o/σ^a_1 as a function of R^a.

Refs. 4,5,6 and 7 corrected as explained previously.

It is remarkable that the line of Brando et al., which was obtained by fitting data above 300 MeV/c, agrees very well also with the data of Brueckner et al. [8], when it is extrapolated down to 175 MeV/c. The same tendency is shown also by new annihilation data below 440 MeV/c obtained by the latter authors [13].

Summarizing, above 400 MeV/c the data are spread and below that they are poor.

In Fig. 2, the line fitting the $\bar{n}p$ data from Ref. 14 is also reported. In the figure the statistical errors on the $\bar{n}p$ data are indicated; the systematic errors are of the same order of magnitude. The $\bar{n}p$ line is superimposed on the $\bar{p}p$ line of Lowenstein et al. above 400 MeV/c.

Considering the large errors (both systematic and statistical), the

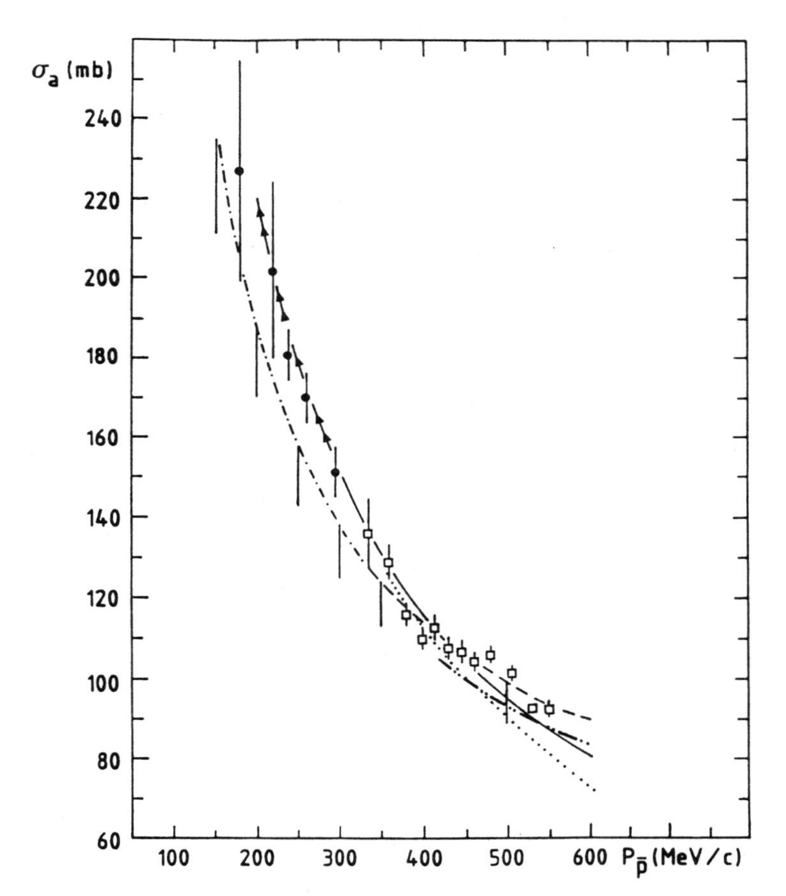

Fig. 2 - $\sigma^a(\bar{p}p)$ and $\sigma^a(\bar{n}p) = \sigma^a(\bar{p}n)$ vs. $\bar{p}$ momentum (see text). $\bar{p}p$ annihilation: (□) Chaloupka et al. [2]; (•) Brueckner et. al. [8]; (-) Brando et al. [6]. The line with arrows is an extrapolation to values of the $\bar{p}$ momentum lower than those of the experiment. (...) Sai et al. [5]; (--) Lowenstein et al. [4]; (..-) Brueckner et al. [7].
$\bar{p}n$ annihilation: (_._) Armstrong et al. [14]. Only the statistical error on the negative side is indicated. The systematic error is similar to the statistical one.

figure suggests $\sigma^a(\bar{n}p)<\sigma^a(\bar{p}p)$ below 400 MeV/c and $\sigma^a(\bar{n}p)\cong\sigma^a(\bar{p}p)$ above with a large degree of uncertainty. The behaviour of R^a obtained from the $\sigma^a(\bar{n}p)$ values of Ref. 14 and the $\sigma^a(\bar{p}p)$ values of Ref. 6 is reported in Fig. 3. The errors displayed are only the statistical ones on $\sigma^a(\bar{n}p)$, but further errors of similar magnitude (sistematic on $\sigma^a(\bar{n}p)$ and those on $\sigma^a(\bar{p}p)$) must be considered.

b) In the Glauber theory analyses of the $\bar{p}$-nucleus interaction, the $\bar{p}$-nucleus scattering amplitude is obtained on the basis of the nucleus structure and of the $\bar{p}$-nucleon scattering amplitude. This can be written in the form

$$f_j(\bar{q}) = \frac{K}{4\pi}\sigma(j)(\rho(j) + i)e^{-1/2\, b(j)q^2}$$

$$j = \bar{p}n,\ \bar{p}p$$

$$\bar{q} = \bar{K}'-\bar{K}$$

$$\rho(j) = \mathrm{Re}f_j(0°)/\mathrm{Im}f_j(0°)$$

If one assumes that the nucleus structure and $\sigma(\bar{p}p)$, $\rho(\bar{p}p)$ and $b(\bar{p}p)$ are known, in principle one can determine $\sigma(\bar{p}n)$ (or $\sigma^a(\bar{p}n)$), $\rho(\bar{p}n)$ an $b(\bar{p}n)$ by fitting $\bar{p}$-nucleus interaction data. Glauber theory seems to be appropriate above 300 MeV/c.

The results of two analyses are reported here. The former [15] obtains $\sigma(\bar{p}n)$ and $\sigma^a(\bar{p}n)$ by utilizing $\sigma(\bar{p}p)$ from Clough et al. [16], $\rho(\bar{p}p)$ and $b(\bar{p}p)$ from best fits to data from different experiments, $\bar{p}$H-2 total cross sections and reaction (=total-elastic) cross sections on He-4, C-12, Ne-20, Ca-40 between 200 and 600 MeV/c. The authors assume $\rho(\bar{p}n)=\rho(\bar{p}p)$ and $b(\bar{p}n)=R\, b(\bar{p}p)$, where $R=\sigma(\bar{p}n)/\sigma(\bar{p}p)$.

The latter analysis [17] utilizes $\sigma(\bar{p}p)$ from Iwasaki et al. [18] and $\rho(\bar{p}p)$ as in the previous analysis and makes a best fit to elastic scattering data on C-12, O-16 and Ca-40 at 300 and 600 MeV/c, considering $\sigma(\bar{p}n)$, $b(\bar{p}n)$ and $\rho(\bar{p}n)$ as free parameters.

The behaviours of R^a obtained through the values of $\sigma^a(\bar{p}n)$ from the previous analyses and of $\sigma^a(\bar{p}p)$ from [6] are shown in Fig. 3.

We see that the values of R^a obtained with different methods between 200 and 600 MeV/c are somewhat different. This depends on the uncertainties on the $\bar{p}p$ and $\bar{n}p$ annihilation cross sections mentioned previously and on those on the $\bar{p}p$ scattering parameters and other quantities assumed as input data of the Glauber calculations. Nevertheless, the different results together seem to suggest that $R^a<1$ and is increasing with the antiproton momentum.

(c) Annihilation events at rest on He-3 and He-4 are analyzed, through an approximated procedure, in Ref. 19. Fits to the π^- multiplicity distributions having R^a as a free parameter lead to $R^a=0.35\pm0.07$ for He-3 and $R^a=0.48\pm0.10$ for He-4. These values show that the $\bar{N}N$ interaction depends strongly on the isospin ($\sigma^a_1/\sigma^a_0=0.21$ and 0.31, respectively).

3. DATA ON R^a_b

The ratio R^a_b was obtained from the analysis of annihilation events on H-2, He-3 and He-4. The main point in these measurements is the identification of the annihilations on n and on p, which differ for the numbers of π^+, π^- and heavy prongs. Considering for the sake of simplicity annihilations at rest, the numbers of π^+, π^- and heavy prongs depend on two processes: the annihilation on a n and on a p and the charge exchange interaction between the π produced and the residual nucleons.

The annihilation reactions are the following:

$\bar{p}n$ annihilation:

$$\left.\begin{array}{l} \bar{p}H\text{-}2 \rightarrow p \\ \bar{p}He\text{-}3 \rightarrow 2p \\ \bar{p}He\text{-}4 \rightarrow \left\{ \begin{array}{l} 2p+n \\ d+p \\ He\text{-}3 \end{array} \right. \end{array} \right\} + h\pi^+ + (h+1)\pi^- + k\pi^0$$

$\bar{p}p$ annihilation:

$$\left.\begin{array}{l} \bar{p}H\text{-}2 \rightarrow n \\ \bar{p}He\text{-}3 \rightarrow \left\{ \begin{array}{l} p+n \\ d \end{array} \right. \\ \bar{p}He\text{-}4 \rightarrow \left\{ \begin{array}{l} p+2n \\ d+n \\ t \end{array} \right. \end{array} \right\} + h\pi^+ + h\pi^- + k\pi^0$$

where h, k=0,1,2, ... $\bar{p}n$ annihilation produces different numbers of π^+ and of π^-; $\bar{p}p$ annihilation produces equal numbers of π^+ and of π^-.

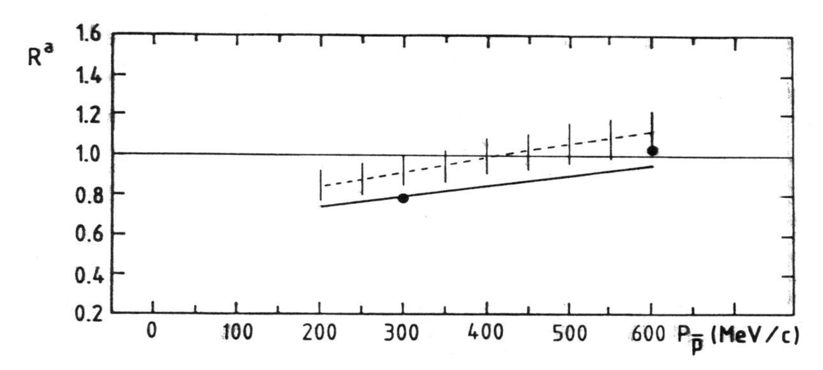

Fig. 3 - Behaviour of R^a vs. $\bar{p}$ momentum. For $\sigma^a(\bar{p}p)$ the data from [6] are used. (--) $\sigma^a(\bar{n}p)$ from [14]. The errors displayed are lower limits (see text).(-) $\sigma^a(\bar{p}n)$ from Glauber analysis of [15]; (●) $\sigma^a(\bar{p}n)$ from Glauber analysis of [17].

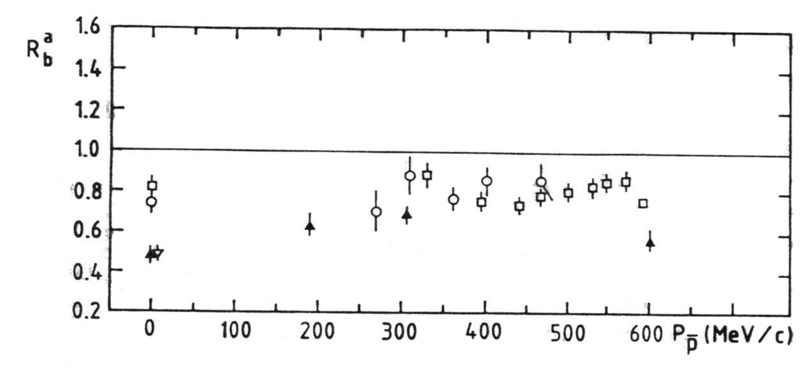

Fig. 4 - Behaviour of R^a_b vs. $\bar{p}$ momentum. (□) $\bar{p}H$-2, Bizzarri et al. [20]; (o) $\bar{p}H$-2, Kalogeropoulos et al. [21]; (▲) $\bar{p}He$-4, Balestra et al. [22]; (▽) $\bar{p}He$-3, Balestra et al. [23]

The original numbers of π and heavy prongs may be changed by the πN charge exchange interaction ($\pi^o n \rightleftarrows \pi^- p$, $\pi^o p \rightleftarrows \pi^+ n$). The main consequence is that some reactions, which arise from annihilations on neutrons (different numbers of π^+ and of π^-) look like annihilations on protons (equal numbers of π^+ and of π^-) and viceversa. However there is a (partial) compensation in the numbers of exchanged events and simple relations exist [22,23] between R^a_b and the ratio (number of $\bar{p}n$ annihilations or $\bar{p}n$ like annihilations)/(number of $\bar{p}p$ or $\bar{p}p$ like annihilations).

The behaviours of R^a_b obtained from bubble chambers (H-2) and a streamer chamber (He-3,He-4) between 0 and 600 MeV/c are shown in Fig. 4. We see that: (1) $R^a_b<1$, that is the $\bar{N}N$ interaction depends on the isospin; (2) at rest, $R^a_b(\text{He-3}) \cong R^a_b(\text{He-4})$; (3) on the average, $R^a_b(\text{H-2}) \neq R^a_b(\text{He})$.

Note that the ratio R^a_b is not affected by systematic errors on the cross sections similar to those affecting R^a. Indeed, in the present analyses, the values of R^a_b are carried out by means of cross section values obtained in the same experiment; so effects dependent on different normalizations are avoided. The same comment holds for the analysis mentioned in Sec. 2 (c).

Also, note that for annihilation at rest on He $R^a \cong R^a_b$ within the errors.

4. COMMENTS

A possible explanation of the difference between $R^a_b(\text{H-2})$ and $R^a_b(\text{He})$ may be as follows. We consider in particular the values at rest.

The annihilation probability depends on the relative motion of the $\bar{p}$ and of the annihilating nucleon. The relative motion at rest depends on the motion of the nucleon inside the nucleus and on the atomic level from which $\bar{p}$ annihilates. We know that in liquid hydrogen annihilations occur mostly from an S atomic state [24] and in gaseus He-4 (1 atm) annihilations occur mostly from P and D atomic states [25]. So we expect that different mixtures of $\bar{N}N$ waves contribute to the annihilation on H-2 and He.

Moreover, the ratio R^a for annihilation in an $\bar{N}N$ S-wave is predicted by a number of potential models to be high (above about 0.7) [26,27]. A similar value is given also by recent analyses of $\bar{N}N$ scattering and antiprotonic hydrogen data [27-29].Some analyses consider also P wave and find low values (0.2, Ref. 28; 0.35, Ref. 29) for R^a in this $\bar{N}N$ state.

So, recalling the value of R^a_b obtained from H-2 (~0.8) and that from He (~0.5), we can say that in the annihilation on H-2 S wave prevailes and in the annihilation on He P and (perhaps) D waves are important.

5. CONCLUSIONS

Measurements on free nucleons and on bound nucleons indicate that below 600 MeV/c R^a and R^a_b are < 1, that is, the $\bar{N}N$ interaction is isospin dependent and the annihilation in the I=0 state is stronger than in the I=1 state. However, at present, the knowledge of R^a from annihilations on free nucleons is somewhat uncertain due to normalization problems in the cross sections and to a poor statistics. A clearer indication of the isospin dependence is given by the annihilation on nuclei.

The different values of R^a_b from H-2 and He are compatible with analyses of $\bar{N}N$ scattering and antiprotonic hydrogen data and show that there is an isospin dependence in the $\bar{N}N$ P state, which is stronger than in the S state.

More details related to the present paper can be found in Ref. 23.

REFERENCES

1. U. Amaldi et al., Nuovo Cimento 46A (1966) 171
2. V. Chaloupka et al., Phys.Lett. 61B (1976) 487
3. W. Brueckner et al., Phys.Lett. 67B (1977) 222
4. D.I. Lowenstein et al., Phys.Rev. D23 (1981) 2788
5. F. Sai et al., Nucl.Phys. B213 (1983) 371
6. T. Brando et al., Phys.Lett. 158B (1985) 505
7. W. Brueckner et al., Phys.Lett. B197 (1987) 463
8. W. Brueckner et al., Phys.Lett. B166 (1986) 113
9. M. Alston-Garnjost, Phys.Rev.Lett. 35 (1975) 1685
10. R.P. Hamilton et al., Phys.Rev.Lett. 44 (1980) 1182
11. K. Nakamura et al., Phys.Rev.Lett. 53 (1984) 885
12. W. Brueckner et al., Phys.Lett. 169B (1986) 302
13. T.A. Shibata, Private Communication
14. T. Armstrong et al., Phys.Rev. D36 (1987) 659
15. L.A. Kondratyuk and M.G. Sapozhnikov, Sov.J.Nucl.Phys.46 (1987) 56
16. A.S. Clough et al., Phys.Lett. 146B (1984) 299
17. G. Bendiscioli, A. Rotondi, P. Salvini and A. Zenoni, Nucl.Phys. A469 (1987) 669
18. H. Iwasaki et al., Nucl.Phys. A433 (1985) 580
19. F. Balestra et al., Nucl.Phys. A474 (1987) 651
20. R. Bizzarri et al., Nuovo Cimento A22 (1974) 225
21. T.E. Kalogeropoulos et al., Phys. Rev. D22 (1980) 2585
22. F. Balestra et al., Nucl.Phys. A465 (1987) 714 and CERN-EP/87-213
23. F. Balestra et al., CERN-EP/88-92. In press in Nucl.Phys.
24. T.B. Day et al., Phys.Rev. 118 (1960) 864
    H. Poth, CERN-EP/86-105
25. F. Reifenroether et al., Phys.Lett. B203 (1988) 9
26. T. Ueda, Prog.Theor.Phys. 62 (1979) 1670
    R.A. Bryan and R. Phillips, Nucl.Phys. B5 (1968) 201
    C. Dover and J.M. Richard, Phys. Rev. C21 (1980) 1466
    J. Côté et al., Phys.Rev.Lett. 48 (1982) 1319
27. L.S. Pinsky, Physics at LEAR with Low Energy Antiprotons, Edited by C. Amsler et al., Harwood Academic Publ. (1988)
28. J. Mahalanabis et al., CERN-TH/87-4833. In press in Nucl.Phys.A
29. I.L. Grach et al., ITEP, N210 (1987) Moscow

LOW ENERGY ANTIPROTON INTERACTION WITH NE NUCLEI

Maria Pia Bussa

for the Bergen-Brescia-Dubna-Frascati-Oslo-Pavia-Torino Collaboration
PS179 experiment at CERN

The results of the analysis of antiproton-Neon annihilation at 600 MeV/c are reported. The experimental apparatus consisted of a self shunted streamer chamber in a magnetic field. The correlation measurements and the strangeness production are reported and discussed in the framework of the more recent theoretical analyses.

INTRODUCTION

In these last years the experimental studies of the interaction of low energy antiprotons on nucleons and nuclei have received a vigorous impulse, due to the LEAR antiprotons beams at CERN. The nuclear physics results are among the most interesting scientific outputs of the first generation of LEAR experiments. For a comprehensive review the reader is referred to the report by Guaraldo at the Villars Conference[1]. Among the aims of the Streamer Chamber PS179 experiment, the following items should be stressed:

- the correlation measurements between final pions and overall multiplicity;
- the strangeness production, in the framework of looking for "unusual" annihilations with their typical signatures.

In this paper the results obtained by the Streamer Group studying $\bar{p}$-Ne annihilations at 600 MeV/c are presented.

SPECIFIC FEATURES OF THE EXPERIMENTAL APPARATUS

The experimental apparatus consisted of a 113 litres self shunted streamer chamber in a magnetic field filled at 1 atm with gas target and exposed to the $\bar{p}$ beams of the LEAR facility at CERN. The details of the apparatus and its performances are reported in ref.2. The ability to measure short nuclear fragment ranges together with the curvature of the fast particles and the very good vertex resolution allow to obtain an accurate identification of the reaction channels and a detailed study of momenta and angles distribution vs. charged tracks multiplicity.

CORRELATION MEASUREMENTS

It is quite possible that the correlations between the numbers of emitted particles of different kind will be more sensitive to the details of the N-A interaction mechanism compared with the inclusive characteristics[3). Unfortunately few correlations have been experimentally measured. The PS179 experiment dealt, in particular, with the correlation between the average number of negative pions and the overall multiplicity M (charged pions plus nuclear fragments).

According to the well established energy transferring mechanisms throught the intranuclear cascade and assuming in first approximation that higher multiplicity events are preferentially produced by higher energy deposition, one might expect a decreasing number of final (survived) pions at high multiplicity. Referring to the overall number of charged pions vs the ejected protons this is, indeed, a classical result of conventional INC models: see, for instance, fig. 2 of ref.4. Streamer Group results seem to contradict this too simple conclusion since $<\pi^->$ exibits a striking increasing behaviour with the multiplicity. However, nothing of "unusual" can be inferred from these data. In fact, a recent version of the INC model of Iljinov et al (ref.3) explains beautifully the experimental result (fig.1a). This version of the model gives a much better description of the elementary $N-\bar{N}$ interaction,includes the effects of the local reduction of the nuclear density during the development of the cascade (trawling effect) and takes into account the de-excitation of the residual nuclei in terms of evaporation and multifragmentation.

The same agreement is obtained by an other recent different model developed by Cugnon et al.[5)] in which the ejection mechanism is assumed to correspond to a clan picture where the ancestors are pions. This simple model retains all the basic premises of INC and handles charge conservation exactly at each step. Good agreement is obtained for multiplicity distributions and in particular for the very strong correlations observed in Ne between $<\pi^->$ and charged prongs multiplicity (fig.1b). The agreement is a direct consequence of charge conservation and has no dynamical content.

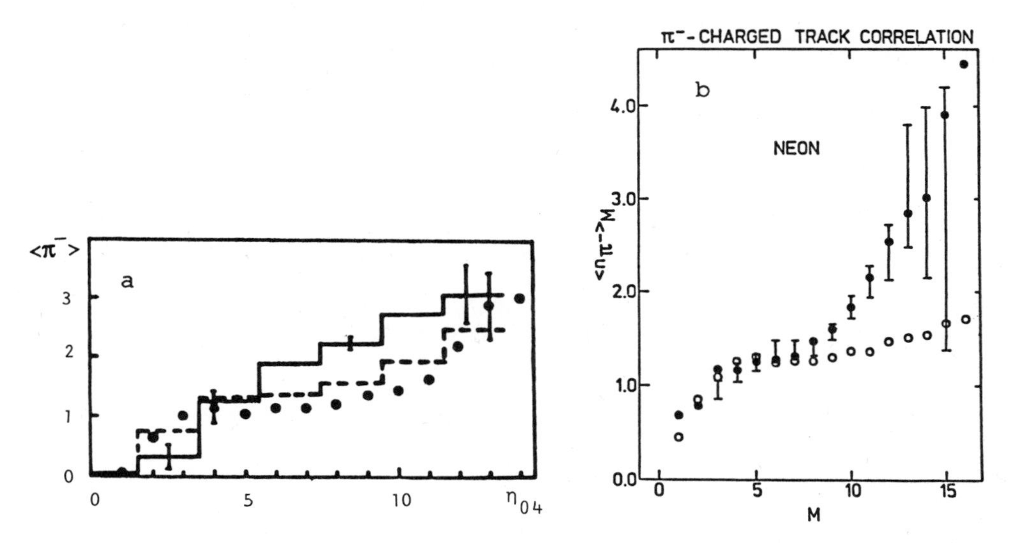

Fig.1a,1b - Correlation between the mean number of negative pions and charged tracks multiplicity in $\bar{p}$-Ne at 607 MeV/c).

a) dots: experimental data; solid and dashed line: intranuclear cascade calculations by Iljinov et al.[3], with and without trawling effect.

b) bars: experimental data; full circles:prediction of the model of Cugnon et al.[5]; open circles: relaxation of the charge conservation.

STRANGE PARTICLE PRODUCTION IN NUCLEAR COLLISIONS

Strange particle production in nuclear collisions has become a topic of interest with the speculation that significant changes in relative and absolute abundance of strange particles (togheter with tails of high momentum protons) could serve as a probe for a quark-gluon plasma formation [6]. Also in a more conventional approach[7] enhanced strangeness production would be signature of annihilations with baryon number B>0.

With the streamer chamber, the inclusive production of V^o from 600 MeV/c antiproton annihilations in Neon was investigated[8]. The $\Lambda^o$and K_s^o were observed through their charged decay modes.

The Λ^o and K_s^o production cross section can be compared with recent results obtained at KEK on Ta at 4 GeV/c (Miyano et al. ref.9),on complex nuclei below 450 MeV/c (Condo et al;, ref.10) and on ^2H at 450-921 MeV/c (Parkin et al., ref.11).

1. R=$\sigma(\Lambda^o)/\sigma(K_s^o)$ behaviour vs. A and energy

The emission probability of Λ^o (1.95±0.43 %) confirms the results obtained on complex nuclei by Condo et al.[10]. Both PS179 and KEK experiments, on medium and medium heavy nuclei, obtain about the same value (about 2.4) for the ratio R=$\sigma(\Lambda^o)/\sigma(K_s^o)$, independently on energy. On light nuclei, according to the recent data on deuterium [11] the R value is less than unity. Preliminary data of PS179 experiment on annihilation on ^4He show a value for R near unity. It is possible to hypothesize a "peripherical-like" component in the production of K_s^o. Then, in a ligther nucleus such a deuteron or ^4He the enhanced surface production would invert the value of the ratio. If this is true, the surface character of annihilation at rest should give a lower value for R. The preliminary annihilation data at rest on Neon seem to confirm this mechanism.

Table 1 K_s^o and Λ^o production cross section . R=$\sigma(\Lambda^o)/\sigma(K_s^o)$ behaviour vs. A and energy.

Reaction	$\sigma(\Lambda^o)$	$\sigma(K_s^o)$	R=$\sigma(\Lambda^o)/\sigma(K_s^o)$
Ne (607 MeV/c) ref.8	12.3 ± 2.8 mb	5.4 ± 1.1 mb	2.3 ± 0.7
Ta (4 GeV/c) ref.9	193 ± 12 mb	82 ± 6 mb	2.4 ± 0.3
^2H (450-921 MeV/c) r.11	0.65 %	2.55 ± 0.12 %	0.25 ± 0.05
^4He (at rest) *)			1.09 ± 0.10
Ne (at rest) *)			1.25 ± 0.19

*) preliminary data from PS179 experiment.

2. Strangeness enhancement in PS179 and KEK experiments

For the $\bar{p}$-Ta experiment[9] at 4 GeV/c the Λ^o production is surprisingly high compared with that obtained by scaling data from $\bar{p}$-p strangeness production and taking into account some simplified geometrical factor.

For the Neon data the same effect is supported, scaling the data from $\bar{p}$-d at 600 MeV/c.(see Table 2)

Table 2 Strangeness enhancement in PS179 and KEK experiments

	KEK			PS179		
	s_{pTa} (mb) (4 GeV/c)	$s_{pp} * A^{2/3}$ (mb)	$E = \frac{\sigma_{pTa}}{\sigma_{pp} * A^{2/3}}$	s_{pNe} (mb) (.6 GeV/c)	$s_{pd} * A^{2/3}$ (mb)	$E = \frac{\sigma_{pNe}}{\sigma_{pd} * A^{2/3}}$
Λ^o+X	193 ± 12	17	11	12.3 ± 2.8	3.1	4.1
K_s^o +X	82 ± 6	60	1.3	5.4±1.1	11.7	0.46

3. Rapidity distributions of Λ^o, K_s^o and associated π^-

From the analysis of rapidity distributions some conclusions on the effective target of nucleons involved in the strangeness production can be drawn.

The weighted average rapidity of Λ^o is $\langle y\rangle$ =(.07±.03) which implies an interacting system of many nucleons. This result is consistent with that deduced for $\bar{p}$ of 4 GeV/c[9]. The weighted average rapidity of K_s^o is y= (.33±.07) indicating a pronounced preference of forward production relative to the $\bar{p}$ beam direction and implying an interacting system ($\bar{p}$,1N) to be compared with a system ($\bar{p}$,3N) deduced at higher energy [9]. We can conclude that in $\bar{p}$ Ne annihilations at 607 MeV/c the data favour the hypothesis that K_s^o and Λ^o production mechanisms are similar to those described by Miyano et al. for $\bar{p}$ -Ta annihilations at 4 GeV/c .

The rapidity distribution of negative tracks ,mostly pions, gives a puzzling result. Both (Λ^o associated and K_s^o associated) π^- rapidity distributions - a parameter not measured at KEK - present different average values compared with those of the associated strange particles: as if pions were produced from totally different effective targets.

The rapidity distribution of negative tracks associated to Λ^o events has an average value $\langle y\rangle$ = 0.31± 0.08 indicating the forward character of these events not correlated with the behaviour of the associated Λ^o distribution,whereas the rapidity distribution of negative tracks associated to K_s^o events is y=-0.01±0.06, which is consistent with a flat production, differently from the forward associated K_s^o .

4. Super cooled quark-matter in 4 GeV/c $\bar{p}$-heavy nuclei annihilation

A different model developed recently by Rafelsky[12] in the frame of the hypothesis of quark matter production in a super cooled condition seems to explain both the strangeness enhancement and the different sources for Λ^{o} and K_{s}^{o} production together with the behaviour of $R=\sigma(\Lambda^{o})/\sigma(K_{s}^{o})$ vs. A.

According to Rafelsky for some magic momentum (3-4 GeV/c) the $\bar{p}$ deposits all the annihilation energy in a narrow forward cone of nuclear matter within the target nucleus. Under these favourable conditions of energy transferring it can be postulated the formation of super cooled quark-matter at rather modest temperature (T=60 MeV).

Both the strangeness enhancement and the absolute strangeness abundance of $\bar{s}s$ pairs are explained by the model.The agreement of experimental data with rapidity distributions computed assuming a unique thermal source for the emitted strange particles at the same low temperature confirms the internal consistency of the hypothesis of a central fireball disintegrating into various particles.

REFERENCES

1) C.Guaraldo, IV LEAR Workshop Proceed.,Villars 1987,p.797.

2) F.Balestra et al.,Nucl.Instr.and Meth.A134 (1985) 30.

3) Ye.S.Golubeva,A.S.Iljinov,A.Botvina,N.Sobolosky,Nucl.Phys.A483(1988)539.

4) P.L.Mcgaughey et al., Phys.Lett. 166B(1986)264.

5) J.Cugnon et al., Phys.Rev C38(1988)795.

6) J.Rafelski,Phys. Lett.,91B (1980) 281 ;J.Rafelski and B.Muller, Phys.Rev.Lett.,48 (1982) 1066.

7) J.Cugnon and J.Vandermeulen,Phys.Lett. 149B (1984) 16.

8) F.Balestra et al.,Phys.Lett.194B (1987) 192.

9) K.Miyano et al.,Phys.Rev.Lett.53 (1984) 1725; KEK Prep. 87-160(1988).

10) G.T.Condo et al.,Phys.Rev.C29 (1984) 1531.

11) S.J.H.Parkin et al.,Nucl.Phys.B277 (1986) 634.

12) J.Rafelski,Proceed. IV LEAR Workshop,Villars (1987).

MEASUREMENT OF ANALYSING POWER OF $\bar{p}C$ ELASTIC SCATTERING AT LEAR

R. Birsa, F. Bradamante, S. Dalla Torre - Colautti, A. Martin,
A. Penzo, P. Schiavon, F. Tessarotto, A. Villari, A.M. Zanetti
INFN, Trieste and University of Trieste, Trieste, Italy

E. Heer, R. Hess, C. Lechanoine - Leluc, D. Rapin,
DPNC, University of Geneva, Geneva, Switzerland

K. Bos, J.C. Kluyver, R.A. Kunne,
NIKHEF-H, Amsterdam, The Netherlands

D.V. Bugg, J.R. Hall,
Queen Mary College, London, Great Britain

(presented by A. Martin)

INTRODUCTION

The polarization parameter $A_{\bar{p}C}$ in antiproton - carbon elastic scattering at small angles has been measured at the Low Energy Antiproton Ring (LEAR) at CERN, as part of the program of experiment PS172[1,2]. The measurement was carried out to investigate the possibility of polarizing the antiprotons by scattering off nuclear targets and of analysing the $\bar{p}$ polarization by means of a carbon polarimeter; these techniques turned out to be very useful in the systematic study of the nucleon - nucleon interactions, because of the large analysing power of carbon for protons at small angles ($\approx$0.4 at 10° for 800 MeV/c incident protons). First calculations using Glauber theory[3] suggested for $\bar{p}C$ elastic scattering a polarization parameter considerably smaller than for proton scattering, but the uncertainty of the calculation, mainly due to lack of precise $\bar{p}p$ and $\bar{p}n$ scattering data, was large, suggesting the measurement of $A_{\bar{p}C}$ at LEAR. A further motivation to perform the experiment, which is independent of the magnitude of the analysing power, was to get information on the antiproton - neutron spin amplitudes, as pointed out by Osland and Glauber for the proton case[4].

In june '84 we performed a first measurement of $A_{\bar{p}C}$ with an extracted $\bar{p}$ beam of 608 MeV/c[5]. Here we present the final results of a new measurement, carried out in august '86 at higher antiproton momenta to look at any energy dependence with improved statistics[6].

Antiproton-Nucleon and Antiproton-Nucleus Interactions
Edited by F. Bradamante *et al.*
Plenum Press, New York, 1990

EXPERIMENTAL METHOD AND SET-UP

The measurement of $A_{\bar{p}C}$ was performed using the standard double scattering technique, i.e. by scattering twice the antiprotons off two carbon targets. The azimuthal distribution of the double scattered antiprotons is given by

$$N^{\pm}(\theta_1,\theta_2,\phi)=N_0^{\pm}(\theta_1,\theta_2)\times[1\pm\varepsilon(\theta_1,\theta_2)\times\cos\phi] ,$$

where $\varepsilon(\theta_1,\theta_2)=A_1(\theta_1)\times A_2(\theta_2)$, A_1 and A_2 are the analysing power in the first and in the second interaction respectively, θ_1 and θ_2 are the scattering angles and ϕ is the angle between the first and the second scattering plane; + and - refer to the sign of θ_1. At first order in $\cos\phi$, experimental asymmetries of the apparatus cancel away alternating the sign of the first scattering angle and evaluating the asymmetry as:

$$\varepsilon(\theta_1,\theta_2) = \langle\cos\phi\rangle^{+}-\langle\cos\phi\rangle^{-}.$$

The experimental set-up is shown in Fig. 1. The first carbon target was located at an intermediate focus and two bending magnets just in front of it allowed to steer the beam in the horizontal plane both to the left and to the right, so that the extracted $\bar{p}$ beam hit the target with an angle θ_1 variable between -9° and $+9^{\circ}$; the beam was monitored continuously by a scintillation counter (B_0) and a multiwire proportional chamber (PC_0). The C2 beam line, 20 m long, was designed to act as a spectrometer. The angular acceptance was ±18 mrad in the horizontal plane and ±36 mrad in the vertical one, the momentum resolution was $\Delta p/p \simeq 0.01$, and a quadrupole in between two bending magnets acted as a field lens so that the line was achromatic. Only elastically scattered antiprotons were thus refocused on the second carbon target around which the experimental apparatus was built. The multiwire proportional chambers PC_1-PC_9 shown in Fig. 1 measured incoming and outgoing particles and a box made up of five scintillation counters surrounded the target to tag annihilation events for the off-line analysis. The asymmetry of the final azimuthal distribution is:

$$\varepsilon_c(\theta_1,\theta_2) = A_{\bar{p}C}(\theta_1)\times A'_{\bar{p}C}(\theta_2) ,$$

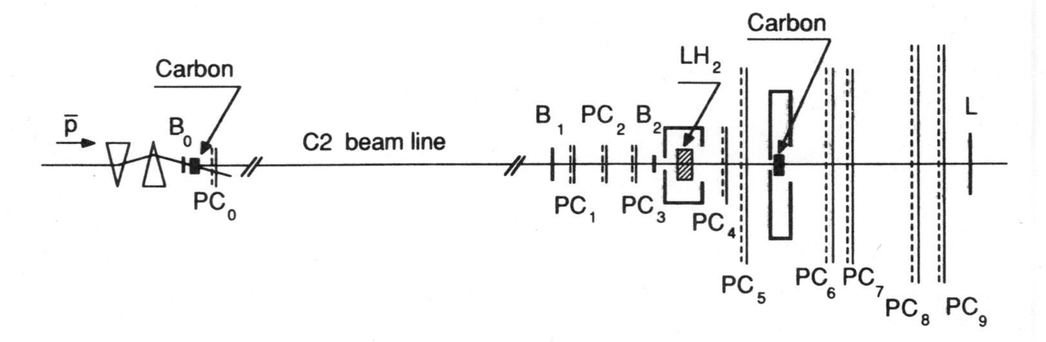

Fig. 1. Schematic layout of the apparatus.

where $A_{\bar{p}C}$ is the polarization parameter in $\bar{p}C$ elastic scattering and $A'_{\bar{p}C}$ is the analysing power in the second $\bar{p}C$ interaction, in general different from $A_{\bar{p}C}$ because of the presence of residual annihilation and inelastic and quasi-elastic scattering events.

Just in front of the second carbon scatterer there was a liquid hydrogen target (LH_2), surrounded by scintillation counters, again for off-line annihilation rejection; using this apparatus we could collect also data with the second scattering on protons, and measure at the same time

$$\varepsilon_p(\theta_1,\theta_2) = A_{\bar{p}C}(\theta_1)\times A'_{\bar{p}p}(\theta_2) ,$$

where $A'_{\bar{p}p}$ is the polarization parameter in $\bar{p}p$ scattering. As we will see, the measurement of ε_p is needed to determine the sign of $A_{\bar{p}C}$.

In August '86 we used this apparatus to measure $A_{\bar{p}C}$ with an extracted $\bar{p}$ beam of 800 and 1100 MeV/c (783 and 1091 MeV/c at the center of the first carbon target and 738 and 1066 MeV/c at the second target respectively), $\theta_1 = \pm 8°$ at 800 MeV/c and $\theta_1 = \pm 8°$ and $\pm 5°$ at 1100 MeV/c. The previous measurement at 608 MeV/c was performed using a very similar set-up.

RESULTS

All the details about data taking and analysis are given in Ref. 6. Here we just stress that a lot of effort was put in studying systematic effects as the measured asymmetries turned out to be very small. This is clear from Fig. 2, where, as an example, the azimuthal distributions after the second $\bar{p}C$ scattering at 800 MeV/c are shown: the number of events with $10.7° < \theta_2 < 12.7°$ are plotted versus ϕ for positive and negative sign of the first scattering angle; the full curves show the expected behavior $N^{\pm}(\phi) = N_0^{\pm}(1 \pm \varepsilon_c \times \cos\phi)$, with ε_c given by a best fit to the data.

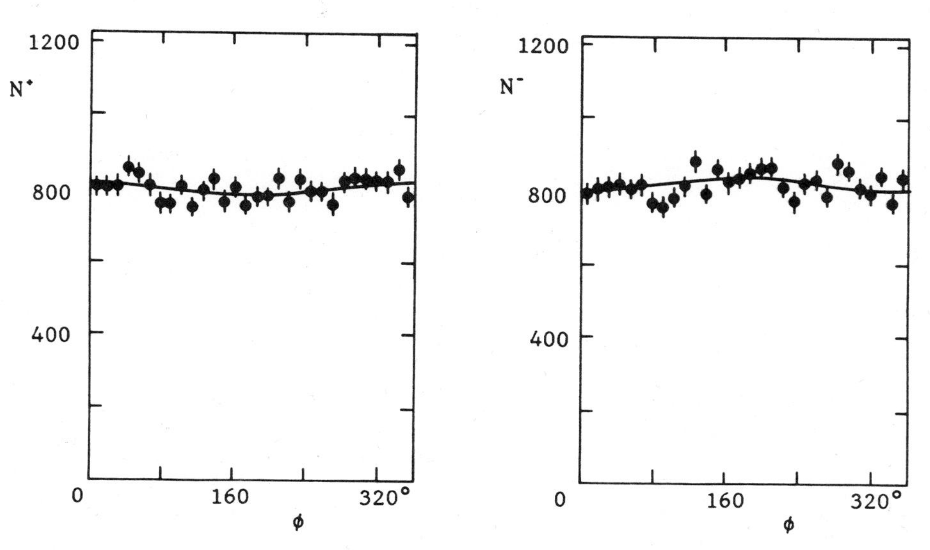

Fig. 2. Typical azimuthal distribution in $\bar{p}C$ scattering at 800 MeV/c; the curves show the expected behavior $N^{\pm}(\phi) = N_0^{\pm}[1 \pm \varepsilon_c \cos\phi]$.

The geometrical acceptance in the second interaction was full up to $\theta_2 \simeq 15°$ for $\bar{p}p$ scattering and $\theta_2 \simeq 17°$ for $\bar{p}C$ scattering. While for $\bar{p}p$ interaction the background (mostly due to scattering on the mylar windows of the LH_2 target and residual annihilations) turned out to be very low ($\simeq$10%) in all the angular range, so that we can assume $A'_{\bar{p}C}=A_{\bar{p}C}$, in $\bar{p}C$ scattering the percentage of elastic events was greater than 85% only in the diffraction peak region ($\theta_2<13°$ at 800 MeV/c and $\theta_2<9°$ at 1100 MeV/c) while at larger angles there were both residual annihilations ($\simeq$7% of the total number of events) and inelastic and quasi-elastic events which can have a different polarization than the elastic ones; so only at small values of the momentum transfer (q_2<165 MeV/c) we can consider $A'_{\bar{p}C}=A_{\bar{p}C}$.

The measured asymmetries for $\bar{p}C$ scattering at 800 and 1100 MeV/c, together with the data at 608 MeV/c, are plotted in Fig. 3 versus the momentum transfer q_2. The curves give the theoretical values of ε_c calculated using Glauber theory[4] and the $\bar{p}$-nucleon amplitudes given by the Dover-Richard (dashed) and Paris (dash-dotted) models. As can be seen, the data at all three energies suggest a small positive value of the asymmetry and there is no evidence for any energy dependence, in agreement with the predictions of theoretical models.

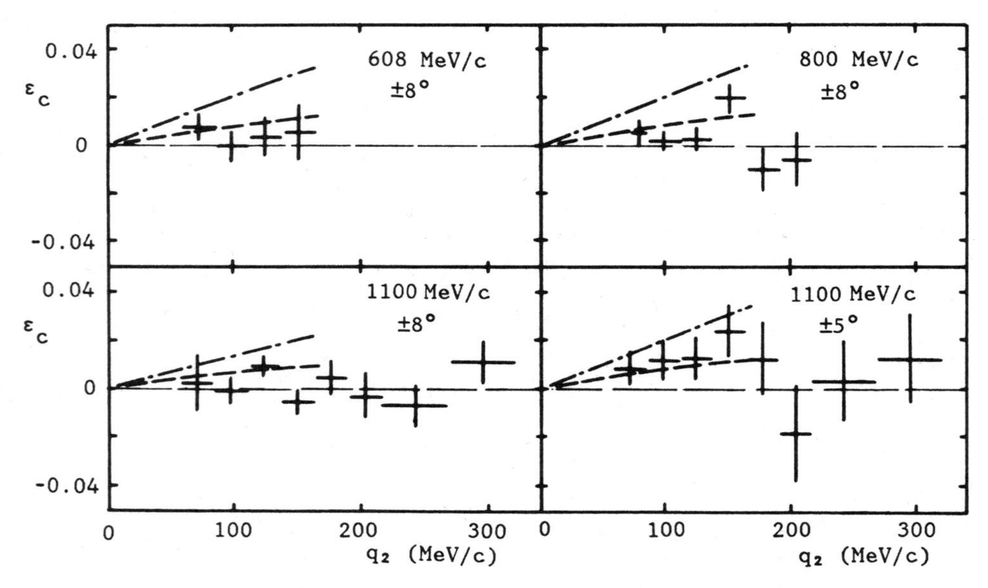

Fig. 3. Measured $\bar{p}C$ asymmetries at 608 Mev/c, 8°, 800 MeV/c,8°, 1100 MeV/c, 5° and 8°.

Assuming at all energies the same linear dependence of $A_{\bar{p}C}$ on the momentum transfer q at small angles, as foreseen by the models ($A_{\bar{p}C}=a_c \times q$), we averaged all our data into a single set of asymmetries using as reference value for the momentum transfer in the first scattering q_1=100 MeV/c. The resulting asymmetries $\langle\varepsilon_c\rangle$ are plotted in Fig. 4; the dashed and dash-dotted curves are again the predictions obtained from Dover-Richard and Paris models. Our data are compatible with the Dover-Richard model ($\Delta\chi^2$=4.4, confidence level (CL) = 0.04) but not with the Paris model ($\Delta\chi^2$=93).

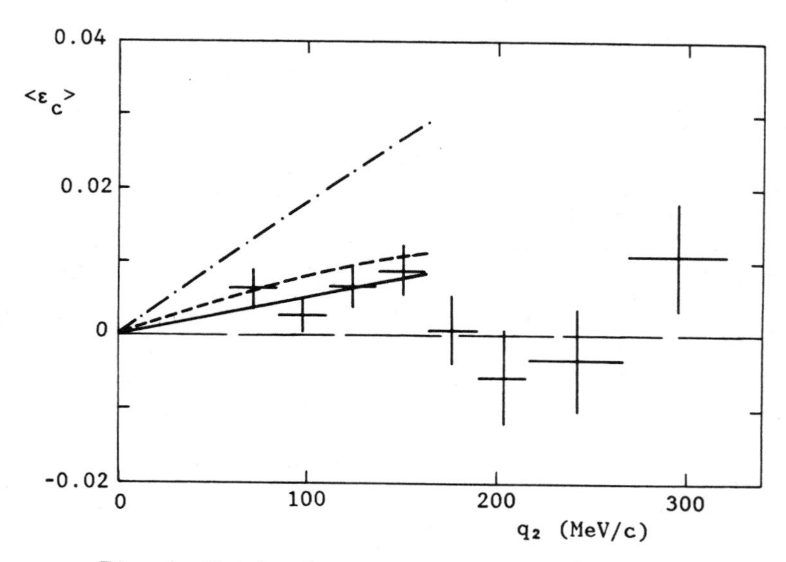

Fig. 4. Weighted averages of $\bar{p}C$ asymmetries.

The full line is the best fit to our data at small momentum transfer ($q_2<165$ MeV/c), assuming $<\varepsilon_c>=a_c^2\times q_1\times q_2$, and gives

$$a_c^2 = 0.52 \pm 0.13\ (\text{GeV/c})^{-2}, \qquad \text{or} \qquad |a_c| = +0.72\pm^{0.09}_{0.10}\ (\text{GeV/c})^{-1}\ .$$

The confidence level for the hypothesis $a_c=0$ is $\simeq 10^{-4}$ ($\Delta\chi^2=16$) so we can conclude that the absolute value of the analysing power in $\bar{p}C$ elastic scattering is small but incompatible with zero.

In the measurement of ε_p, the statistical errors were larger than for the carbon data; again no energy dependence or deviation from a linear dependence of $A_{\bar{p}p}$ on the momentum transfer in the considered range has been seen ($A_{\bar{p}p}=a_p\times q$). We averaged the measured values of the asymmetry using the same procedure than for $\bar{p}C$ data; the final values of $<\varepsilon_p>$ are plotted in Fig. 5.

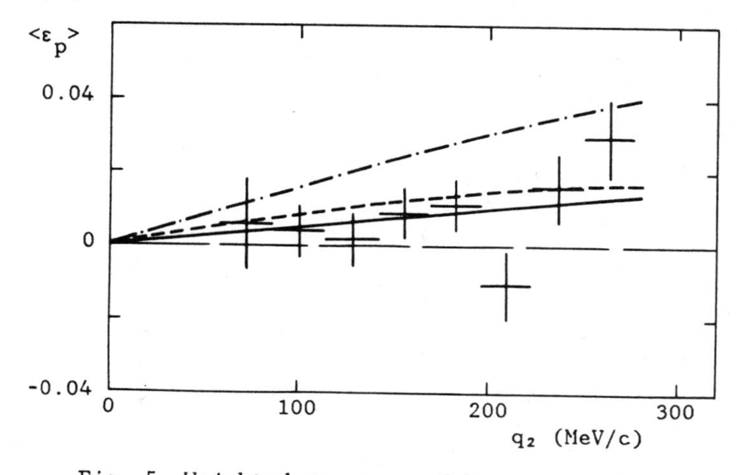

Fig. 5. Weighted averages of $\bar{p}p$ asymmetries.

The curves give the predictions from Dover - Richard and Paris models; the full line is the best fit to our data assuming $\langle\varepsilon_p\rangle = A_{\bar{p}C}(q_1)\times A_{\bar{p}p}(q_2) = a_c a_p q_1 q_2$ and gives

$$a_c a_p = +\ 0.48 \pm 0.17\ (\text{GeV/c})^{-2}\ .$$

a_p has been evaluated from the measurements with polarized target performed by our collaboration[7]: the small angle data at 783 and 1089 MeV/c give $a_p = +\ 0.83 \pm 0.04\ (\text{GeV/c})^{-1}$. Using for $|a_c|$ the value obtained from the $\bar{p}C$ data, one gets

$$|a_c|\, a_p = 0.60 \pm {}^{0.08}_{0.09}\ (\text{GeV/c})^{-2}.$$

This result is perfectly consistent with the estimated value of $a_c a_p$ if $a_c > 0$ ($\Delta\chi^2 = 0.45$, CL=0.50) while the hypothesis $a_c < 0$ gives $\Delta\chi^2 = 36$ and must be rejected. Thus the final result is

$$a_c = +\ 0.72 \pm {}^{0.09}_{0.10}\ (\text{GeV/c})^{-1}\ .$$

ESTIMATE OF $A_{\bar{p}n}$

From the measurement of $A_{\bar{p}C}$ we can extract informations on $A_{\bar{p}n}$, the polarization parameter in antiproton - neutron elastic scattering. Assuming also for $A_{\bar{p}n}$ a linear dependence on the momentum transfer at small angles ($A_{\bar{p}n} = a_n \times q$), and equal $\bar{p}p$ and $\bar{p}n$ total cross sections, from Glauber theory one gets in first approximation that $a_n = 2a_c - a_p$. From our measurement of a_c and a_p we get:

$$a_n = +0.61 \pm {}^{0.18}_{0.21}\ (\text{GeV/c})^{-1}.$$

This is the first measurement of the polarization parameter in $\bar{p}n$ elastic scattering at small angles, and within errors is in agreement with the estimates of the potential models, which predict for $\bar{p}n$ a polarization somewhat smaller than for $\bar{p}p$ elastic scattering.

REFERENCES

1. R. Birsa et al., Study of spin effects in $\bar{p}p$ total cross-section and elastic scattering at LEAR, Letter of Intent CERN/PSCC/79-55, PSCC/I6(1979).
2. J. Bailey et al., $\bar{p}p$ total cross-section and spin effects in $\bar{p}p \to K^+K^-, \pi^+\pi^-, \bar{p}p$ above 200 MeV/c, Proposal CERN/PSCC/80-77, PSCC/S 16.
3. G. Alberi et al., Antiproton-carbon elastic scattering, Proc. 5th European Symposium on Nucleon-Antinucleon Interactions, Bressanone, Italy, 1980 (CLUEP, Padua, 1980) p.51.
4. P. Osland and R.J. Glauber, Polarization in elastic proton-nucleus scattering, Nucl. Phys. A 326 (1979) 255.
5. R. Birsa et al., Polarization at small angles in elastic $\bar{p}p$ and $\bar{p}C$ scattering at 550 MeV/c, Phys. Lett. B 155 (1985) 437.
6. A. Martin et al., Polarization at small angles in antiproton-carbon elastic scattering at LEAR energies, CERN report CERN-EP/88-35, accepted for publication in Nucl. Phys. A.
7. R.A. Kunne et al., Asymmetry in $\bar{p}p$ elastic scattering, Phys. Lett. B 206 (1988) 557.

LEAR IN THE NINETIES

Pierre Lefèvre

PS Division
CERN
CH-1211 Geneva 23

INTRODUCTION

The LEAR machine has been described on several occasions, in particular last year during the Second Course of this Erice School [3]. It will not be repeated here, but we will give some background information to show the time scale of the successive improvements in the machine.

Historical and financial aspects

The first idea for LEAR was presented at the International Accelerator Conference held in Protvino (USSR) in 1977, and again in the same year during the Cern Workshop on Intermediate Energy Physics.

A conceptual design report issued in 1979, anticipated the design report in 1980. A 12,5 MSF project, authorised for 1980-1983, led to the first p in LEAR in 1982 and the first $\bar{p}$ pre-run for physics in 1983. A 2,5 MSF consolidation project (1983-1985) was attributed to improve the performance of the first generation of experiments (mainly to ensure more than 4000 hours of physics instead of 2000 hours in the original design, and to improve the main power supplies in the low energy domain).

In 1985 a second 4 year consolidation project of about 2 MSF was approved in order to improve the operation and controls and to adapt the 16 bit digital function generators and auxiliary power supplies to the precision required for low energies and for momentum scanning.

The ultra low energy domain (below 100 MeV/c) can not be opened operationally without a third consolidation programme (2,5 MSF will be asked for the coming years).

LEAR has always fulfilled the experimental programme because enough machine study time and money was obtained to satisfy the approved experiments and to anticipate the new demands for physics (options).

This situation is the case of the majority of the present second generation of LEAR experiments (ACOL era), but some of them will suffer some delay concerning the required performances if the necessary priorities and money are not attributed in due course.

Antiproton-Nucleon and Antiproton-Nucleus Interactions
Edited by F. Bradamante *et al.*
Plenum Press, New York, 1990

One aim of this presentation is to identify the sensitive points.

Overview of main machine characteristics

The antiprotons are produced on a target (AP) at 3,5 GeV/c (from PS 26 GeV protons), collected (ACOL) and accumulated (AA) at 3,5 GeV/c.

They are unstacked and transferred to the PS where they are decelerated to 600 MeV/c, then transferred to LEAR. After injection and cooling in LEAR they are either accelerated to a momentum between 600 MeV/c and 2 GeV/c (LEAR High Energy Cycle) or decelerated to a momentum between 600 MeV/c and 105,5 MeV/c (T = 5,8 MeV) (Low Energy Cycle). The machine is preparing a lower momentum deceleration scheme to 61.3 MeV/c (T = 2 MeV) and we estimate that the future lowest possible energy is close to T = 1 MeV (40 MeV/c).

To be complete we note that LEAR can also be fed by 300 MeV/c (T = 50 MeV) protons or H^-. A new H^- feeding scheme is in preparation if the old Linac I becomes unavailable for LEAR (after transformations linked to the project of Pb ion production for SPS).

The machine's basic parameters have been summarized in the Erice Workshop [1.1], and the classical scheme shown last year [3.1].

For the purpose of this school we will just recall two of the particularities of LEAR.

The first one is the phase space stochastic cooling systems acting in the transverse and longitudinal planes [1.2].

A "Fast-cooling" acts on fixed momentum flat tops : at injection (600 MeV/c), and for the deceleration cycle on strategic flat tops (309, 200, 105 MeV/c). Applied during 3 to 10 minutes the aim is to restore on each intermediate flat top the emittances adiabatically blown up during deceleration; the minimum 95% emittances (surface/π) obtained are :

$E_H \sim E_V \sim 10\ 10^{-6}$ rad.m

$\Delta p/p \sim 10^{-3}$

Cooling time τ_C = 1 min for N = 10^9 particles

In addition a "Slow-cooling" is used to maintain the beam qualities during the long extraction process (15 min. to 5 hours). Each of these systems is composed of two subsystems of different lengths to cope with the large velocity differences between the low energy domain and the higher one.

The second LEAR process we would like to emphasise is the ultra-slow stochastic extraction [1.3]. The principle is to apply RF noise in different forms and characteristics to the longitudinal beam distribution, on a harmonic of the revolution frequency around 16 MHz in the following order :

- a shaping noise (several seconds) produces a rectangular distribution; it is followed by the necessary transverse adjustments for a typical one third integer (3QH = 7) resonant extraction (horizontal bumps, QH jump) and chromaticity adjustments $\xi_V \sim 0$, $\xi_H \sim 0{,}6$ to satisfy the condition of alignment on the same transverse plane separatrix of the particles of different energies;

- a sweeping noise, swept into the distribution, during the extraction process (15 min to 5 h), causes the diffusion of particles into the resonance;

- a fixed noise on the resonance, called the chimney, reduces the ripple and enables a duty factor of 80%.

Present Performance

The ACOL aim which will be achieved in 88 is to produce $10^7 \bar{p}/s$; the present achievement is $10^{10} \bar{p}/h$ per production cycle out of six in the supercycle, developments are being done in the PS complex to guarantee these performances :

- 1 GeV injection Booster-PS
- 10 bunches recombination
- adiabatic debunching in ACOL
- faster cooling in ACOL

As far as LEAR is concerned, the machine can already consume, in dedicated mode, all the ACOL production (remember that this figure depends on the number of cycle available for antiproton production). In parallel mode (LEAR physics shares the antiprotons with the high energy $Sp\bar{p}S$ Collider) LEAR receives 15% of the production (but with more production cycles available). A typical transfer batch from AA ranges from $3\ 10^8$ to $2\ 10^{10}\ \bar{p}$ (for a higher intensity transfer a LEAR multibatch filling has to be developed).

- LEAR extraction spill $10^5\ \bar{p}/s$ to $10^7 \bar{p}/s$
- Spill length 15 min to 5 h.
- Repetition time 1h10 to 1h30 (depending on cooling time) for 1h spill length.

All the present performances are already better than announced in the Design Report (as presented in 1982, [1.1]) in the 100 MeV/c to 2 GeV/c range if we accept some delays in the realisation of 100 MeV/c full performances (expected in 1989).

The total transmission estimated from AA unstacking to LEAR extracted and transported to the experiment is around 50% (with exception to 100 MeV/c operation); it means that the extracted emittances are no longer a problem for the approved experiments as summarized in the following table.

Extraction Momentum	600 - 1500	300 - 200 (100)**
E_H 10^6 rad.m *	2	4
E_V " *	5	10
$\Delta p/p$ $\pm\ 10^{-3}$	0,5	1

LEAR Emittances for 95% of the beam
* (phase space surface/π)
**(expected 1989).

We have to note that for a hig ι spill rate ($>10^6 \bar{p}/s$) there are some counting problems (statistics) and some 1 MHz microstructure (under study). For high circulating intensity the cooling is less efficient and more time consuming as the extraction efficiency seems to decrease (under study).

STUDIES AND MACHINE IMPROVEMENTS

Already in achievement

The PS-LEAR transfer line efficiency has been raised to better than 90% with a new optics to enlarge the acceptance and a set of 7 pick-up electrodes H and V to control the dipolar trajectories.

A new set of micro VAX workstations operating through an interactive database allows new editing energy scanning and surveyance programmes.

A fast extraction facility to feed Penning Traps in parallel with other operations has been developed; the beam is bunched (at 105 MeV/c) at the harmonic 4 or 8, one bunch is then fast extracted (with full aperture kicker) followed by debunching and slow extraction of the remaining beam at same or lower energy. The same fast extraction can be achieved after a slow extraction at an higher energy.

Under development

Additional sextupoles as poleface windings in the edge blocks of the main bending magnets are under study with the following aims :

- increase the chromaticity corrections for high energy;
- provide two additional families of sextupoles for low energy resonance compensation;
- increase the dynamical acceptance, reduced by the non-linearities
- reduce the losses and perturbation during the extraction process for large momentum deviation particles :

$$\Delta Q_V \sim 700 \cdot \left(\frac{\Delta p}{p}\right)^2 .$$

The same poleface windings will include additional vertical dipoles for a better control of the closed orbit in the straight sections. They will improve the machine vertical acceptance, in particular for injection, permit the adjustment of the beam in the electron cooling and jet target sections (this last point is of great importance to operate the horizontal jet target) expected in January 89, but a more realistic installation date for this equipment is the January 90 machine shutdown.

The jet target operation requires for physics a continuous scanning in energy. The small steps are realised, with fixed machine magnet fields, by moving the beam across the horizontal jet by the momentum stochastic cooling system ; the transverse cooling also follows this energy change. For that purpose a high energy (> 600 MeV/c) continuously adjustable stochastic cooling new system is developed with synthetised delays notch filters and hopefully digital delays.

For the low energy stochastic cooling the pick-ups are connected in travelling wave mode, and cryogenic amplifiers are used to improve the signal to noise ratio. Below 100 MeV/c this improvement is not sufficient and we have installed prototypes of helix and meander couplers for momentum and transverse stochastic cooling respectively. For these low energy systems, digital delays and line filters could solve the problem of very long delay cables and strongly improve the flexibility of the system.

The electron cooling is an indispensable tool to operate a low momentum jet target (< 400 MeV/c) for a fast enough compensation of the multiple scattering blow up. The ICE prototype electron cooler has been modified by a CERN-KfK Karlsruhe collaboration, and installed in 1987 in LEAR and pre-tested with protons. The principle of possible low energy electron cooling in LEAR has been demonstrated. Now a considerable amount of work has to be done for integration of that equipment in the LEAR operation. Several years will be needed to change equipments (HV supply, new collector), to rebuild part of the mechanics, add solenoid compensation, in order to reach the necessary reliability and to operate at variable energy and electron intensity. It will anyway be ready to operate a low energy jet target, but we have strong hope to use it as soon as possible to facilitate the low energy deceleration process and in that respect we hope to find the necessary resources. With these improvements the electron cooling can realistically be operational in 1991.

Special quadrupoles set up for options

A low beta section can be provided in SS2 for the jet target if necessary. A limited one (β decreased from 5m to 1 or 2m) can be done almost free of charge, using the existing quadrupoles and the existing previous 250A unipolar correcting dipole power supplies and could be operational in 1989.

For a large β reduction (0,1 or 0,2m) up to 2GeV/c, 4 new 1000A, 10^{-5} accuracy power supplies have to be bought and this is not foreseen.

A dispersion suppression (in the long straight sections) is in preparation for 1989 with 4 new ±50A trimming power supplies able to make $D_H \sim D'_H \sim 0$ (with a reduced acceptance). It can help for injection, jet targets and $H^-\bar{p}$ corotating beams.

Solenoidal field compensation. A small solenoid (0,1Tm) can be easily compensated (electron cooling), a medium one (1Tm at 1 GeV/c) however is more difficult (under study for jetset detector second generation or for spin splitter experiment) and first indications are that it seems hopeless to try to compensate strong solenoidal fields (above 1.6Tm at 1 GeV/c).

ULTRA LOW ENERGIES IN THE NINETIES

Who are the users of antiproton beams below 100 MeV/c? The answer to this preliminary question is not so easy.

The first experiment PS 189 [1.7, 2.2] uses an RF Spectrometer for the antiproton inertial mass measurement and requires 20 MeV/c (T = 200 KeV) beam of 100 ms duration (slow extraction mode). A second one, PS200 [4] aims to measure the antiproton gravitational mass and asks for 50 KeV or 100 KeV antiproton to be injected into a Penning trap (250 to 500 ns bunch duration) in a fast extraction mode.

In addition to the previous two approved experiments, other users may be interested.

PS196 [2.3] the second antiproton inertial mass measurement

experiment using a Penning trap is satisfied with fast extracted 105 MeV/C particles, but could probably appreciate lower energy particles for degradation to increase their density at the trap entrance.

Other experiments on the floor (PS175, 177, 201, 194 etc) or new experiments are potential users of very low energy antiprotons with long spills in the ultra slow extraction mode. They have to solve the problem of the material in the beam in front of their experiment for the detection devices and the thin window (vacuum isolation). A clean windowless experiment, with differential pumping is very expensive if we want to protect efficiently the LEAR 10^{-12} Torr good vacuum, but could be of large interest for the future as an evident method to preserve and use the very good LEAR beam density.

Deceleration in LEAR

The official goal is to decelerate to T = 2 MeV (61.3 MeV/c). Such a beam has (in 1988) circulated in the machine. The extraction with reduced performance will be available in 1989, and we aim at full performances in 1990 (for 10^9 $\bar{p}$ decelerated).

Summary of the qualities of a fast extracted bunched beam (10^8 particle by fast extracted batch) :

$E_H \sim E_V$ = 5 to 10 10^{-6} rad.m (surface/π , 95%)

$(\Delta p/p)_{total}$ = 2 to 4 10^{-3} (95%)

ΔW_{total} = 8 to 16 KeV

Δl_{bunch} = 500 to 250 ns

A slow extraction is also foreseen (spill length 15 to 30 min) and a medium slow extraction (spill length 100 ms),

the following qualities are expected :

$E_H \sim 5\ 10^{-6}$ rad.m

$E_V \sim 10^{-6}$ rad.m

$(\Delta p/p)_{total} \sim 10^{-3}$ to 2 10^{-3}

As the machine is density limited by the space charge and the intrabeam scattering, there is an Emittance/Intensity limitation and one <u>can not</u> expect miracles from a strong cooling without drastic intensity reduction.

What is the lowest possible momentum in LEAR? For the future we have some hope to be able to decelerate the beam to T = 1 MeV (~40 MeV/c) but the 2 MeV is a necessary step, and an answer to this possibility can not be expected before 1991; It is not yet sure that, due to the previous limitations, the internal deceleration will be favorable compared to external ones.

The other possibility is to extend the deceleration in the machine towards 200 KeV (20 MeV/c) to statisfy the PS189 demand is not fully excluded, and will be very difficult, time consuming and requires several intermediate steps for study. We do not recommend to rely on this possibility to feed approved experiments in a reasonable time scale. We are convinced of the necessity to present two alternatives, a cheap one, the degradation and a clean one, the external post deceleration.

Degradation

PS196 has his own degradation foils at the entrance of its Penning trap to slow down the 105 MeV/c beam.

PS200 has recently adopted this method at least for the first stage. They plan to incorporate a degrading foil before the last electrostatic deceleration at the entrance of the cooling trap.

For PS189 we prepare Be and Al foils for tests of degradation from 5 MeV in 1988 and 2 MeV in end 1989. These measurements are useful because of some uncertainties in the programme codes in this domain of energy with antiparticles. The transmission from degrader into the Penning trap is good enough to permit the experiment, we have some hope that it can also be the case for the RF spectrometer in spite of its reduced acceptance, at least for the debugging stage of the experiment. We have in mind to improve the degraded density and then the transmission by achromatic optics and by a bunching before degrading and debunching after degrading scheme. It could perhaps permit to have enough transmitted particles to make possible the data taking stage of the experiment with adequate statistics.

Postdeceleration

Studies and proposals have been made since 1982 Erice Workshop, but they have always been considered as too expensive.

We have first considered a synchrotron decelerator from 5 MeV as included in the proposal (Elena) [1.4], then other types of synchrotrons has been proposed. Finally they have been judged too complicated and expensive, intensity and density limited (low energy beam dynamics), not very favorable for the time structure and finally hardly compatible with the present south hall layout.

A RF quadrupole (RFQ) linear structure starting from 5 or 2 MeV has been studied both by us and by the Los Alamos (PS200) team. It is a rather expensive solution which convinced PS200 to adopt the degradation technics. For PS189 it has a major inconvience, the spill time required (100 ms) obliged us to consider a DC type RFQ, large enough to reduce the for power density and with a huge and very expensive RF power (power supplies and amplifiers) system. We have some hope with a 4 rods Frankfurt University type RFQ to have a cheaper, simpler and less power consuming system. This solution is under study, but we are convinced that a reasonable solution can only be found for use of a RFQ for PS189 if they accept a short (< 1ms) spill, leading to a classical pulsed RFQ solution.

A cyclotron has also been considered as a general purpose post decelerator. It is a classical RF cyclotron as proposed in Erice (1982) [1.5] but with lower energy (from 2 MeV to 200 KeV). A common study with the users is underway, it is promising for time structure, easily adaptable to lower available energy from LEAR, but on the other hand the mechanics, RF and axial ejection scheme are rather complicated, requiring sophisticated field connections, and lead to complication for axially extraction at 200 KeV and consequently at a rather large radius.

Another promising approach of a cyclotron decelerator is to use the

SIN Superconducting Cyclotron trap used by PS175, equiped with axial pulsed extraction, modified injection and reduced pressure (0,1mb) as proposed also in Erice 82 [1.6].

This very interesting possibility of a general purpose tool is actively studied by the authors [2.4, 2.5] and they will soon produce a proposal.

JET TARGETS IN LEAR

The Jetset experiment PS202.

The jetset experiment PS202 [3.3] will be installed in 1989 (Jet target) and in 1990 (detector) as it has been approved (stage 1), without solenoid nor low beta but equiped with a thin vacuum chamber. As already mentioned the Continuous Scanning operation will be ready.

Polarised target

Another proposal also in the pipeline is the Heidelberg polarised target (FILTEX) [5]. It requires more from the machine: a low momentum electron cooler (expected already in 1991), a high intensity multibatch injection scheme and a low beta section.

OPTIONS

Corotating beams

$H^-\bar{p}$ corotating beams are still an open option. For that purpose the H^- injection and the H^- stripping mechanisms are under study and a publication in preparation.

A dispersion adjustment and suppression in straight section 1 requiring trimming power supplies for the main quadrupole is in preparation. In 1990 we could be ready to make the first tests of protonium in flight ($p\bar{p}$) production rate, if the physics demands it [2.1] and if machine study time is scheduled.

Two problems remain, the present H^- source (2mA) has to be changed or improved to produce 10 to 20mA. The second problem is linked to the CERN lead ion project, definitively killing the H^- Linac 1 facility for LEAR. A replacement has to be found and is under study. One solution could be to recuperate the tank 1 of the Linac 1 (if it becomes available) and use it with the present source and H^- RFQ as a 10 MeV H^- facility for LEAR. The lower injection energy scheme (137 MeV/c) requires a lot of studies (injection, life time, stripping, cooling) perhaps reserving some surprises.

Spin splitter [3.4]

We will not discuss the proposal of Spin Splitter techniques to polarise in situ the LEAR beam, as it will be discussed in another session.

Minicollider

The $p\bar{p}$ mini collider option seems to be abandoned because of too low luminosity ($L = 10^{29}\ cm^{-2}$ sec) and a too long bunch length (5m).

SUPERLEAR

An overview of a possible machine has been presented in the

September 86 PSSC session [3.2]. It is a 2 GeV/c to 10 or 15 GeV/c compact machine (L = 120 to 150m) with 6T to 10T superconducting bendings and 15T/m normal quadrupoles or 30T/m superconducting quadrupole.
Such a machine designed for high intensity ($N \geq 10^{12}\bar{p}$) is equipped with strong stochastic and electron cooling devices. The expected maximum luminosity is $L = 10^{32} cm^{-2}s^{-1}$ for jet targets and $3\ 10^{30}$ to $10^{31} cm^{-2}s^{-1}$ for $p\bar{p}$ collisions. The jet target limit is given by the $10^7\bar{p}/s$ production rate matched to the $(dN/dt) = L.\sigma loss$ consumption (we have taken $\sigma p\bar{p}$interaction = 100 mb in the 2 to 10 GeV/c region).

For the collider the limit is given by the beam-beam detuning effect; we have chosen $\Delta Qbb < 5\ 10^{-3}$ (classical for Hadron colliders) leading to $L = 3\ 10^{30}$. To increase the luminosity to 10^{31} we think of very strong cooling system (cooling time lower than 1 minute) to withdraw the beam beam limit to the $\Delta Qbb \sim 0,025$ region.

We are sure to be able to propose an adequate machine (within the previous limits) if the physics proposed is considered as interesting.

LEAR COMPLEX

References

[1] Workshop on Physics at Lear with Low-energy cooled antiproton, 1982, Ettore Majorama Centre for Scientific Culture, Erice, Italy. Edited by U. Gastaldi and R. Klapisch, 1984, Plenum Press New York.

[1.1] p. 15, P. Lefèvre, Construction of the Lear facility : Status report.

[1.2] p. 27, D. Möhl, Phase space cooling techniques and their combination.

[1.3] p. 49, R. Cappi, R. Giannini, W. Hardt, Ultraslow extraction.

[1.4] p. 633, H. Herr, A Small decelerator ring for extra low energy antiprotons (Elena).

[1.5] p. 643, P. Mandrillon, The Medicyc Cyclotron as a temporary decelerator for Lear.

[1.6] p. 819, P. Blüm and L.M. Simons, The extraction of Low Energetic Antiprotons out of a cyclotron trap.

[1.7] p. 761, C. Thibault, High precision p-$\bar{p}$ mass comparison.

[2] International school of Physics with Low Energy Antiprotons, Fundamental Symmetries (1st 1986, Erice, Italy). Edited by P. Block, P. Pavlopoulos and R. Klapisch 1987, Plenum Press, New York.

[2.1] p. 307 U. Galstaldi, Low and High Resolution Protonium Spectroscopy at LEAR in ACOL time.

[2.2] p. 77, A. Coc et al, Present Status of the Radiofrequency mass spectrometer.

[2.3] p. 59, G. Gabrielse, Penning traps, masses and antiprotons.

[2.4] p. 89, L.M. Simons, The cyclotron trap as decelerator and ion source.

[2.5] p. 85, G. Torelli, Stochastic cooling in a trap.

[3] International School of Physics with low energy antiprotons on Spectroscopy on light and heavy quarks (2nd 1987, Erice, Italy) edited by U. Gastaldi, R. Klapisch and F. Close, 1989 Plenum Press, New York.

[3.1] p. 359, P. Lefèvre, LEAR, past, present and near future.

[3.2] p. 373, R. Giannini, P. Lefèvre and D. Möhl, A high luminosity 10 GeV proton-antiproton storage ring with Superconducting magnets.

[3.3] p. 341, K. Kirseborn, Spectroscopy at LEAR with an interval gas jet target.

[3.4] p. 339, A Spin splitter for antiproton in LEAR, Y. Onel, A. Penzo, R. Rossmanith.

[4] A measurement of the grantational acceleration of the antiproton, CERN/PSCC/86-2.

[5] Measurement of Spin-dependence in p$\bar{p}$ interaction at low momenta. CERN/PSCC/85-80, (PSCC/p92), CERN/PSCC/86-12, (PSCC/p92 Add1).

Participants of the Third Course of the International School of Physics with Low Energy Antiprotons on Antiproton-Nucleon and Antiproton-Nucleus Interactions, held June 10-18, 1988, in Erice, Sicily, Italy.

PARTICIPANTS

M. Agnello — INFN - Sezione di Torino
Via Pietro Giuria 1
I-10125 Torino
Italy

C.J. Batty — Rutherford Appleton Laboratory
Chilton near Didcot
Oxfordshire OX1 0QX
England

G. Bendiscioli — Dipartimento di fisica nucleare
Università di Pavia
Via Bassi 6
I-27100 Pavia
Italy

F. Bradamante — Dipartimento di Fisica
Università di Trieste
Via A. Valerio 2
I-34127 Trieste
Italy

S.J. Brodsky — S L A C
Stanford University
P.O. Box 4349
Stanford
California 94305
USA

M.P. Bussa — INFN - Sezione di Torino
Via Pietro Giuria 1
I-10125 Torino
Italy

J. Carbonell — Institut des Sciences Nucléaires
Université de Grenoble I
53 Avenue des Martyrs
F-38026 Grenoble cedex
France

A. Coc — CERN - EP
CH-1211 Geneva 23
Switzerland

P. Deneye | Institut de Physique
Sart-Tilman
Université de Liège
B-4000 Liège 1
Belgium

S. Furui | Institut für Theor. Kernphysik
Universität Bonn
Nussallee 14-16
D-5300 Bonn 1
FRG

U. Gastaldi | CERN - EP
CH-1211 Geneva 23
Switzerland

A.M. Green | Research Institute for
Theoretical Physics
University of Helsinki
Siltavuorenpenger 20
SF-00170 Helsinki
Finland

J. Haidenbauer | Institut für Kernphysik
Kernforschungsanlage GmbH
Postfach 1913
D-5170 Jülich 1
FRG

N. Hamann | Fakultät für Physik
Universität Freiburg
Hermann-Herderstrasse 3
D-7800 Freiburg
FRG

D. Hertzog | Nuclear Physics Laboratory
University of Illinois
23 Stadium Drive
Champaign
Illinois 61820
USA

A.S. Jensen | Institute of Physics
University of Aarhus
Ny Munkegade
DK-8000 Aarhus C
Denmark

V.A. Karmanov | Lebedev Institute of Physics
USSR Academy of Sciences
Leninsky Prospekt 53
117 924 Moscow
USSR

R. Klapisch | CERN - DG
CH-1211 Geneva 23
Switzerland

A.D. Krisch
The Harrison M. Randal Laboratory of Physics
The University of Michigan
Ann Arbor
Michigan 48109
USA

R. Kunne
D P N C
Université de Genève
CH-1211 Geneva 4
Switzerland

P. Lefèvre
CERN-PS
CH-1211 Geneva 23
Switzerland

G.Q. Liu
Research Institute for Theoretical Physics
University of Helsinki
Siltavuorenpenger 20 C
SF-00170 Helsinki
Finland

E. Lodi-Rizzini
INFN - Facoltà ingegneria
Università di Brescia
Viale Europa 39
I-25100 Brescia
Italy

F. Malek
Institut des Sciences Nucléaires
Université de Grenoble I
53 Avenue des Martyrs
F-38026 Grenoble cedex
France

L. Mandrup
Institute of Physics
University of Aarhus
Ny Munkegade
DK-8000 Aarhus C
Denmark

G. Margagliotti
Dipartimento di Fisica
Università di Trieste
Via A. Valerio 2
I-34127 Trieste
Italy

A. Martin
Dipartimento di Fisica
Università di Trieste
Via A. Valerio 2
I-34127 Trieste
Italy

G. Matthiae
Dipartimento di Fisica
Università di Roma II
Via O. Raimondo
I-00173 Roma (La Romanina)
Italy

B. Minetti

Dipartimento di Fisica
Politecnico di Torino
C.so Duca degli Abruzzi 24
I-10129 Torino
Italy

M. Morando

Dipartimento di Fisica
Università di Padova
Via Marzolo 8
I-35100 Padova
Italy

A. Morel

Service de Physique Théorique
CEN de Saclay
Orme des Merisiers
F-91191 Gif-sur-Yvette
France

F. Murgia

INFN - Sezione di Cagliari
Via Ada Negri 18
I-09100 Cagliari
Italy

D. Möhl

CERN-PS
CH-1211 Geneva 23
Switzerland

N. Nägele

IMEP - Wien
Boltzmanngasse 5
A-1090 Wien
Austria

F. Perrot

CERN - EP
CH-1211 Geneva 23
Switzerland

M. Pignone

INFN- Sezione di Torino
Corso M. d'Azeglio 46
I-10125 Torino
Italy

K.V. Protasov

Lebedev Institute of Physics
USSR Academy of Sciences
Leninsky Prospekt 53
117 924 Moscow
USSR

J. Rafelski

Department of Physics
The University of Arizona
Tucson
Arizona 85721
USA

A. Raimondi

INFN - Sezione di Cagliari
Via Ada Negri 18
I-09100 Cagliari
Italy

D. Rapin

DPNC
Université de Genève
CH-1211 Geneva 4
Switzerland

J.-M. Richard

Institut des Sciences Nucléaires
Université de Grenoble I
53 Avenue des Martyrs
F-38026 Grenoble cedex
France

P. Salvini

INFN - Sezione di Pavia
Via Bassi 6
I-27100 Pavia
Italy

W. Schweiger

Institut für Theor. Fakultät
Bergische Universität
Postfach 100 127
D-5600 Wuppertal 1
FRG

I.S. Shapiro

Lebedev Institute of Physics
USSR Academy of Sciences
Leninsky Prospekt 53
117 259 Moscow
USSR

T.-A. Shibata

Institut für Kernphysik
Universität Mainz
Postfach 3980
D-6500 Mainz 1
FRG

D. Stoll

Institut für Theor. Physik III
Glückstrasse 6
D-8520 Erlangen
FRG

A. Zenoni

INFN - Sezione di Pavia
Via Bassi 6
I-27100 Pavia
Italy

INDEX